Advances in Soft Computing 39

Editor-in-Chief: J. Kacprzyk

Advances in Soft Computing

Editor-in-Chief

Prof. Janusz Kacprzyk
Systems Research Institute
Polish Academy of Sciences
ul. Newelska 6
01-447 Warsaw
Poland
E-mail: kacprzyk@ibspan.waw.pl

Further volumes of this series can be found on our homepage: springer.com

Marek Kurzynski, Edward Puchala, Michal Wozniak, Andrzej Zolnierek (Eds.)
Computer Recognition Systems, 2005
ISBN 978-3-540-25054-8

Abraham Ajith, Yasuhiko Dote, Takeshi Furuhashi, Mario Köppen, Azuma Ohuchi, Yukio Ohsawa (Eds.)
Soft Computing as Transdisciplinary Science and Technology, 2005
ISBN 978-3-540-25055-5

Barbara Dunin-Keplicz, Andrzej Jankowski, Andrzej Skowron, Marcin Szczuka (Eds.)
Monitoring, Security, and Rescue Techniques in Multiagent Systems, 2005
ISBN 978-3-540-23245-2

Frank Hoffmann, Mario Köppen, Frank Klawonn, Rajkumar Roy (Eds.)
Soft Computing Methodologies and Applications, 2005
ISBN 978-3-540-25726-4

Mieczyslaw A. Klopotek, Slawomir T. Wierzchon, Kryzysztof Trojanowski (Eds.)
Intelligent Information Processing and Web Mining, 2005
ISBN 978-3-540-25056-2

Abraham Ajith, Bernard de Bacts, Mario Köppen, Bertram Nickolay (Eds.)
Applied Soft Computing Technologies: The Challenge of Complexity, 2006
ISBN 978-3-540-31649-7

Mieczyslaw A. Klopotek, Slawomir T. Wierzchon, Kryzysztof Trojanowski (Eds.)
Intelligent Information Processing and Web Mining, 2006
ISBN 978-3-540-33520-7

Ashutosh Tiwari, Joshua Knowles, Erel Avineri, Keshav Dahal, Rajkumar Roy (Eds.)
Applications and Soft Computing, 2006
ISBN 978-3-540-29123-7

Bernd Reusch, (Ed.)
Computational Intelligence, Theory and Applications, 2006
ISBN 978-3-540-34780-4

Miguel López-Díaz, María ç. Gil, Przemysław Grzegorzewski, Olgierd Hryniewicz, Jonathan Lawry
Soft Methodology and Random Information Systems, 2006
ISBN 978-3-540-34776-7

Ashraf Saad, Erel Avineri, Keshav Dahal, Muhammad Sarfraz, Rajkumar Roy (Eds.)
Soft Computing in Industrial Applications, 2007
ISBN 978-3-540-70704-2

Ashraf Saad, Erel Avineri, Keshav Dahal,
Muhammad Sarfraz, Rajkumar Roy (Eds.)

Soft Computing in Industrial Applications

Recent and Emerging Methods and Techniques

Springer

Editors

Dr. Ashraf Saad
Department of Computer Science
Armstrong Atlantic State University
11935 Abercorn Street
Savannah, Georgia 31419-1997
USA
E-mail: ashraf@cs.armstrong.edu

Dr. Erel Avineri
Centre for Transport & Society
Faculty of the Built Environment
University of the West of England
Frenchay Campus
Coldharbour Lane
Bristol BS16 1QY
UK
E-mail: Erel.Avineri@uwe.ac.uk

Dr. Keshav Dahal
MOSAIC Research Group
University of Bradford
Department of Computing
Bradford BD7 1DP
UK
E-mail: K.P.Dahal@Bradford.ac.uk

Dr. Muhammad Sarfraz
Information & Computer Science Department
King Fahd University of Petroleum & Minerals
KFUPM #1510
Dhahran 31261
Saudi Arabia
E-mail: sarfraz@ccse.kfupm.edu.sa,
sarfraz@kfupm.edu.sa

Prof. Rajkumar Roy
Decision Engineering Centre
Manufacturing Department
Cranfield University
Bedford MK43 OAL
UK
E-mail: r.roy@cranfield.ac.uk

Library of Congress Control Number: 2007923718

ISSN print edition: 1615-3871
ISSN electronic edition: 1860-0794
ISBN-10 3-540-70704-2 Springer Berlin Heidelberg New York
ISBN-13 978-3-540-70704-2 Springer Berlin Heidelberg New York

Springer is a part of Springer Science+Business Media

springer.com

Printed in Germany

Typesetting: by the authors and SPS using a Springer LaTeX macro package

Printed on acid-free paper SPIN: 11585275 89/SPS 5 4 3 2 1 0

Preface

On behalf of all members of the International Technical Program Committee of the 11th Online World Conference on Soft Computing in Industrial Applications (WSC11), we would like to extend our sincere welcome to you. The conference continues a tradition started over a decade ago by the World Federation of Soft Computing (WFSC) to bring together researchers interested in advancing state of the art in the field. Continuous technological improvements since then continue to make this online forum a viable gathering format for a world class conference.

The program committee received a total of 63 submissions, of which 61 papers qualified for peer review by the International Program Committee. Each paper was then reviewed by at least three referees, culminating in the acceptance of 30 papers for publication. Authors of all accepted papers were then notified to prepare and submit their final manuscripts and conference presentations. This resulted in a total of 28 final submissions by 73 authors that comprise the six sessions of the conference program. Based on the reviewers' reports, the authors provided revised versions of the papers – all of them are featured in this book. Also featured is an invited paper based on a keynote presentation. The authors of several outstanding papers have been invited to submit significantly revised and extended versions of their papers to the Applied Soft Computing Journal.

We extend our sincere thanks to all authors and to all members of the International Program Committee for their clear and unwavering commitment to the success of WSC11. Reflecting the worldwide nature of WSC11, authors, members of the program committee and the conference organizers are from over 20 countries and five continents. We also extend our thanks to our keynote speaker, Dr. Pieter Mosterman of the MathWorks for his contributed talk.

November 29, 2006

Ashraf Saad
General Chair of WSC11
Savannah, Georgia, USA

Erel Avineri
Program Chair of WSC11
Bristol, UK

Message from the WSC11 General Chair and Program Chair

It is our pleasure to officially announce the start of the conference. The official WSC11 web site has been relocated since August to the following URL: http://www.cs.armstrong.edu/wsc11/. Please make the necessary changes to any web pages that you maintain with reference to the conference. That will increase the chances of search engines pointing to the correct WSC11 web site.

An opening note has been posted to the conference web site along with the final pdf version of all accepted papers. With regard to the presentation of papers and the keynote, we will be able to support (for the first time in WSC's history) real-time presentations via audio conferencing. This is made possible through a kind three-week trial offer (for the duration of the conference) of Elluminate (http://www.Elluminate.com), a Java-based (http://java.sun.com/products/javawebstart/) webinar environment. In return, we will provide feedback about the use of this web-based conferencing tool in support of our worldwide conference. In order to get an idea of the use of this tool, please visit the following URL: https://sas.elluminate.com/m.jnlp?sid=1125&password=M.161974A26FAAF95DB6C50F2C6CFF05 where an image version of the opening note is currently posted for testing purposes.

Therefore, we request from each correspondence author to email us back by Friday, September 22, with his/her availability to make a 25-30 minutes presentation during the upcoming two weeks (Sep 25-Oct 6). Please provide us with 2-3 possible times, and indicate your local time zone as it relate to GMT (e.g., EST in the US is GMT-5, while Brazil should be GMT-4). A presenter will need a Java-enabled computer, with a reasonable high quality connection to the Internet, and which is also equipped with a speaker and a microphone (or a headset). We will schedule all presentations and upload into Elluminate the presentation slides that have been submitted in August. A final schedule of presentations will be posted and emailed to all by Monday, September 25. All interested participants will then be able to connect to a presentation at the scheduled time, up to a maximum of 30 seats per session. We will expect session chairs to attend as many of the presentations of their sessions as possible.

It is indeed an exciting development for us to be able to support a synchronous mode of interaction for WSC11 given our global community. We also hope to witness a strong level of participation in the sessions by researchers from all four corners of the globe.

September 18, 2006

Ashraf Saad
General Chair of WSC11
Savannah, Georgia, USA

Erel Avineri
Program Chair of WSC11
Bristol, UK

WSC11 Organization and International Program Committee

General Chair

Ashraf Saad, Armstrong Atlantic State University**, USA
** Formerly with the Georgia Institute of Technology

Program Chair

Erel Avineri, University of the West of England, Bristol, UK

Advisory Board

Hisao Ishibuchi, Osaka Prefecture University, Japan
Rajkumar Roy, Cranfield University, UK
Ajith Abraham, Chung-Ang University, Korea
Mario Köppen, Fraunhofer IPK, Berlin, Germany

International Co-chairs

Lakhmi Jain, University of South Australia, Australia
Serge Popov, Kharkiv University of Radio Electronics, Ukraine
Muhammad Sarfraz, King Fahd University of Petroleum and Minerals, Saudi Arabia
Ashitosh Tiwari, Cranfield University, UK

Publicity Chair

Keshav Dahal, University of Bradford, UK

International Technical Program Committee

Janos Abonyi, University of Veszprem Folyamatmérnöki Tanszék, Hungary
Bart Baesens, Catholic University of Leuven, Belgium
Valeriu Beiu, United Arab Emirates University, UAE
Sugato Bagchi, IBM Research, USA
Soumya Banerjee, BITS Mesra, India
Christian Blum, Universitat Politecnica de Catalunya, Spain
Ulrich Bodenhofer, Software Competence Center, Austria
Andrea Bonarini, Politecnico de Milano, Italy

Oscar Castillo, Instituto Tecnológico de Tijuana, Mexico
Siam Charoenseang, King Mongkut's University of Technology, Thailand
Leandro Coelho, Pontifical Catholic University of Parana, Brazil
Carlos A. Coelho, CINVESTAV, Mexico
Oscar Cordon, University of Granada, Spain
Gaspar Cunha, University of Minho, Potugal
Suash Deb, National Institute of Science & Technology, India
Guy De Tré, Ghent University, Belgium
Mauro Dell'Orco, University of Bari, Italy
Giuseppe Di Fatta, University of Konstanz, Germany
Katrin Franke, Fraunhofer IPK, Germany
Aureli Soria Frisch, Universitat Pompeu Fabra, Spain
Xiao-Zhi Gao, Helsinki University of Technology, Finland
Takeshi Furuhashi, Nagoya University, Japan
Crina Grosan, Babes-Bolyai University, Romania
Roderich Gross, Universite Libre de Bruxelles, Belgium
Hani Hagras, University of Essex, UK
Ioannis Hatzilygeroudis, University of Patras, Greece
Ayanna Howard, Georgia Institute of Technology, USA
Yaochu Jin, Honda Research Institute Europe, Germany
Uri Kartoun, Ben Gurion University of the Negev, Israel
Okyay Kaynak, Bogazici University, Turkey
Frank Klawonn, University of Applied Sciences, Germany
Joshua Knowles, University of Manchester, UK
Andreas König, Technische Universitat Kaiserslautern, Germany
Renato Krohling, University of Dortmund, Germany
Reza Langari, Texas A&M, USA
Luis Magdalena, Universidad Politecnica de Madrid, Spain
Max Manfrin, Universite Libre de Bruxelles, Belgium
Christophe Marsala, Universite P. et M. Currie, France
Patricia Melin, Instituto Tecnológico de Tijuana, Mexico
Sanaz Mostaghim, ETH-Zurich, Switzerland
Mehmet K Muezzinoglu, University of Louisville, USA
Lakshmi Narasimhan, The University of Newcastle, Australia
Detlef D Nauck, British Telecom, UK
Nadia Nedjah, State University of Rio de Janeiro, Brazil
Andreas Nuernberger, Universität Magdeburg, Germany
Jae C. Oh, Syracuse University, USA
Sankar K. Pal, Indian Statistical Institute, India
Vasile Palade, Oxford University, UK
Gerardo Rossel, Universidad Abierta Interamericana, Argentina
Yos Sunitiyoso, University of the West of England, Bristol, UK
Vicenc Torra, AI Research Institute, CSIC, Spain
Edward Tunstel, Jet Propulsion Lab/NASA, USA
Marley Vellasco, Pontifical Catholic University of Rio de Janeiro, Brazil
Christian Woehler, DaimlerChrysler AG, Germany
Berend Jan van der Zwaag, University of Twente, The Netherlands

Contents

Hybrid Dynamic Systems in an Industry Design Application

Pieter J. Mosterman and Elisabeth M. O'Brien

The MathWorks, Inc., 3 Apple Hill Drive, Natick, MA 01760, USA
{pieter_j_mosterman,elisabeth.obrien}@mathworks.com

Abstract. The term *hybrid dynamic system* is a term for a mathematical system that combines behavior of a continuous nature with discontinuous changes. Such systems are often formed by the underlying computational representation of models used in the design of control and signal processing applications, for example in the automotive and aerospace industries. This paper outlines the benefits of Model-Based Design and illustrates how many different formalisms may be essential in model elaboration, such as time-based block diagrams, state transition diagrams, entity-flow networks, and multi-body diagrams. The basic elements of the underlying hybrid dynamic system computational representation are presented and it is shown how these elements combine to form different classes of behaviors that need to be handled for simulation.

Keywords: Model-Based Design; Hybrid Dynamic Systems; Hybrid Systems; Multi-Formalism Modeling; Embedded Control Systems; Networked Embedded Systems.

1 Introduction

Model-Based Design improves the design workflow of engineered sytems by employing computational models. In the embedded control systems realm, these models often are designed using Simulink® [20]. An embedded control system typically consists of a controller and a plant, where the plant is a physical system that is controlled to operate according to desired behavior.

The elements of Model-Based Design, illustrated in Fig. 1, can be summarized as:

- Executable specifications from models allow immediate feedback on the behavior of a specification, as opposed to documented behavior that often is misinterpreted.
- Design with simulation supports a faster exploration of the design space as opposed to constructing physical prototypes.
- Automatic code generation reduces the tedious and error-prone process of translating a design into a specification for the software engineers and manually writing the corresponding computer code.
- Test and verification can be performed in a much earlier stage in the design as a computational model is available with access to all internal variables, including those that may be difficult to obtain on a physical prototype.

A. Saad et al. (Eds.): Soft Computing in Industrial Applications, ASC 39, pp. 1–16, 2007.
springerlink.com

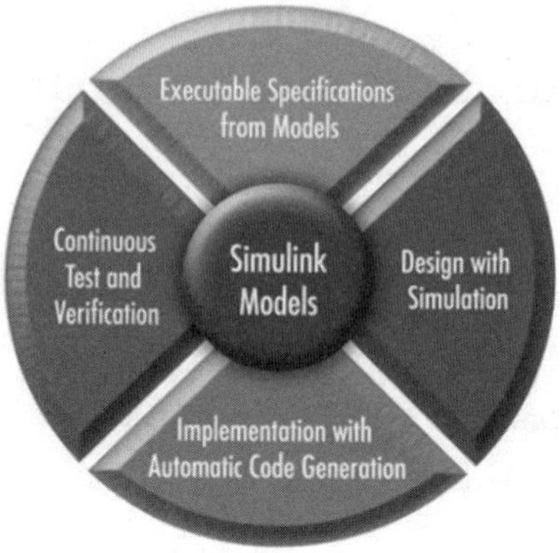

Fig. 1. Model-Based Design elements leverage Simulink® models

The adoption of Model-Based Design has enterprise-wide implications [1]. For example, the extensive use of models throughout the design process has created the desire to facilitate model reuse. This reuse, in turn, requires design tools that support the exchange of models between engineering teams. For example, to obtain a high-fidelity plant model, a SolidWorks [21] computer-aided design (CAD) model of the geometry can be exported into a SimMechanics [18, 25] multibody model of the dynamics. Thus, modeling effort is reused as models are shared across teams.

A controller model may initially be a discrete state based model that is then extended to include implementation effects such the validation of input data. This approach implies that an execution engine for a supporting tool set such as Simulink and Stateflow® [22] has to efficiently handle both a data driven approach as well as an event driven approach. Because of the widely differing execution semantics that different models may employ, execution engines are required to be versatile and powerful such that efficient algorithms, tailored to the needs of a specific model, can be invoked.

An important distinction in execution semantics can be made between those that require continuity of variables, possibly in higher derivatives, and those that allow discrete changes. Combining those two execution semantics results in *hybrid dynamic systems*, or *hybrid systems* for short (e.g., [2, 10, 23]).

The modeling formalisms that capture the discrete part of a hybrid system often are state transition diagrams [9], for example, a high-level language such as statecharts [7] may be employed. Statecharts are state transition diagrams that include language features such as hierarchy, parallelism, and event broadcasting.

The modeling formalisms that capture the continuous part of a hybrid system often are designed for plant modeling, i.e., the modeling of physics [5, 8], and they typically rely on differential equations, possibly combined with algebraic constraints.

The combination of state transition diagrams and differential and algebraic equations may be desired if, for example, there are widelydiffering time scales at which physical phenomena occur. In such a situation, it may be beneficial to abstract fast continuous behavior into a discrete change. The slower continuous behavior is then modeled by differential equations, while the discrete behavior may be modeled by a state transition diagram [12].

For example, a nonelastic collision between two bodies can be modeled in detail by accounting for dissipation effects that occur from when the bodies initiate contact to when they achieve the same velocity. Alternatively, detailed behavior from the dissipative effects can be disregarded, and the velocities can be instantaneously set to be equal.

This paper provides the elements that constitute a hybrid dynamic system. Complications and idiosyncrasies in the behavior of such hybrid dynamic systems and an ontology of mode transition behavior are presented. It is illustrated how instantaneous changes in variables, in combination with the inequalities that define mode switching, can lead to rich and complex mode transition behavior [13].

Section 2 provides a more detailed introduction to Model-Based Design. Section 3 illustrates the use of Model-Based Design for a power window control system, which concretely shows a number of different modeling formalisms that are employed throughout the design. Section 4 introduces the underlying computational representation across different modeling formalisms as a hybrid dynamic system and discusses the characteristics of such a system. Section 5 presents the conclusions of this work.

2 Model-Based Design

The benefits of Model-Based Design are manifold and mostly stem from the use of computational technologies. In addition, rather than isolated usages of computational models, it is important that a tool infrastructure is available to move a model through the design stages while elaborating it along the way.

2.1 Why Model-Based Design?

Model-Based Design uses an executable specification, which facilitates communication across engineering groups and enables rapid design iterations which greatly decreases development time. This approach contrasts with a more traditional approach in which the specification typically consists of a paper document. The document needs to be shared among many engineers or groups of engineers, and is often miscommunicated or distributed copies are not kept up to date.

The model that results from an executable specification is not only the repository for all of the information about the concept and design but also the design implementation. Once the specification has been made executable, Model-Based Design enables the exploitation of simulation so that the design space can be searched for an optimal design efficiently. Moreover, this search may now be automated.

Following simulation, implementation is achieved through automatic code generation. Transforming a paper specification of a design into software such as C-code is an error-prone process. Automatic code generation can reduce both design and hand-coding errors while substantially alleviating the tediousness of the coding task.

Model-Based Design further enables unambiguous communication between everyone involved in the overall design, within one company and across companies, such as between suppliers and the original equipment manufacturer (OEM). When everyone works off the same model, or at least an elaborated form of a core model, they can speak the same language and communicate more effectively.

Another key benefit of Model-Based Design is early test and verification. If a model is available early on in the design process, and it is executable, it is possible to design the tests to ensure that the final product complies with the original requirements based on the model. Therefore, design testing can be performed early on in the design process, as opposed to having to wait until the physical product has become available.

As a result, Model-Based Design eliminates the need for physical prototypes in the early design phases. Their use can be deferred much longer than in a traditional design approach, which decreases the reworking of a prototype because it has already been tested in much greater detail in a computational setting.

2.2 Practicing Model-Based Design

Model-Based Design relies heavily on model elaboration, as shown in Fig. 2. On the left of the diagram is the core control algorithm, which is often designed using synthesis techniques based on simplified plant models, such as low-order linear versions of more complex plant models.

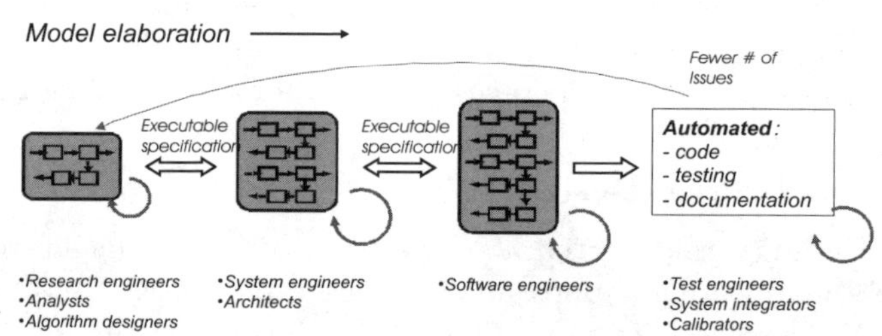

Fig. 2. Model elaboration

Once the core control algorithm has been derived, it is handed to the system engineers who embed it into an overall system. At this point, data validation, input/output (I/O) functionality, redundancy management, and testing functionality will be included.

The next step is implementation, in which the control algorithm needs to be coded in C, Ada, or any other desired target language, to embed the control algorithm into a physical environment as software that executes on a hardware target. This step is typically done by software engineers. Operating system issues may arise here; for example, computations that have been designed for the algorithm as well as for the system must fit into the computational resources available. The algorithm may need to fit onto a number of microprocessors; there may be high priority tasks, low priority tasks, and different sample rates, which are all coded into tasks or multiple tasks; and it is necessary to verify and validate that the system still operates according to specification.

Finally, the system must be integrated with other systems that have been built. This requirement leads to the notion of "systems of systems." Using an automobile power window as an example, it may be necessary to validate that the window operates properly in concert with the electrical system by not drawing electrical power when the engine is started. This is achieved by combining and integrating the system of systems, as well as calibrating it to make sure that it operates properly.

Model elaboration, then, is the process of moving the model through a number of phases where increasing detail is included. This facilitates communication between the engineering teams responsible for the separate phases. As mentioned previously, data validation and analysis need to be performed, I/O and interfaces need to be established, and redundancy management all need to be included in the design. With Model-Based Design–and its use of executable models–testing happens every time a model is simulated, and thus is an integrated aspect of the design process. This integration enables continuous testing and validation that the model satisfies the requirements and is working according to specifications.

3 A Power Window

To provide a concrete example of the use of Model-Based Design, the design of a power window (see Fig. 3) is outlined. The power window is an example of Model-Based Design for embedded control system development from concept through to implementation. It illustrates the use of different modeling formalisms that have different models of computation, the combination of which results in a hybrid dynamic system.

3.1 System Requirements

Electronics are used in automobiles to control various functions such as the opening and closing of windows and sun-roof, adjusting the mirrors/headlights, and locking and unlocking the doors. These systems are subject to stringent operating constraints, as failure may result in dangerous and possibly life-threatening situations. Therefore, careful design and analysis is mandatory before deployment.

Some quantitative requirements for the control of a power window may be as follows:

- The window must be fully opened and closed within 4 s.
- If the down or up command is issued for at least 200 ms and at most 1 s the window has to be fully opened or closed, respectively (auto-up/auto-down).

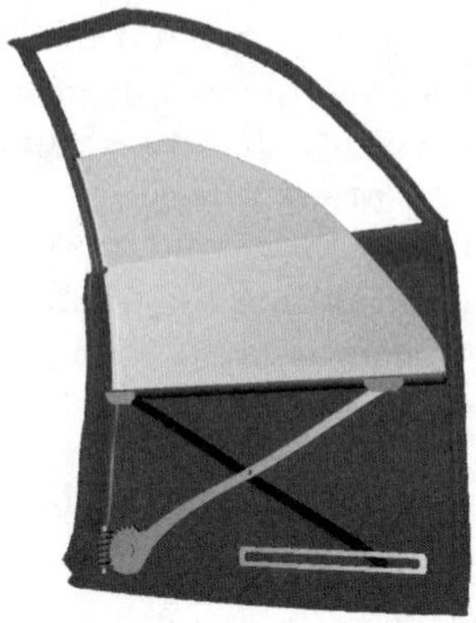

Fig. 3. An automobile power window

- After a command is issued, the window must start moving within 200 ms.
- The force exerted in the presence of an object should be less than 100 N.
- When an object is present, the window should be lowered by approximately 10 cm.

3.2 Discrete Event Control

The core control algorithm is of a discrete event nature and best modeled by using a statechart. The statechart contains the basic states of the power window system: up, auto-up, down, auto-down, rest, and emergency. It models the state transitions between these states and accounts for the precedence of driver commands over the passenger commands. It also includes emergency behavior that is to be activated when an object is detected to be present between the window and the door frame while moving up. In the emergency state, the window is moved down by 10 cm.

While in the state in which the driver command is neutral, the passenger is in control and can command the window up or down. Figure 4 shows part of the Stateflow chart that switches between neutral, up, and down states, as commanded by the passenger, *passengerNeutral*, *passengerUp*, and *passengerDown*,

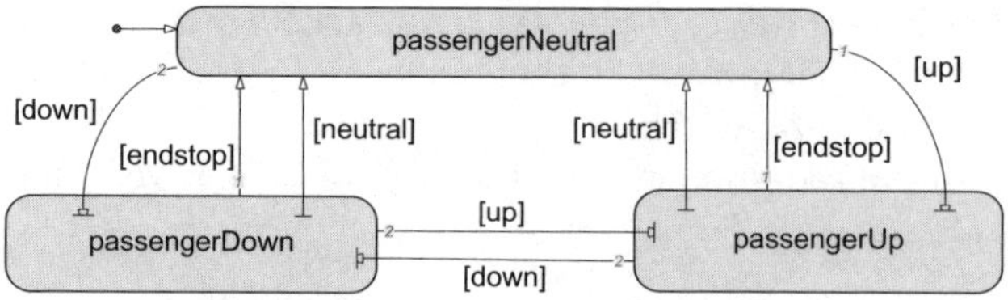

Fig. 4. A state transition diagram

respectively. The transitions between states are based on conditions *down*, *up*, *neutral*, and *endstop*. The statechart is executed periodically at a 10 ms rate and the conditions are evaluated at this rate. If one of them is true, the corresponding transition is taken, where the order of evaluation is explicitly shown by the numbers on the state transition arrows.

The *passengerDown* and *passengerUp* states contain subcharts that implement the auto-up and auto-down state transition logic.

Simulink enables testing the design with a variety of test vectors as inputs to the state machine. A model coverage report permits verification that the design is completely excited with the test vectors that have been employed, thereby showing that the design is void of hidden functionality. The generated report documents which transitions have been excited and which have not.

3.3 The Emergency Rollback

Further fulfillment of the requirements results in increased design complexity. Once the discrete event control has been designed and verified, it can be coupled to the continuous time plant model shown in Fig. 5 to ensure the window is retracted 10 cm upon detecting an object. The plant model contains two integrators. One computes velocity from the acceleration that results from the actuation force. The other computes the window position from its velocity. Viscous friction is modeled by the gain block that feeds back a friction force to be subtracted from the actuation force.

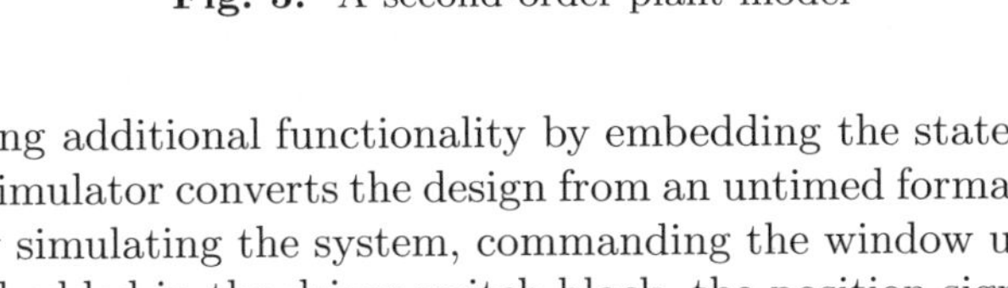

Fig. 5. A second-order plant model

Implementing additional functionality by embedding the statechart in a continuous time simulator converts the design from an untimed formalism to a timed formalism. By simulating the system, commanding the window up by switching the switch embedded in the driver switch block, the position signal can be analyzed to verify that the 10 cm requirement is satisfied.

3.4 Verifying the 100 N Force Limit

After an initial analysis of the discrete event control and continuous dynamics, a detailed plant model can be used to evaluate performance in a more realistic implementation. Models at such a level of detail are best designed in the power domain, i.e., as energy flow. This approach is facilitated by several domain specific blocksets.

Using a tool for modeling physical systems such as SimMechanics allows inertias, joints, and bodies to be used as basic elements of the modeling formalism. For example, Fig. 6 shows a SimMechanics model of the scissor-type lift mechanism that is used to move the window up and down and that is shown in Fig. 3. On the one end, a DC motor drives one of the two levers that constitute the scissor-like mechanism. Driving torque provided by the DC motor causes the worm part of a worm gear to rotate which, in turn, causes the lever to rotate.

The SimMechanics model shows the torque coming from the DC motor as an inport block on the left. This torque is used to actuate a rotational joint, which represents the worm part with inertia modeled by the *worm* rigid body block. The worm connects to the main gear through a gear ratio as modeled by the *worm gear* block. Both the worm and the main lever rotate with one degree of freedom relative to the door. The *main gear* lever body attaches to the bottom of the window by a *rotate & slide* joint, to ensure the attachment can move to the left or right as the window moves up and down. The angle of the main lever is measured, in this case for visualization purposes.

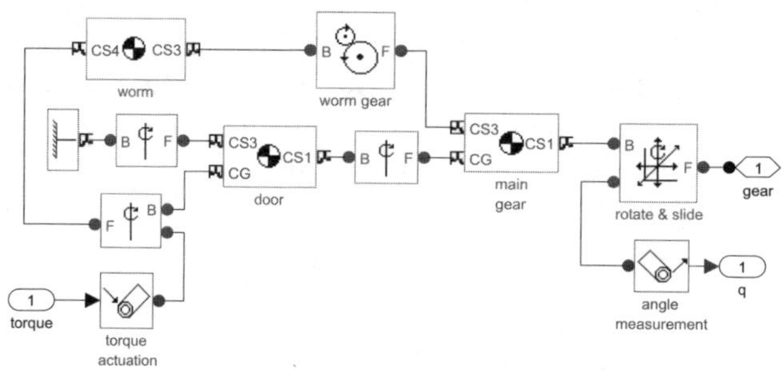

Fig. 6. A multibody diagram of the lift mechanism

Note that there is no direction associated with the connection of joints and bodies. Instead, a joint carries two variables: force and velocity. The modeler does not have to determine if the *main gear* block computes the force or the velocity. This is automatically derived by the compiler. Similarly, the DC motor model is designed using SimPowerSystems [19], and contains undirected connections in the electrical domain.

At this point in the design it becomes clear that the armature current drawn by the DC motor is the only available measurement. The control system as derived earlier now has to be modified to accommodate an input different from the window position. Instead, when the armature current is more than 1.7 A, an object is detected.

In Fig. 7 the force exerted by the window during a simulation of the window moving up is shown. At approximately 2.7 s, an obstacle is detected and the window is retracted by 10 cm. As shown in Fig. 7, the force, indeed, remains below 100 N, as per the requirement. An assertion check can be inserted so that if the window exerts a force above 100 N the simulation will stop. This step is done by way of adding a *check static bound* block. Note that the force does fall below -100 N when the direction of motion is reversed, which does not violate the requirement and is safe because it is irrelevant how forcefully the window is being pulled down.

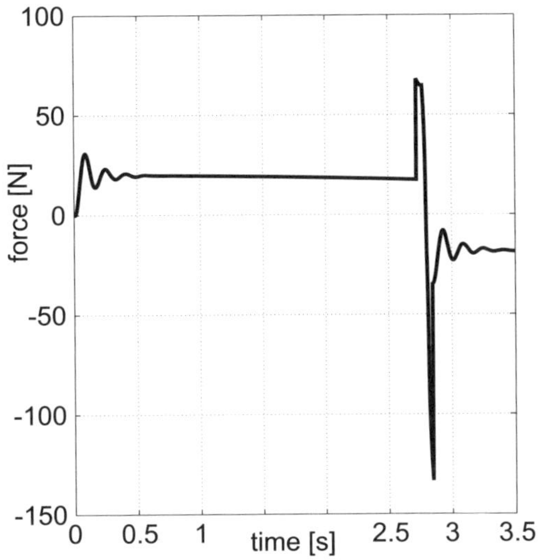

Fig. 7. Simulation of the force exerted by the window

3.5 Further Model Elaboration

Further model elaboration may include architectural elements such as the use of a controller area network (CAN) [4] bus to communicate the user command as input using some switch hardware to the hardware that controls the window movement. Communication is achieved by packaging the commands entered through the window control switches into a network frame that is sent to the window controller, which unpacks the frame to retrieve the command value. The CAN bus is modeled as an entity-flow network using SimEvents™ [17]. Part of this model is depicted in Fig. 8. It shows a write port that sends a prepared frame to a transmit buffer. A flow controller connects to a gating block to release frames for actual transmission. Once released, a frame is copied so as to make it available on the communication channel and to queue it so the channel state can be determined.

Network traffic is often best modeled as irregularly spaced in time. To efficiently simulate such behavior, a discrete event simulator typically employs an event calendar that captures the times when an event occurs [3]. Simulation then progresses in time by simply updating the current time with the time at which the earliest event on the calendar occurs. In some applications, this update may take place in the order of a hundred thousand times over the course of one simulation run. Numerical integration schemes are not required, which enhances the efficiency of the simulation significantly and allows handling a large number of discrete events.

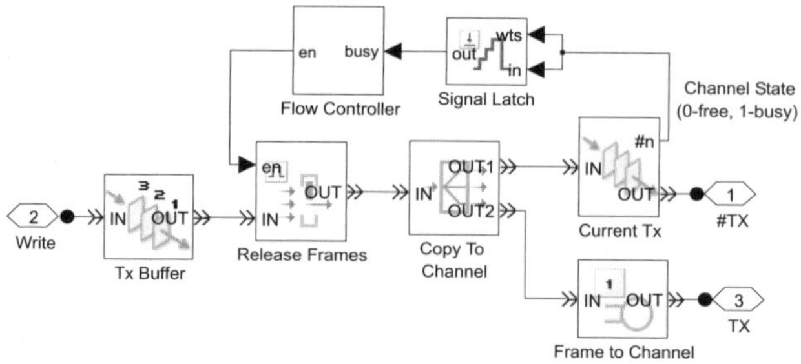

Fig. 8. An entity-flow network

Using SimEvents for the design of event driven systems allows convenient modeling of how packets of information on the network move and how other network traffic may affect the performance of the system. Because the CAN bus is shared and driver commands are put on the bus, the speed at which commands are retrieved by the control system, which moves the window, is affected by other network traffic.

Additional communication effects can be added until a sufficient level of detail is achieved. Controller code can then be automatically generated for any specific target platform, and coverage analysis tools can be used to ensure that the model is generating the desired output.

4 Hybrid Dynamic Systems

As illustrated by the power window design example in Section 3, many different modeling formalisms are typically employed in the design of an engineered system. In the case of the power window design, these formalisms include state transition diagrams (Fig. 4), time-based block diagrams (Fig. 5), multibody diagrams (Fig. 6), and entity-flow networks (Fig. 8).

4.1 Elements of a Hybrid Dynamic System

The semantics of the formalisms used are rather different from each other, varying from mechanical primitives to discrete states and transition elements, and are based on widely differing models of computation. For example, whereas time-based block diagrams may be used to model ordinary differential equation behavior, state transition diagrams may be used to capture finite state machine behavior. Similarly, multibody diagrams may be based on differential and algebraic equations, while entity-flow networks may rely on discrete-event models of computation.

An important aspect of these different models of computation is whether state behavior is allowed to exhibit discontinuous changes or whether state behavior must be continuous, possibly with further constraints on higher order derivatives.

A mathematical system that contains both classes of behavior is often referred to as a *hybrid dynamic system*, or a *hybrid system* for short. In this paper, state variable behavior with continuity constraints corresponds to differential equation behavior as captured by, for example, time-based block diagrams (Fig. 5) and multibody diagrams (Fig. 6); state behavior that may be discontinuous corresponds to discrete event behavior as captured by, for example, state transition diagrams (Fig. 4) and entity-flow networks (Fig. 8).

To illustrate a hybrid dynamic system, consider a model of the dynamics of the power window in Fig. 3, presented in Fig. 9. Here, the window is modeled as a rigid body that moves in the vertical direction. When the window moves between the bottom and the top of the door frame, the window movement is determined by the net force acting on it, which derives from the actuator force combined with the frictional force and gravity.

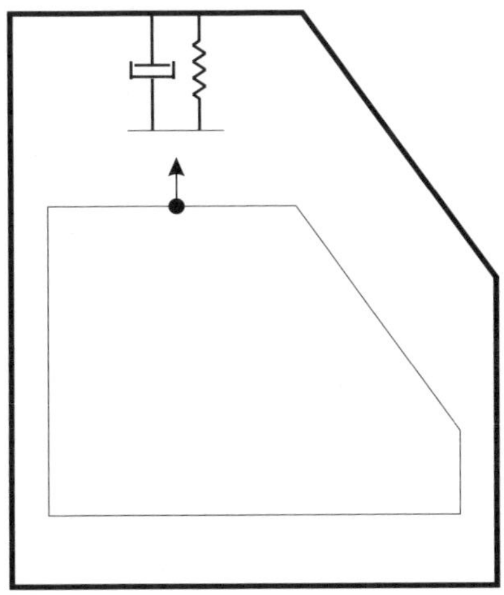

Fig. 9. A power window system model

As illustrated in Fig. 9, the top of the door frame can be modeled as a stiff spring-damper system. This system acts as an additional force when the window reaches the top of the door frame. The force is composed of a viscous (damping) force and a displacement (spring) force. The spring-damper force builds up very quickly to balance the combination of the actuator force, the frictional force, and gravity. When a balance of forces is achieved, the window stops moving.[1]

The window behavior can now be schematically captured by the state spaces in Fig. 10. The two state spaces correspond to the two *modes* of operation of the window. In Fig. 10(a), the behavior of the window when it is between the

[1] Note that typically the actuator force will be turned off when the window reaches the top of the door frame. Such feedback aspects are not considered here to avoid unnecessary complexity in the illustrative behavior.

bottom and the top of the door frame is shown, which is called the *free* mode. In this mode, the actuator force causes a positive velocity, v, according to which the window starts to move up and increase its position, x. In Fig. 10(b), the behavior of the window when it is moving against the top of the door frame is shown, which is called the *stuck* mode. In this mode, the door frame, modeled by the spring-damper system, exerts a rapidly increasing force to bring the window movement to a halt.

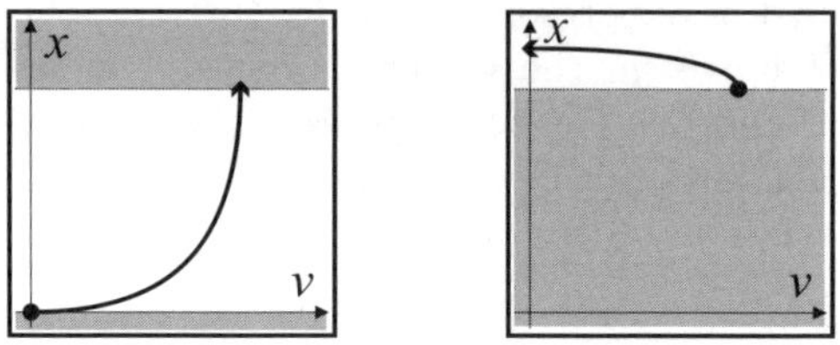

(a) Free (b) Endstop

Fig. 10. Modes of behavior for a power window

The state space behavior in Fig. 10 shows a number of important elements that are present in hybrid dynamic systems [14, 15]:

- Differential equations determine the window behavior in continuous time. For example, for the power window, $F_{net} = m_{window}\dot{v}$, i.e., the net force, F_{net}, acting on the window with mass m_{window} corresponds to the window acceleration $\dot{v}$, where the dot operator is used to express differentiation with respect to time.
- Inequality constraints determine where the differential equations are operational. This is called the *operational area* (or *patch* [6]). For the power window, the differential equations for the *free* mode are operational when $x > x_{bottom}$ and $x < x_{top}$, with x being the window position and x_{bottom} and x_{top} the values corresponding to the bottom and top of the door frame.
- A mode transition function determines which mode is active. For the power window, the mode transition function captures the change from *free* to *stuck* when $x \geq x_{top}$. The mode transition function is often provided as a state transition diagram.

4.2 Further Model Abstraction

For many analysis and synthesis tasks, abstractions are applied to the model to obtain a simplified representation. The abstractions applied determine the level of detail to capture versus the level of computational complexity suited for the algorithms employed.

For example, the state space behavior in Fig. 10 may be simplified, as shown in Fig. 11. Here, the differential equation behavior that couples velocity and position in the *free* mode is partitioned into two piecewise linear modes of operation, shown in Fig. 11(a) and Fig. 11(b). Though this may reduce the complexity of the mathematics involved in computing the up movement, it requires the derivation of inequalities to properly define the operational areas.

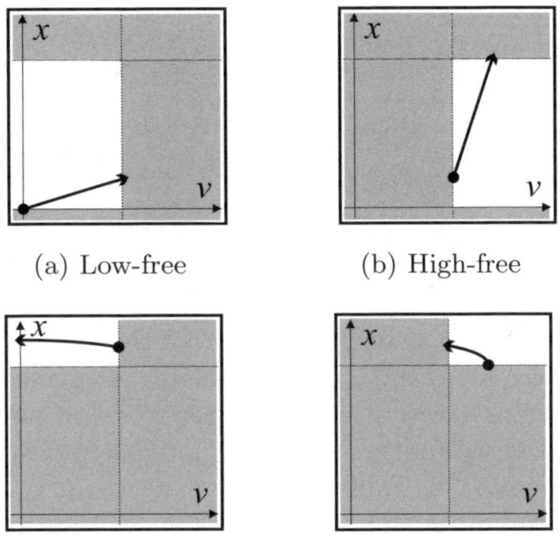

Fig. 11. Simplified continuous-time model of the power window behavior

The model can be further simplified by removing the stiff behavior caused by the spring-damper system in order to quickly reduce the window velocity to 0, shown in Fig. 11(c) and Fig. 11(d) for the low and high velocity partitioning. Instead, the window velocity may be immediately set to 0, resulting in a nonelastic collision model. This instantaneous change in velocity is shown in Fig. 12 as a line with a double arrow head. An important observation is that the instantaneous change covers two modes: it is initiated in the *high-stuck* mode and terminates in the *low-stuck* mode. This exemplifies that the instantaneous change may exit the operational area. In general, the point in the state space where an instantaneous change leaves the operational area indicated by the open circle in Fig. 12(b), is difficult to determine.

The simplification in Fig. 12 illustrates another important element of hybrid dynamic systems, the *admissible space*. Note that, in general, a system of differential and algebraic equations may contain variables that are operated on by a time differentation, but that are not state variables. These variables are sometimes referred to as *generalized state variables* [24]. Referring to Fig. 12 of the power window example, even though the window position and velocity are two generalized state variables, the window velocity is required to be 0, leaving only one degree of freedom, or state, for the dynamic behavior, i.e., the window position.

The space that represents the degree of freedom is called the *admissible space*. In Fig. 12 it is the line at which the velocity is 0, indicated by the thick line. In Fig. 11, the admissible space is the entire state space, indicated by the thick border.

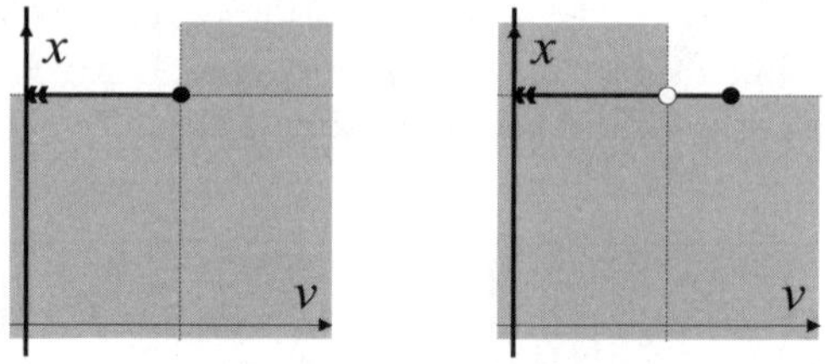

(a) Low-stuck (b) High-stuck

Fig. 12. Abstraction classes for endstop models

4.3 Mode Transition Sequences

In Fig. 12(b) the admissible space lies outside of the operational area, and thus this mode has to be departed immediately when it is reached. In general, an important characteristic of hybrid dynamic systems is that one mode change may immediately be followed by another mode change without any continuously evolving behavior in between. This is illustrated by the scenario in Fig. 13. Once the window reaches the top of the door frame, it changes from the *high-free* mode to the *high-stuck* mode. Before another mode of continuously evolving behavior is arrived at, a consecutive mode change moves the hybrid dynamic system into the *low-stuck* mode.

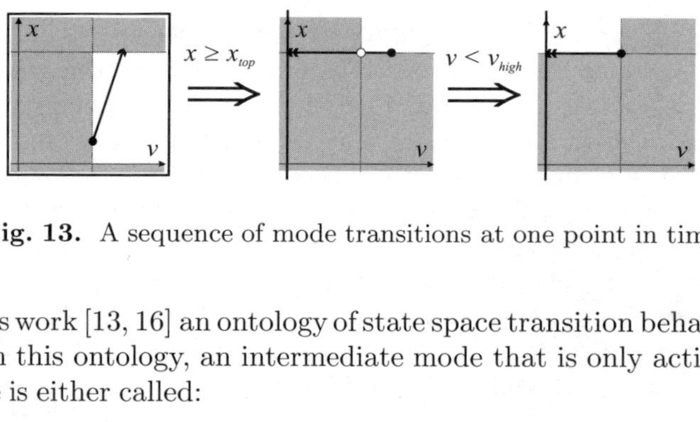

Fig. 13. A sequence of mode transitions at one point in time

In previous work [13, 16] an ontology of state space transition behavior has been developed. In this ontology, an intermediate mode that is only active at a given point in time is either called:

- A *pinnacle*, which causes a change in the state. This situation happens when the admissible space is outside of the operational area, and the mode is entered with a state outside of the admissible space.
- A *mythical mode*, which has no effect on the state. This situation happens when the mode is entered with a state outside of the operational area and within the admissible space.

To support computational simulation, these different classes of behavior have to be properly handled. Details on approaches and algorithms are discussed elsewhere [11, 14].

5 Conclusions

Model-Based Design is increasingly adopted in industry to aid in the design of engineered systems. The use of computational models offers a variety of advantages over the use of paper documents and physical prototypes. An important aspect of computational models is that they typically can be executed so the behavior of a design can be studied by means of simulation.

This paper has given an overview of Model-Based Design and introduced some of the benefits that can be derived from it. A concrete example has been given by illustrating elements of the design of a power window control system. This example motivated the need to support widely differing formalisms such as state transition diagrams, time-based block diagrams, entity-flow networks, and multibody diagrams.

The execution semantics of each of these formalisms are very different and require different technology for simulation. A general classification can be made in terms of behavior that is continuous in time and behavior that may be discontinuous. Combining formalisms with elements in both classes leads to *hybrid dynamic systems*.

The basic elements of a hybrid system and an overview of hybrid dynamic system behavior in geometrical terms was given.

Acknowledgments

MATLAB, Simulink, Stateflow, Handle Graphics, Real-Time Workshop, and xPC TargetBox are registered trademarks and SimBiology, SimEvents, and SimHydraulics are trademarks of The MathWorks, Inc. Other product or brand names are trademarks or registered trademarks of their respective holders.

References

[1] Paul Barnard. Graphical techniques for aircraft dynamic model development. In *AIAA Modeling and Simulation Technologies Conference and Exhibit*. Providence, Rhode Island, August 2004. CD-ROM.

[2] Maria Domenica Di Benedetto and Alberto L. Sangiovanni- Vincentelli, editors. Hybrid Systems: Computation and Control, volume 2034 of Lecture Notes in Computer Science. Springer- Verlag, March 2001.

[3] Randy Brown. Calendar queues: A fast 0(1) priority queue im- plementation for the simulation event set problem. Communi- cations of the ACM, 31(10):12201227, 1988.

[4] CAN specification. Technical Report, 1991. Robert Bosch GmbH.

[5] F.E. Cellier, H. Elmqvist, and M. Otter. Modelling from physical principles. In W.S. Levine, editor, The Control Handbook, pages 99107. CRC Press, Boca Raton, FL, 1996.

[6] John Guckenheimer and Stewart Johnson. Planar hybrid sys- tems. In Panos Antsaklis, Wolf Kohn, Anil Nerode, and Shankar Sastry, editors, Hybrid Systems II, volume 999, pages 202225. Springer-Verlag, 1995. Lecture Notes in Computer Science.

[7] David Harel. Statecharts: A visual formalism for complex sys- tems. Science of Computer Programming, 8:231274, 1987.

[8] D.C. Karnopp, D.L. Margolis, and R.C. Rosenberg. Systems Dynamics: A Unified Approach. John Wiley and Sons, New York, 2 edition, 1990.

[9] Zvi Kohavi. Switching and Finite Automata Theory. McGraw- Hill, Inc., New York, 1978.

[10] Nancy Lynch and Bruce Krogh, editors. Hybrid Systems: Com- putation and Control, volume 1790 of Lecture Notes in Com- puter Science. Springer-Verlag, March 2000.

[11] Pieter J. Mosterman. An overview of hybrid simulation phe- nomena and their support by simulation packages. In Frits W. Vaandrager and Jan H. van Schuppen, editors, Hybrid Systems: Computation and Control, volume 1569, pages 164177. Lecture Notes in Computer Science; Springer-Verlag, March 1999.

[12] Pieter J. Mosterman. HyBrSim a modeling and simulation environment for hybrid bond graphs. Journal of Systems and Control Engineering, 216:3546, 2002. special issue paper.

[13] Pieter J. Mosterman. Mode transition behavior in hybrid dy- namic systems. In Proceedings of the 2003 Winter Simulation Conference, pages 623631, New Orleans, LA, December 2003. invited paper.

[14] Pieter J. Mosterman. Hybrid dynamic systems: Modeling and execution. In Paul A. Fishwick, editor, Handbook of Dynamic System Modeling, chapter 15, pages 15-115-23. CRC Press LLC, Boca Raton, FL, 2007.

[15] Pieter J. Mosterman and Gautam Biswas. A hybrid modeling and simulation methodology for dynamic physical systems. SIM- ULATION: Transactions of The Society for Modeling and Sim- ulation International, 178(1):517, January 2002.

[16] Pieter J. Mosterman, Feng Zhao, and Gautam Biswas. An on- tology for transitions in physical dynamic systems. In AAAI98, pages 219224, July 1998.

[17] SimEvents. SimEvents Users Guide. The MathWorks, Natick, MA, March 2006.

[18] SimMechanics. SimMechanics Users Guide. The MathWorks, Natick, MA, March 2006.

[19] SimPowerSystems. SimPowerSystems Users Guide. The Math- Works, Natick, MA, March 2006.

[20] Simulink. Using Simulink. The MathWorks, Inc., Natick, MA, March 2006.

[21] SolidWorks. Introducing SolidWorks. SolidWorks Corporation, Concord, MA, 2002.

[22] Stateflow. Stateflow Users Guide. The MathWorks, Natick, MA, March 2006.

[23] Frits W. Vaandrager and Jan H. van Schuppen, editors. Hy- brid Systems: Computation and Control, volume 1569 of Lecture Notes in Computer Science. Springer-Verlag, March 1999.

[24] George C. Verghese, Bernard C. Levy, and Thomas Kailath. A generalized state-space for singular systems. IEEE Transactions on Automatic Control, 26(4):811831, August 1981.

[25] Giles D. Wood and Dallas C. Kennedy. Simulating mechani- cal systems in simulink with simmechanics. Technical Report 91124v00, The MathWorks, Inc., Natick, MA, 2003.

Part I

Soft Computing in Computer Graphics, Imaging and Vision

Object Recognition Using Particle Swarm Optimization on Fourier Descriptors

Muhammad Sarfraz and Ali Taleb Ali Al-Awami

[1] Department of Information and Computer Science, King Fahd University of Petroleum and Minerals, Dhahran 31261, Saudi Arabia
sarfraz@kfupm.edu.sa

[2] Department of Electrical Engineering, King Fahd University of Petroleum and Minerals, Dhahran, 31261 Saudi Arabia
aliawami@kfupm.edu.sa

Abstract. This work presents study and experimentation for object recognition when isolated objects are under discussion. The circumstances of similarity transformations, presence of noise, and occlusion have been included as the part of the study. For simplicity, instead of objects, outlines of the objects have been used for the whole process of the recognition. Fourier Descriptors have been used as features of the objects. From the analysis and results using Fourier Descriptors, the following questions arise: What is the optimum number of descriptors to be used? Are these descriptors of equal importance? To answer these questions, the problem of selecting the best descriptors has been formulated as an optimization problem. Particle Swarm Optimization technique has been mapped and used successfully to have an object recognition system using minimal number of Fourier Descriptors. The proposed method assigns, for each of these descriptors, a weighting factor that reflects the relative importance of that descriptor.

Keywords: curve fitting, NURBS, approximation, simulated evolution, algorithm.

1 Introduction

Fourier descriptors [1, 2, 14], like Moment descriptors [9], have been frequently used as features for image processing, remote sensing, shape recognition and classification. Fourier Descriptors can provide characteristics of an object that uniquely represent its shape. Several techniques have been developed that derive invariant features from Fourier Descriptors for object recognition and representation [1-5, 14]. These techniques are distinguished by their definition, such as the type of data exploited and the method for deriving invariant values from the image Fourier Descriptors.

Granlund [1] introduced Fourier descriptors using complex representation in 1972. This method ensures that a closed curve will correspond to any set of descriptors. The Fourier descriptors have useful properties [3, 4]. They are invariant under similarity transformations like translation, scaling and rotation. The objects having these kind of transformations can be easily recognized using some recognition algorithms with Fourier descriptors as invariant features. For example, the Fourier descriptors, of the boundary [11-13], for recognizing closed contours is proposed in [5]. However,

A. Saad et al. (Eds.): Soft Computing in Industrial Applications, ASC 39, pp. 19–29, 2007.
springerlink.com

despite its success in some applications, it has certain limitations. Occlusion is the severe shape distortion, when the shape gets distorted, the Fourier descriptors don't work well for recognition [6-8].

This paper has used Fourier descriptors, with different combinations, for the recognition of objects captured by an imaging system which may transform, make noise or can have occlusion in the images. An extensive experimental study, similar to the moment invariants [9], has been made using various similarity measures in the process of recognition. These measures include Euclidean Measure and Percentage error. Comparative study of various cases has provided very interesting observations which may be quite useful for the researchers as well as practitioners working for imaging and computer vision problem solving. Although the whole study has been made for bitmap images, but it can be easily extended to gray level images.

From the analysis and results using Fourier Descriptors, the following questions arise: What is the optimum number of descriptors to be used? Are these descriptors of equal importance? To answer these questions, the problem of selecting the best descriptors has been formulated as an optimization problem. Particle Swarm Optimization technique has been mapped and used successfully to have an object recognition system using minimal number of Fourier Descriptors. The goal of the proposed optimization technique is to select the most helpful descriptors that will maximize the recognition rate. The proposed method will assign, for each of these descriptors, a weighting factor that reflects the relative importance of that descriptor.

The outline of the remainder of the paper is as follows. Getting of bitmap images and their outline is discussed in Sections 2 and 3 respectively. Section 4 deals with the study of Fourier descriptors. The concepts of similarity measures are explained in Section 5. Algorithm for object recognition problem has been devised in Section 6. Detailed experimental study and analyses are made in Section 7 whereas Section 8 deals with interesting observations during the experimental study. Finally, Section 9 concludes the paper as well as touches some future work.

2 Fourier Theory

To characterize objects we use features that remain invariant to translation, rotation and small modification of the object's aspect. The invariant Fourier descriptors of the boundary [11-13] of the object can be used to identify an input shape, independent on the position or size of the shape in the image.

Fourier transform theory has played a major role in image processing for many years. It is a commonly used tool in all types of signal processing and is defined both for one and two-dimensional functions. In the scope of this research, the Fourier transform technique is used for shape description in the form of Fourier descriptors. The shape descriptors generated from the Fourier coefficients numerically describe shapes and are normalized to make them independent of translation, scale and rotation.

The Fourier transform theory can be applied in different ways for shape description. In this research, the procedure has been implemented in such a way that the boundary of the image is treated as lying in the complex plane. So the row and column co-ordinates of each point on the boundary can be expressed as a complex number. For details, the reader is referred to [5, 14].

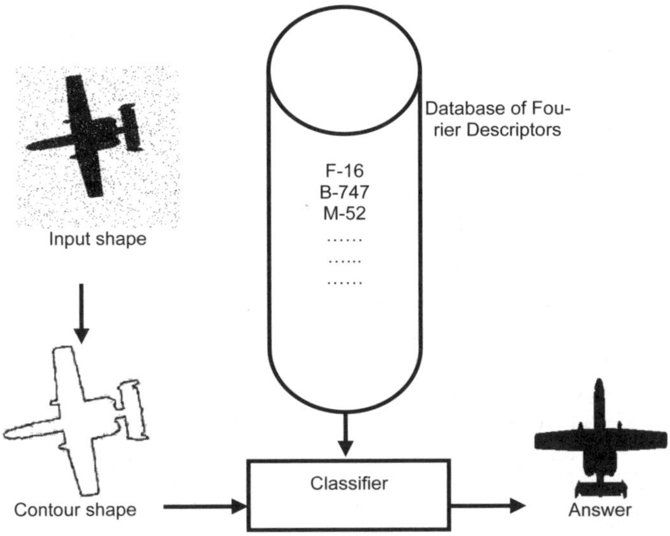

Fig. 1. Pictorial Description of the method

3 Similarity Measures

This paper implements two different simple classifiers that calculate different similarity measures of the corresponding Fourier descriptors of the input shape and each of the shapes contained in the database as shown in Figure 1. The similarity measures, attempted for experimental studies, are Euclidean Distance (ED) and Percentage Error (PE).

Given two sets of descriptors, how do we measure their degree of similarity? An appropriate classification is necessary if unknown shapes are to be compared to a library of known shapes. If two shapes, A and B, produce a set of values represented by $a(i)$ and $b(i)$ then the distance between them can be given as $c(i) = a(i) - b(i)$. If $a(i)$ and $b(i)$ are identical then $c(i)$ will be zero. If they are different then the magnitudes of the components in $c(i)$ will give a reasonable measure of the difference. It proves more convenient to have one value to represent this rather than the set of values that make up $c(i)$. The easiest way is to treat $c(i)$ as a vector in a multi-dimensional space, in which case its length, which represents the distance between the objects, is given by the square root of the sum of the squares of the elements of $c(i)$.

The similarity measures, attempted for experimental studies, are as follows:

1. $\sqrt{\sum_{i=1}^{n}\left(a(i)-b(i)\right)^2}$ (Euclidean Distance (ED))
2. $\sum_{i=1}^{n}\left|\frac{a(i)}{b(i)}\right|$ (Percentage Error (PE))

In this study, n is the number of FDs considered, $a(i)$ is the ith FD of the template image, and $b(i)$ is the ith FD of the test image. A tolerable threshold ρ is selected to decide a test object recognized. This threshold is checked against the least value of the selected similarity measure.

4 Results and Analysis

The recognition system is tested by generating the test objects by translating, rotating, scaling, adding noise, and adding occlusion to the model objects contained in a database of different sizes. The test objects were randomly rotated, translated, and scaled. Some were considered without scale of their model sizes. About 100 test objects were used for each of the experiments for testing similarity transformation.

The salt & pepper noise [15-16] of different densities is added to the objects for generating the noisy test objects. Median filter was used in the experiment to filter the noise, so that the noise remains on the boundary of the object. Median filtering is a type of neighborhood processing that is particularly useful for removing 'salt and pepper' noise from an image. The median filter [15-17] considers each pixel in the image and it looks at its nearby neighbors to decide whether or not it is representative of its surroundings. Instead of simply replacing the pixel value with the *mean* of neighboring pixel values, it replaces it with the *median* of those values. The median is calculated by first sorting all the pixel values from the surrounding neighborhood into numerical order and then replacing the pixel being considered with the middle pixel value.

As would be seen in the experiments, FDs are not promising for the recognition of occluded objects. Around 20% occlusion was added into the objects to make tests. We split the experiments into different categories explained in the following paragraphs.

The first series of experiments has been made to view results for different combinations of the Fourier Descriptors. Various experiments can be seen presenting different scenarios of the combination of Fourier Descriptors, similarity measures, and nature of data used.

The recognition system is tested by generating the test objects by translating, rotating, and scaling and adding noise to the model objects contained in a database of size 60. The test objects were randomly rotated, scaled and translated. Sixty test objects were used for each of the experiments for testing similarity transformation, 16 test objects were used for noisy objects with similarity transformations, and 60 test objects were used for occluded objects. The salt & pepper noise of density 10% is added to the objects for generating the noisy test objects. Median filter was used in the experiment to filter the noise, so that the noise remains on the boundary of the object.

The procedures taken to analyze and test the system are as follows:

1. **The base case:** That is, the Fourier descriptors FD 1-6 (highlighted in Table 1) are used as features and the Euclidean distance is considered for comparison. The percentage of recognition recorded in case of just similarity transformations is 83.3%. In case of similarity transformations with noise, it is about 93.75%. It is worth noting that in the latter case, only translation is considered as a similarity transformation, i.e., the test images are not rotated or scaled. The recognition rate of occluded objects is only 8.3%, which is very low. The Fourier Descriptors for an object in

case of similarity transformations, noise and occlusion are experimented. It was found, by computation, that the Fourier descriptors do not change much in case of similarity transformations and noise. However, the occlusion caused the change in the values of descriptors.

Table 1. Recognition rates for different numbers of Fourier descriptors using Euclidean distance

Number of FDs Used	**4**	**6 Base Case**	**11**	**18**	**22**	**29**
Transformations	71.67%	83.33%	93.33%	90%	93.33%	95%
Noise	75%	93.75%	93.75%	93.75%	93.75%	93.75%
Occlusion	5%	8.33%	20%	18.33%	23.33%	23.33%

2. **Fourier Descriptors and Euclidean distance:** Experiments are made to obtain the recognition rates of transformed, noisy, or occluded images considering different numbers of FDs using Euclidean distance. The recognition rates for using different numbers of FDs, ranging from 1 to 40, are computed. Some sample results are tabulated in Table 1. For example, using 11 FDs improves the recognition rate of transformed images to 93.33% and improves the recognition rate of occluded images to 20%.

   From these results, it can be concluded that a good compromise between recognition performance and computational cost is achieved using 11 FDs. That is, increasing the number of FDs beyond 11 does not help much as the maximum recognition rate obtained for transformed images using up to 40 FDs is 95%. Another observation is that the maximum recognition rate is achieved by using 29 FDs. Thus, using more FDs does not improve the recognition performance further.

Table 2. Recognition rates for different numbers of Fourier descriptors using percentage of errors

Number of FDs Used	**4**	**6 Base Case**	**9**	**16**	**22**	**29**
Transformations	70%	80%	86.67%	75%	68.33%	68.33%
Noise	87.5%	81.25%	81.25%	81.25%	81.25%	81.25%
Occlusion	8.33%	11.67%	13.33%	8.33%	6.67%	11.67%

3. **Fourier Descriptors and percentage of errors:** Similarly, the recognition rates of transformed, noisy, or occluded images using the sum of percentage of error (PE) have been obtained. The recognition rates for using different numbers of FDs, ranging from 1 to 40, are summarized. Some of these numerical results are tabulated in Table 2. It can be seen that, using PE with FDs results in less efficient performance than using ED. Moreover, increasing the number of FDs does not

necessarily guarantee a better performance. From Table 2, it can be observed that the best recognition rate of transformed images is achieved using 9 FDs. However, larger number of FDs gives rise to a lower performance.

5 Optimization of the Feature Vector Using PSO

The problem of selecting the best descriptors can be formulated as an optimization problem. The goal of the optimization is to select the most helpful descriptors that will maximize the recognition rate and assign for each of these descriptors a weighting factor that reflects the relative importance of that descriptor.

Since the problem of selecting the best descriptors can be formulated as an optimization problem, one need to define an objective function. The objective function, in this case, is made up of the following two terms:

- the recognition rate,
- the number of useful descriptors.

In other words, it is required to maximize the recognition rate using the minimum number of descriptors. The function of the optimization algorithm is to assign a weight w_i for every descriptor so that the objective function is optimized, where w_i belongs to [0,1].

The mathematical formulation of the objective function is defined as follows:

$$J = -H + \alpha * \min(PE),$$

where

- H is the number of hits (the number of correct matches),
- PE is the percentage of errors of all the training images for a given set of weights,
- α is a factor that is adjusted according to the relative importance of the min(PE) term.

In most of the simulations, $\alpha = 0.7$ is experienced as best case. The first term makes PSO search for the weights that result in the highest recognition rate and the second term makes the PSO reach to the highest recognition with the minimum number of descriptors.

5.1 Particle Swarm Optimization

A novel population based optimization approach, called Particle Swarm Optimization (PSO) approach, has been used. PSO was introduced first in 1995 by Eberhart and Kennedy [18]. This new approach features many advantages; it is simple, fast and can be coded in few lines. Also, its storage requirement is minimal.

Moreover, this approach is advantageous over evolutionary algorithms in more than one way. First, PSO has memory. That is, every particle remembers its best solution (local best) as well as the group best solution (global best). Another advantage of PSO is that the initial population of the PSO is maintained, and so there is no need for applying operators to the population, a process which is time- and memory-storage-consuming. In addition, PSO is based on "constructive cooperation" between particles, in contrast with the other artificial algorithms which are based on "the survival of the fittest."[19]

PSO starts with a population of random solutions. Each particle keeps track of its coordinates in hyperspace which are associated with the fittest solution it has achieved so far. The value of the fitness (*pbest*) is stored. Another best value is also tracked, which is the global best value. The global version of the PSO keeps track of the overall best value, and its location, obtained thus far by any particle in the population, which is called (*gbest*).

PSO consists of, at each step, changing the velocity of each particle toward its pbest and gbest. Acceleration is weighted by a random term, with separate random numbers being generated for acceleration toward *pbest* and *gbest*.

Several modifications have been proposed in the literature to improve the PSO algorithm speed and convergence toward the global minimum. One of the most efficient PSO versions uses a time decreasing inertia weight, which leads to a reduction in the number of iterations. The performance of this modified algorithm depends on the method of tuning the inertia weight. The most straightforward way to adjust the inertia weight is to force it to linearly decrease with the number of iterations [20]. As an alternative to inertia weight technique, constriction factors have been proposed [21]. Constriction-factor-based PSO has been proved to be superior upon inertia-weight-based PSO in terms of the number of iterations to reach the best solution.

PSO starts with a population of random solutions "particles" in a D-dimension space. The ith particle is represented by:

$$X_i = (x_{1i}, x_{i2}, \ldots, x_{iD}). \tag{1}$$

Each particle keeps track of its coordinates in hyperspace which are associated with the fittest solution it has achieved so far. The value of the fitness for particle *i* (*pbest*) is also stored as:

$$P_i = (p_{1i}, p_{i2}, \ldots, p_{iD}). \tag{2}$$

The global version of the PSO keeps track of the overall best value (*gbest*), and its location, obtained thus far by any particle in the population.

PSO consists of, at each step, changing the velocity of each particle toward its *pbest* and *gbest* according to equation (4). The velocity of particle *i* is represented as:

$$V_i = (v_{1i}, v_{i2}, \ldots, v_{iD}). \tag{3}$$

Acceleration is weighted by a random term *w*, with separate random numbers being generated for acceleration toward pbest and gbest. The position of the ith particle is then updated according to equation (5).

$$v_{id} = w * v_{id} + c_1 * Rand(\ ) * (p_{id} - x_{id}) + c_1 * Rand(\ ) * (p_{gd} - x_{id}) \tag{4}$$

$$x_{id} = x_{id} + v_{id} \tag{5}$$

where

$$p_{id} = pbest \text{ and } p_{gd} = gbest$$

A simplified method of incorporating a constriction factor is represented in equation (6), where K is a function of c_1 and c_2 as illustrated by equation (7). Eberhart and Shi [21] experimented the performance of PSO using an inertia weight as compared with PSO performance using a constriction factor. They concluded that the best approach is to use a constriction factor while limiting the maximum velocity V_{max} to the dynamic range of the variable X_{max} on each dimension. They showed that this approach provides performance superior to any other published results.

$$v_{id} = K * \left[v_{id} + c_1 * Rand(\ ) * (p_{id} - x_{id}) + c_1 * Rand(\ ) * (p_{gd} - x_{id}) \right], \qquad (6)$$

$$K = \frac{kk}{\left| 2 - \phi - \sqrt{\varphi^2 - 4\phi} \right|}; \ kk = 2,\ \phi = c_1 + c_2,\ \phi > 4. \qquad (7)$$

5.2 Test Results Using PSO

The PSO algorithm can be found in the current literature at many places. Therefore, it is not necessary to describe the steps of the PSO algorithm here. However, the reader is referred to [18-19] if interested to see the PSO algorithm details.

The proposed PSO–based approach was implemented using a MATLAB library built by the authors. In our implementation, the inertia weight w is linearly decreasing from 0.9 to 0.4, c_1 and c_2 are selected as 2, a population of 20 particles is used, and the maximum number of iterations is 400. The search process stops whenever the maximum number of iterations is exceeded.

Table 3. Optimized weights for different numbers of Fourier descriptors

Experiment No.	**1**	**2**	**3**	**4**	**5**
Training set*	X	X	X	O	X, O, N
No. of FDs considered	11	11	6	6	11
Optimized Weights obtained	0.0009 0.0035 0.0024 0.0328 0.0000 0.0156 0.0394 0.0651 0.0000 0.0000 0.0000	0.1033 0.4004 0.4368 0.5681 0.6245 0.9140 0.8876 0.2022 0.3879 0.4992 0.5281	0.0005 0.0048 0.0041 0.0079 0.0296 0.0152	0.0002 0.0020 0.0009 0.0015 0.0001 0.0036	0.0143 0.0195 0.0138 0.1742 0.1083 0.3051 0.2216 0 0.1515 0.5409 0.2698

Several experiments have been attempted to use PSO to search for the optimum descriptor weights. These experiments are summarized in Tables 3 and 4. In these tables, "No. of FDs considered" means the number of Fourier descriptors considered in the optimization process. For example, if this number is *F*, the PSO is supposed to search for *F* weights, a single weight for a single FD, that maximize the recognition rate with the minimum number of descriptors. Hence, if only *f* FDs can do the job, PSO will assign nonzero weights for *f* FDs and it will assign zero weights for $F - f$ FDs. As a result, the number of descriptors used in the recognition process is only *f* out of *F*.

Table 4. Total number of optimized weights used for different numbers of Fourier descriptors and the recognition rate

Experiment No.		**1**	**2**	**3**	**4**	**5**
Training set*		X	X	X	O	X, O, N
No. of FDs Considered		11	11	6	6	11
No. of FDs Used		7	11	6	6	10
Recognition Rate	**X**	93.33%	95%	95%	90%	98.33%
	N	93.75%	93.75%	87.5	87.5%	87.5%
	O	25%	23.33	20%	20%	25%

*X = transformed objects, O = occluded objects, N = noisy objects

Table 3 demonstrates the computed optimized weights for different numbers of Fourier descriptors. Table 4 displays about the total number of optimized weights used for different numbers of Fourier descriptors and the recognition rate achieved. In the first experiment when a database of 60 transformed objects, in Tables 3 and 4, was considered, one can see a much better recognition results than using even 7 un-weighted FDs in Table 2.

In the second experiment, the results are better than using 11 un-weighted FDs. One can see, in Table 2, a similar recognition rate is obtained using 29 un-weighted FDs.

Experiment 3, when considered for 6FDs, shows generally, much better results than using 6 un-weighted FDs. However, in experiment 4, using occluded images only in the training, does not help improve the recognition rate of occluded images.

Experiment 5 for a mixed data set of transformed, noisy and occluded objects has produced wonderful results of 98.33% recognition rate. Such an achievement has not been obtained by any number of un-weighted FDs.

6 Conclusion and Future Work

This work has been reported to make a practical study of the Fourier descriptors to the application of Object Recognition. The implementation was done on a P-IV PC using MATLAB. The ultimate results have variations depending upon the selection of number of FDs, similarity transformations, noise, occlusion, and data size. The variety of similarity measures and different combinations of FD features, used in the process, make a difference to the recognition rate. The results have been tested using up to 40 FDs, and different size of databases. Different combinations of these parameters implied different results.

Two similarity measures, including ED, and PE, provided different recognition results. The images used are all bitmapped images, further investigations are being done with some more complex images.

It can be seen that, using PE with FDs results in less efficient performance than using ED. Moreover, increasing the number of FDs does not necessarily guarantee a better performance.

The images that have to be recognized but failed to be recognized by most of the FD combinations are to be analyzed further. This leads to the theory of optimization to find out appropriate features or attributes in the image that made it difficult to be recognized. The methodology of PSO has been utilized successfully for this purpose. Using PSO, to find the most suitable descriptors and to assign weights for these descriptors, improved dramatically the recognition rate using the least number of descriptors.

In future, the authors would like to treat the problem as multi-objective optimization method. They would like to maximize the recognition rate while minimizing the number of descriptors.

Acknowledgments

The authors are thankful to the anonymous referees for the valuable suggestions towards the improvement of this manuscript. This work has been supported by the King Fahd University of Petroleum and Minerals against the Project # ICS/OPTICAL TEST/337.

References

1. G. H. Granlund, Fourier Preprocessing for hand print character recognition, IEEE Trans. Computers, Vol C-21, Febr. 1972, pp. 195-201.
2. A Project led by Julien Boeuf and Pascal Belin, and supervised by Henri Maître: http://www.tsi.enst.fr/tsi/enseignement/ressources/mti/descript_fourier/index.html.
3. O. Betrand, R. Queval, H. Maître, Shape Interpolation by Fourier Descriptors with Application to Animation Graphics, Signal Processing, June 1981, 4:53-58.
4. H. Maître, Le traitement des images, ENST, December 2000, pp. 70-72.
5. C.T. Zahn, R.Z. Rhoskies, Fourier descriptors for plane closed curves, IEEE trans. Compu. 21 (1972) 269-281.

6. Thomas Bernier, Jacques-Andre landry, A new method for representing and matching shapes of natural objects, Pattern Recognition 36 (2003), 1711-1723.
7. N. Ansari, E.J. Delp, Partial Shape Recognition: a landmark based approach, IEEE Trans. PAMI 12 (1990), 470-183.
8. J. Zhang, X. Zhang, H. Krim, G.G. Walter, Object representation and recognition in shape spaces, Pattern Recognition 36(5), 2003, pp. 1143-1154.
9. M. Sarfraz, Object Recognition using Moments: Object Recognition using Moments: Some Experiments and Observations: Geometric Modeling and Imaging – New Advances, Sarfraz, M. and Banissi, E. (Eds.), ISBN-10: 0-7695-2604-7, IEEE Computer Society, USA, 2006, pp. 189-194.
10. John W. Gorman, O. Robert Mitchell, Frank P. Kuhl, Partial shape recognition using dynamic programming, IEEE Transactions on pattern analysis and machine intelligence, Vol.10(2), March 1988.
11. G. Avrahami and V. Pratt. Sub-pixel edge detection in character digitization. Raster Imaging and Digital Typography II, pp. 54-64, 1991.
12. Hou Z. J., Wei G. W., A new approach to edge detection, Pattern Recognition Vol. 35, pp. 1559-1570, 2002.
13. N. Richard, T. Gilbert, Extraction of Dominant Points by estimation of the contour fluctuations, Pattern Recognition Vol. 35, pp. 1447-1462, 2002.
14. M. Sarfraz, Object Recognition using Fourier Descriptors: Some Experiments and Observations: Computer Graphics, Imaging and Visualization – Techniques and Applications, Banissi, E., Sarfraz, M., Huang, M. L., and Wu, Q. (Eds.), ISBN: 0-7695-2606-3, IEEE Computer Society, USA, 2006, pp. 281-286.
15. Rafael Gonzalez, Richard Woods and Steven Eddins, Digital Image Processing Using Matlab, Prentice Hall, 2003.
16. R. Jain, R. Kasturi, B. Schunk, Machine Vision, McGraw Hill, 1995.
17. http://www.cee.hw.ac.uk/hipr/html/median.html.
18. J. Kennedy, R. Eberhart, Particle swarm optimization, Proc. IEEE Intl. Conf. Neural Networks, 4, Nov/Dec 1995, pp. 1942 –1948.
19. R. Eberhart and J. Kennedy, A new optimizer using particle swarm theory, Proc. the Sixth Intl. Symposium on Micro Machine and Human Science, MHS '95, 4-6 Oct 1995, pp. 39 -43.
20. Y. Shi, R. Eberhart, A modified particle swarm optimizer, The 1998 IEEE Intl. Conf. on Evolutionary Computation Proc., IEEE World Congress on Computational Intelligence, 4-9 May 1998, pp. 69 – 73.
21. R.C. Eberhart, Y. Shi, Comparing inertia weights and constriction factors in particle swarm optimization, Proceedings of the 2000 Congress on Evolutionary Computation, Vol. 1, 16-19 July 2000, pp. 84 -88.

Gestix: A Doctor-Computer Sterile Gesture Interface for Dynamic Environments

Juan Wachs[1], Helman Stern[1], Yael Edan[1],
Michael Gillam[2], Craig Feied[2], Mark Smith[2], and Jon Handler[2]

[1] Department of Industrial Engineering and Management, Ben-Gurion University of the Negev, Be'er-Sheva, Israel, 84105
{juan, helman, yael}@bgu.ac.il
[2] Institute for Medical Informatics, Washington Hospital Center, 110 Irving Street, NW, Washington, DC, 20010
{feied,smith,handler,gillam}@medstar.net

Abstract. In this paper, we design a sterile gesture interface for users, such as doctors/surgeons, to browse medical images in a dynamic medical environment. A vision-based gesture capture system interprets user's gestures in real-time to navigate through and manipulate an image and data visualization environment. Dynamic navigation gestures are translated to commands based on their relative positions on the screen. The gesture system relies on tracking of the user's hand based on color-motion cues. A state machine switches from navigation gestures to others such as zoom and rotate. A prototype of the gesture interface was tested in an operating room by neurosurgeons conducting a live operation. Surgeon's feedback was very positive.

Keywords: hand gesture recognition, medical databases, browsing, image visualization, sterile interface.

1 Introduction

Computer information technology is increasingly penetrating into the hospital domain. It is important that such technology be used in a safe manner to avoid serious mistakes leading to possible fatal incidents. Keyboards and mice are today's principle method of human – computer interaction (HCI). Unfortunately, it has been found that a common method of spreading infection involves computer keyboards and mice in intensive care units (ICUs) used by doctors and nurses [1]. Kiosks using touch screens [2] introduced recently into hospitals, to provide patient information, bring no guarantee to stop the spread of bacteria (such as an outbreak of SARS). When an epidemic crisis erupts access to information is absolutely critical, and kiosk users may forego the washing of hands in the interest of speed.

By the early 1990's scientists, surgeons and other experts were beginning to draw together state of the art technologies to develop comprehensive frameless image-guidance systems for surgery, such as the StealthStation [3]. This is a free-hand stereotactic pointing device, which transmits its position via attached light emitting diodes (LEDs), and converts this position in to the corresponding location in the image space of a high-performance computer monitor. Also, touch-screens are a

A. Saad et al. (Eds.): Soft Computing in Industrial Applications, ASC 39, pp. 30–39, 2007.
springerlink.com

popular means of interaction. As in traditional POS (point of sale) environments, one style of touch screen does not work in all healthcare environments. In a hospital, different departments will insist on different touch screen characteristics. Medical offices want large screens, with large buttons, to help reduce training time [4]. In a setting like an operating room (OR), touch screen displays must be sealed to prevent the buildup of contaminants, and should also have smooth surfaces for easy cleaning with common cleaning solutions.

Many of these deficiencies may be overcome by introducing a more natural human computer interaction (HCI) mode into the hospital environment. The basis of human-human communication is speech and gesture including facial expression, hand and body gestures and eye gaze. Some of this concepts where exploited in systems for improving medical procedures and systems. In FAce MOUSe [5], a surgeon can control the motion of the laparoscope by face gestures. Hand gestures for mouse functions for doctor-computer interfaces appeared in Graetzel et al. [6]. Zeng et al. [7] use finger position to obtain data on breast palpations. Other systems [8] suggest a teleoperated robotic arm using hand gestures for multipurpose tasks. Wheelchairs are guided by hand gestures in [9]. In [10] a Gesture Pendant is used to control home devices. In this paper we explore only the use of hand gestures, which can in the future be further enhanced by other modalities. Gesture capture is vision based and used to manipulate windows and objects, especially images, within a graphical user interface (GUI).

In this paper we explore only the use of hand gestures. In particular we propose a doctor-computer interface system based on the recognition of gestures in sterile dynamic environments such as operation rooms. Much of the research on real-time gesture recognition has focused exclusively on dynamic or static gestures. In this work, we consider hand motion and posture simultaneously. This allows for much richer and realistic gesture representations. Our system is user independent without the need of a large multi-user training set. Operation of the gesture interface was tested in a hospital environment in real-time. In this domain the non-contact aspect of the gesture interface avoids the problem of possible transfer of contagious diseases through traditional keyboard/mice user interfaces.

System specifications, architecture and methodology are presented in Section 2, as well as, description of the Gibson 3D data browser used as our domain of application. In Section 3 image processing operations using color-motion fusion for segmentation of the hand from the background are described. Section 4 provides details of the tracking module, its mapping into navigational gestures, and state machine switching between other gestures such as zoom and rotates. Example results including implementation in a neurosurgeon operating environment is described in section 5. Final conclusions are provided in section 6.

2 System Overview

2.1 System Specifications

Some structural characteristics of a gesture interaction model for a medical environment are presented in [11], and extended for the OR domain in [6]. For the correct design of a hand gesture interaction system for doctors/surgeons, the following specifications should be considered: (1) Real-time interaction - during surgery the surgeon can watch a computer monitor to see the position of hand gesture command. (2) Fatigue - gestural commands must be concise and rapid to minimize effort. (3) Intuitiveness – gestures should be cognitively related to the command or action it represents, (4) Unintentionality - most systems capture every motion of the user's hand, and as a consequence unintentional gesture may be interpreted by the system. The system must have well-defined means to detect the correct intention of the gesture. (5) Robustness - the system should be capable to segment hand gestures from complex backgrounds containing; object motion, variable lighting and reflected color, (6) Easy to learn – doctors/ surgeons are time pressed individuals, so long training times should be avoided. (7) Unencumbered – doctors/ surgeons may wear gloves and frequently hold instruments, so additional devices attached to the hand, such as data gloves, colored or infrared markers must be avoided. The above considerations should improve computer usability. A method to include Psycho-Physiological factors in the design of a hand gesture interface is given in [12].

2.2 Architecture and Methodology

A web-camera placed above a screen (Fig. 1) captures a sequence of images of the hand. The hand is tracked by a tracking module which segments the hand from the background using color and motion cues. This is followed by black/white (BW) thresholding and various morphological image processing operations. The location of the hand in each image is represented by the 2D coordinates of its centroid. A two layer architecture is shown in Fig 2. The lower level, Gestix, provides tracking and recognition functions, while the higher level, Gibson, manages the user interface.

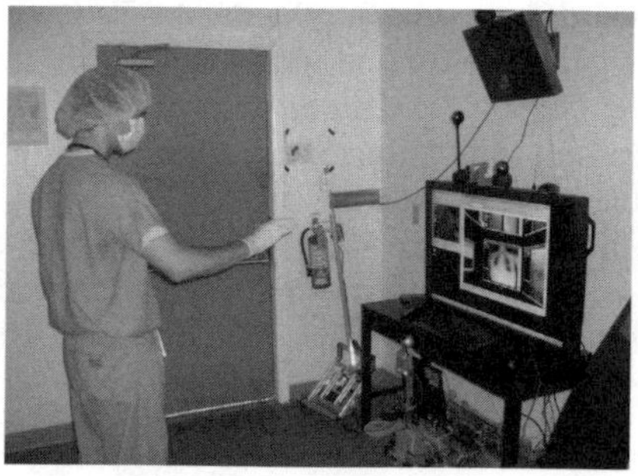

Fig. 1. Gesture capture

2.3 The Gibson Data Browser

The Gibson image browser is 3D visualization medical tool that enables examination of images, such as CT scans and X-rays. To interface the gesture recognition routines

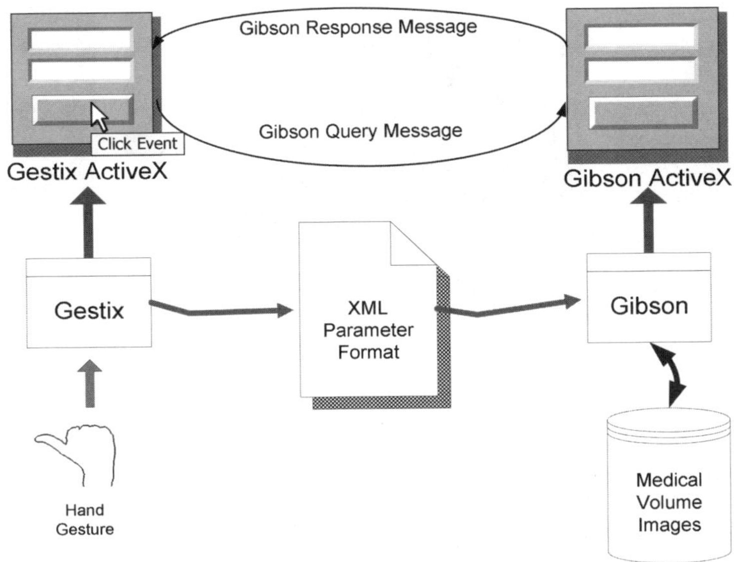

Fig. 2. Architecture of Gestix-Gibson Interface

with the Gibson system, information such as, the centroid of the hand, its size representing zoom, and orientation for rotation angle, is used to enable screen operations in the Gibson GUI. Both the gesture interface and the Gibson image browser are embedded in ActiveX controls which are communicated using messages and windows events. The messages between Gestix and Gibson are based on customized windows events (GibsonQuery and the GibsonResponse) and are intended to browse and manipulate images in the database. The current status of the hand detection module (Gestix) is sent to the Gibson control in XML format as metadata.

3 Segmentation

The CAMSHIFT [13] algorithm is used to track and recognize gestures. Within the CAMSHIFT module, a probability distribution image comprised of pixels representing hand colors is created from a 2D hue-saturation skin color histogram [14]. This histogram is used as a look-up-table to convert the acquired camera images of the hand into corresponding hand pixels, a process known as back projection, see Fig 3(a-b). The initial 2D histogram is generated in real-time by the user in the 'calibration' stage of the system, for more details see [15].

The grayscale image obtained from the RGB channels is smoothed using a Gaussian filter. The absolute difference between two consecutives images is computed, and thresholded to convert to a BW motion image Morphological operations are used clean the image, removing holes and small noise. (see Fig 3(c-d)).

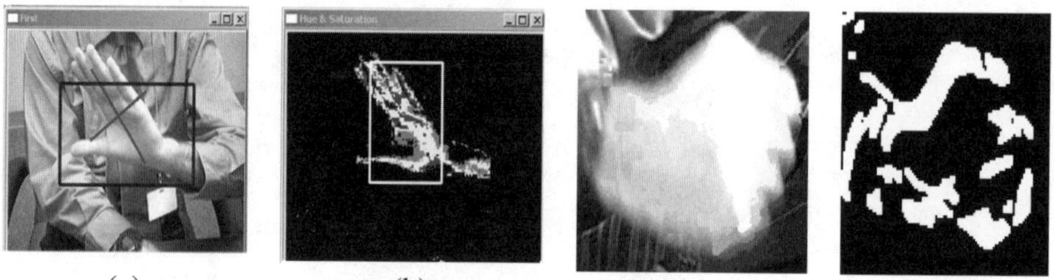

(a) (b) (c) (d)

Fig. 3. (a) User hand image. (b) Back-projected image (color cue). (c) Motion gesture. (d) Image differencing (motion cue).

As a result of the color cue we have an intensity image p_k, representing the skin color probability at frame k, and a second BW image used as a motion indicator ϕ_k, obtained from the motion cue at frame k. At frame k, I_k is the fused intensity image [16] according to (1).

$$I_k(i, j) = \alpha_k \min\{1, p_k(i, j).d\} * \phi_k(i, j) + (1 - \alpha_k) * p_k(i, j) \qquad (1)$$

Here d is an amplifying factor (d=1.3 for best performance), and α is a motion assessment variable which increases and decreases for large small amounts of motion. Motion indication reinforcement is introduced to overcome the weak ability of color only handle extreme color changes and noise from light variations. Motion only cannot be fully trusted because of the resultant halo effect, reflections and cast shadows. Also, color fusion avoids the defect of motion only, which detects not only the hand but the entire body movement.

4 Hand Tracking and Operation Modes

The finite state machine (Fig. 4) is used to illustrate the operational architecture of gesture system. Gesture operations are initiated by a calibration procedure in which a skin color model of the users hand is constructed. Control between dynamic gestures used for browsing through images and pose gestures (used for rotation and zoom) are affected by mode switch gestures. Superimposed over the image is a rectangular frame. The area inside the frame is called the "neutral area". Movements of the hand across the boundary of the rectangle constitute directional browser commands. When a doctor decides to perform a specific operation on a medical image, he/she places the hand in the ‘neutral area’ momentarily, and an attention window event is called. The spatio-temporal information and other attributes of the posture are sent to a “mode detector” to determine whether a zoom or rotation pose gesture is presented.

4.1 Directional Navigation

When a doctor/surgeon wishes to browse the image database, he/she moves the hand rapidly out from a ‘neutral area’ toward any of four directions, and then back to the neutral area. This movement is referred to as a ‘flick’ gesture.

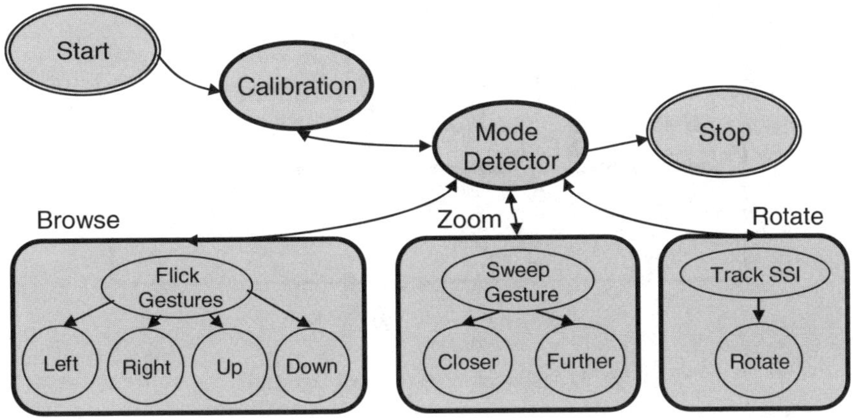

Fig. 4. State machine for the gesture-based medical browser

Interaction is designed in this way because the doctor will often have his hands in the 'neutral area' without intending to control the Gibson data browser. Only when a flick gesture is moved towards one of the four quadrants (left, right, up, down), is the image cylinder moved in the direction of the flick.

4.2 Zoom Mode

Once the zoom-mode is activated the size of image is changed according to the detected area of the hand as it moves toward and away from the screen. To go back to the normal mode, the hand is moved out from the neutral area to any of the 4 directions. The "zoom mode" is activated, when the hand is in the neutral area, by an abrupt rotation (sweep gesture) of the wrist counter clock wise, from 90° to 180°.

4.3 Rotation

The rotation operation is helpful when the doctor wants to rotate the image to a desired angle. To pursue this goal, the physician/surgeon places a sterilized straight instrument (SSI) in the fist of the hand, and holds it in the range of .25 m to 2 m from the camera. When the area of the tracking window becomes smaller than some threshold, the rotation mode is activated. When in rotation mode, the angle to which the medical image is rotated is determined by the angle between the SSI output and the horizontal axis. The architecture to detect the SSI segment in an image is shown in Fig. 5.

To quickly eliminate unlikely candidate lines far from the hand an expanded (2.5 times) window around the tracking window is cropped out of the image, I_{org}. This image is input to a canny edge detector using two thresholds t_1=50 and t_2=210, and a 3x3 mask, to obtain an edge image, I_{edge}. This is followed by the probabilistic Hough Transform, referred to as pHT [17], with distance resolution=1, angle resolution of 1°,

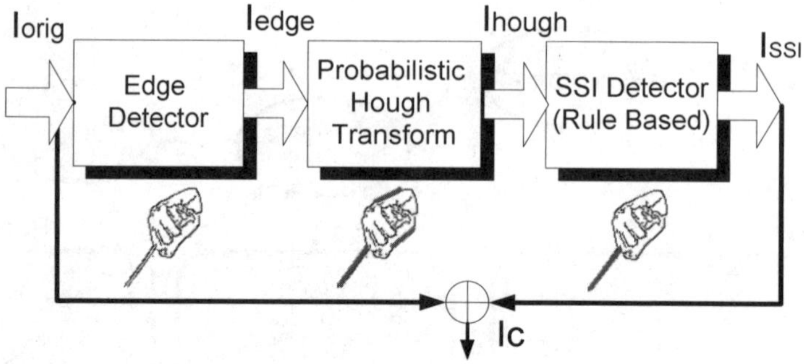

Fig. 5. Architecture of the SSI Algorithm

threshold of 30, minimum line length of 30, and the maximum allowable gap between collinear line segments of 2. The pHT algorithm finds the set of straight line segments in the image longer than some minimum. The entire set of line segments is represented in the image $I_{hough..}$ These segments are good candidates to be the sides of the SSI; however, they may also represent other straight lines such as: doors, windows, tables, wrist, etc. The best segment to represent the SSI is selected by a SSI detector which uses prior knowledge to design a set of crisp rules which is applied to the set of line segments. Using the end points of each line i segment returned from pHT. The closest end point of a line segment i to the hand is determined by finding the end point closest to the centroid of the hand tracking window. This end point is designated as the start end point $S_f(i)$, and the other as the end point, $E_f(i)$. With both endpoints identified it is possible to find the rotation angle of each line segment $\Phi_f(i)$, as measured CCW from the horizontal of the current frame f. To select among all line segments that one most likely to represent the SSI, a rule based evidence test is conducted. This rule based algorithm is described below.

SSI Detector Algorithm (Rule Based)

At frame f, for each candidate line segment i, represented by ($S_f(i)$, $E_f(i)$, $\Phi_f(i)$) found in the image the length of all line segments and perform the following tests: (1) Is i the longest of the lines?, (2) Is the difference between the angle of the major axis of the tracking ellipse of the hand and $\Phi_f(i)$ is small (within a given ε value), (3) Is the change of the angle of the ith line segment with respect to its angular position in the preceding frame small, i.e.; Is $(\Phi_f(i)-\Phi_{f-1}(i))<\varepsilon$)?, (4) Draw a line from the centroid of the tracking window to the far endpoint of the instrument. Find the shortest distance from the close endpoint of the instrument to this line traced. Is this distance small? The test is comprised of a set of queries for each line segment. If enough responses are positive, then enough evidence is accumulated to infer that the line segment represents the SSI. The test looks for a line segment that is long, aligned with the major axis of the detected hand tracking ellipse and has not changed significantly from the last SSI detected, and is pointing to the center of the hand. This last test (test 4) is designed to eliminate all line segments except those radiating out of the center of the hand. Of the set of segments passing test 4 only those aligned with the major axis of the hand will be selected by 3. Every positive response adds one vote to the total votes of a candidate line. The line with the highest number of votes is selected as the segment representing the SSI. Considering I_{SSI} as the image with the single detected line segment, it is overlaid on I_{org} obtain the composite image I_c. An example of I_c is shown in Fig. 6(d). An alternative method for detecting the SSI is to use a Fuzzy Inference System by replacing the above test questions by fuzzy rules. For example, for the question "is i the longest of the lines?"; one may define a linguistic variable line length and membership functions for the terms; short, medium and long, and a linguistic variable, degree of line i being a SSI with terms; not likely, likely, very likely, most likely. Then the following fuzzy rule may be defined: If **<line i>** is< **long>** then the **<degree of line i being a sterile instrument>** is< **very likely>**. A composite premise may be defined to generate a SSI fuzzy acceptance test rule. This method, however, is left for future work.

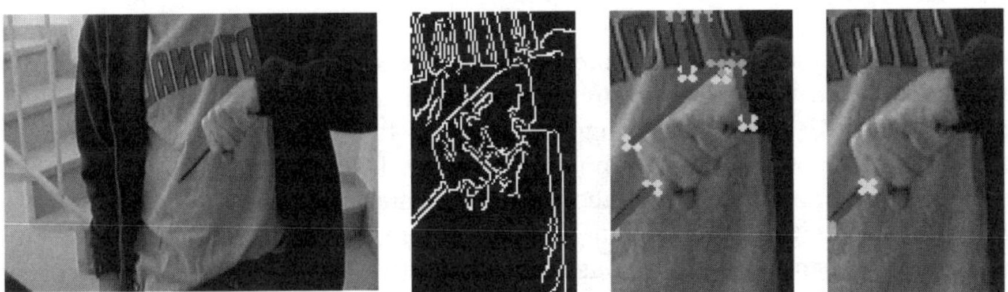

(a) (b) (c) (d)

Fig. 6. The SSI algorithm applied to a single frame. (a) Original image. (b) After canny edge detector. (c) Output of the pHT. (d) Output of the SSI detector.

5 System Implementation

An example of a rotation gesture with a complex background is shown in Fig, 6(a). The pHT output, Fig. 6.(c) consists of five candidate line segments. Three of them violate rules 3 and 4, because they are short, their angles are significantly different from the previously tracked line segment, and the angles are far different from the camshift tracking window. As for the remaining two candidates, those with a positive slopes, only the bottom segment is directed toward the centroid of the hand and is this selected. An example of detection and tracking is shown in Fig. 7, where SSIs were found for 80% of the frames of a 1min, 13 sec video. The system was also tested in a hospital environment during a live neurosurgeon operation where surgeons browsed MRI images using the gesture system.

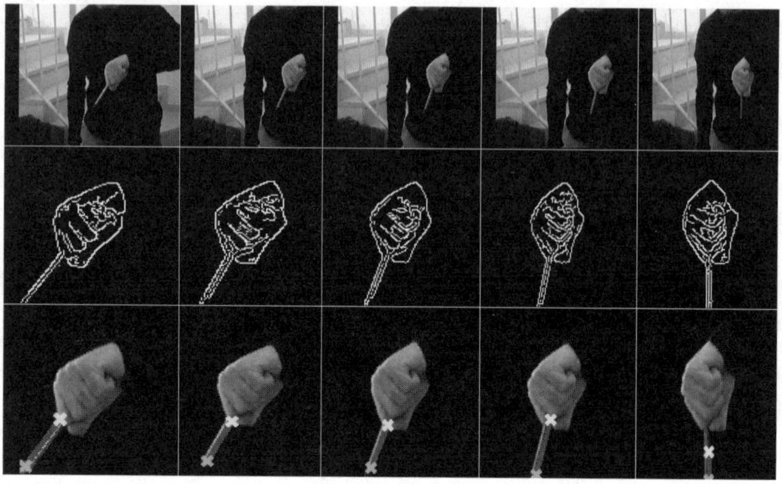

Fig. 7. Sequence of successful detection and tracking of an SSI

6 Conclusions

A vision-based system that can interpret user's gestures in real-time to manipulate windows and objects within a medical data visualization environment is presented. A hand segmentation procedure using color-motion fusion extracts binary hand blobs from each frame of an acquired image sequence. Dynamic navigation gestures are translated to commands based on their relative positions on the screen. Static gesture poses are identified to execute non-directional commands, such as zoom and rotate. The gesture recognition system was implemented in a sterile medical data-browser environment (named Gibson). The system was tested in a hospital environment during a live neurosurgeon operation where surgeons browsed MRI images using the gesture system. Future work includes replacement of the rotation gesture to operate with the hand palm only, and the development of two handed gestures to achieve increased accuracy for the zoom and rotation gestures. In addition, the development of a fuzzy inference system will be conducted for detecting the hand held sterile instrument.

Acknowledgement

This work was partially supported by the Paul Ivanier Center for Robotics Research and Production Management, Ben-Gurion University of the Negev.

References

1. M. Schultz, J. Gill, S. Zubairi, R. Huber, F. Gordin, "Bacterial contamination of computer keyboards in a teaching hospital," Infect Control Hosp. Epidemiol., vol. 4, no. 24, pp. 302-303, 2003.
2. D. Nicholas, P. Huntington, P. Williams, P. Vickery, "Health information: an evaluation of the use of touch screen kiosks in two hospitals". Health Information Librarian Journal. 2001 Dec, vol. 18, no. 4, pp. 213-9.

3. K. R. Smith, K. J. Frank, R.D. Bucholz, "The NeuroStation- a highly accurate, minimally invasive solution to frameless stereotatic neurosurgery. Comput Med Imaging Graph, no. 18, pp. 247-256, 1994.
4. H. Colle, K. Hiszem, "Standing at a kiosk: Effects of key size and spacing on touch screen numeric keypad performance and user preference," Ergonomics, no. 18, pp. 1406-1423, 2004.
5. A. Nishikawa, T. Hosoi, K. Koara, D. Negoro, A. Hikita, S. Asano, H. Kakutani, F. Miyazaki, M. Sekimoto, M. Yasui, Y. Miyake, S. Takiguchi, and M. Monden. "FAce MOUSe: A Novel Human-Machine Interface for Controlling the Position of a Laparoscope," IEEE Trans. on Robotics and Automation, vol. 19, no. 5, pp. 825-841, 2003.
6. C. Graetzel, T.W. Fong, S. Grange, and C. Baur, "A non-contact mouse for surgeon-computer interaction," Technology and Health Care, vol. 12, no. 3, 2004, pp. 245-257.
7. T J. Zeng, Y. Wang, M.T. Freedman and S.K. Mun, "Finger tracking for breast palpation quantification using color image features", SPIE Optical Engineering, vol. 36, no. 12, pp. 3455-3461, Dec. 1997.
8. J. Wachs, H. Stern, Y. Edan, U. Kartoun, "Real-Time Hand Gesture Using the Fuzzy-C Means Algorithm", In Proc. of WAC 2002, Florida, June 2002.
9. Y. Kuno, T. Murashima, N. Shimada, and Y. Shirai, "Intelligent Wheelchair Remotely Controlled by Interactive Gestures," In Proc. of 15th Intl Conf. on Pattern Recognition, vol.4, pp.672-675, 2000
10. T. Starner, J. Auxier, D. Ashbrook, and M. Gandy. "The Gesture Pendant: A Self-illuminating, Wearable, Infrared Computer Vision System for Home Automation Control and Medical Monitoring". In Fourth Intl Symp. on Wearable Computers, pp. 87-94, 2000.
11. T. Baudel, and M. Beaudouin-Lafon. "CHARADE: Remote Control of Objects using Free-Hand Gestures," Communications of the ACM. vol. 36, no. 7, pp. 28-35, 1993.
12. H. Stern, J. Wachs, Y. Edan, "Optimal Hand Gesture Vocabulary Design Using Psycho-Physiological and Technical Factors," 7th International Conference Automatic Face and Gesture Recognition, FG2006. Southampton, UK, April 10-12 2006
13. G. R. Bradski, "Computer vision face tracking for use in a perceptual user interface," Intel Technical Journal, pp. 1-15, 1998.
14. D. Comaniciu and P. Meer, "Robust Analysis of Feature Spaces: Color Image Segmentation," CVPR'97, pp. 750-755.
15. J. Wachs, H. Stern, Y. Edan, M. Gillam, C. Feied, M. Smith, J. Handler, "A Real-Time Hand Gesture Interface for Medical Visualization Applications,", in the 10th Online World Conference on Soft Computing in Industrial Applications, September 19th - October 7th, 2005.
16. H. Stern, B. Efros, "Adaptive Color Space Switching for Tracking under Varying Illumination", Journal of Image and Vision Computing, vol. 23, no. 3, 2005, pp. 353-364.
17. N Kiryati, Y Eldar, and AM Bruckstein. "A probabilistic Hough Transform," Pattern Recognition, vol. 24, no. 4, pp. 303-316, 1991.

Differential Evolution for the Registration of Remotely Sensed Images

I. De Falco[1], A. Della Cioppa[2], D. Maisto[1], and E. Tarantino[1]

[1] Institute of High Performance Computing and Networking, National Research Council of Italy (ICAR–CNR)
Via P. Castellino 111, 80131 Naples, Italy
{ivanoe.defalco,ernesto.tarantino}@na.icar.cnr.it

[2] Natural Computation Lab – DIIIE, University of Salerno, Via Ponte don Melillo 1, 84084 Fisciano (SA), Italy
adellacioppa@unisa.it

Abstract. This paper deals with the design and implementation of a software system based on Differential Evolution for the registration of images, and in its testing by means of a set of bidimensional remotely sensed images on two problems, i.e. mosaicking and changes in time. Registration is carried out by finding the most suitable affine transformation in terms of maximization of the mutual information between the first image and the transformation of the second one, without any need for setting control points. A comparison is effected against a publicly available tool, showing the effectiveness of our method.

Keywords: Differential evolution, image registration, remote sensing, affine transformation, mutual information.

1 Introduction

Registration is a fundamental task in image processing used to match two or more pictures taken with different methods or at different times. During the years several techniques have been developed for various applications [1, 2, 3]. A field which is highly interested is that of remote sensing imagery [4, 5, 6, 7]. The goal of the paper consists in the design and implementation of a software system for the registration of images, and in its testing by means of a set of 2–D satellite images. Among the methods proposed in literature, the one based on the use of an affine transformation [8] to "align" at best the two images to be registered appears of interest. Thus the problem becomes that of finding the best among all the possible transformations, each of which is represented by a set of real parameters. Evolutionary Algorithms (EAs) [9, 10] are successfully used to face several multivariable optimization tasks and their use has been introduced in Image Registration as well, in particular in the medical [11, 12] and in the remote sensing [13, 14, 15, 16] areas. Therefore we employ them to find the optimal combination of the parameter values involved in the affine transformation. Specifically, Differential Evolution (DE) [17] has been implemented. DE is a version of an EA which has proven fast and reliable in many applications [18].

A. Saad et al. (Eds.): Soft Computing in Industrial Applications, ASC 39, pp. 40–49, 2007.
springerlink.com

Here, differently from other approaches [4,16,19] (including DE [20]) in which a set of control points must be provided and the fitness function takes into account how well those points are mapped, we examine DE ability to perform automatic image registration without any *a priori* knowledge.

Paper structure is as follows: Section 2 describes the image registration problem and defines the affine transformation and the mutual information. Section 3 contains DE basic scheme and illustrates the application of our system to the image registration task. Section 4 reports on the two remote sensing problems faced, i.e mosaicking and changes in time, and shows the results achieved by our tool and its comparison against a publicly available registration tool, *Im-Reg* [21,22]. Finally Section 5 contains conclusions and future works.

2 Image Registration

Registration is often necessary for integrating information taken from different sensors, or finding changes in images taken under disparate conditions or diverse viewing angles. Depending on the application, the goals of registering images may be quite different. In remote sensing two problems are typically faced, i.e. *Mosaicking* and *Change Discovery*. The former deals with spatially aligning two images of neighboring areas taken at the same time so as to obtain a larger image, whereas the latter consists in firstly aligning two images of about the same area but taken at different times, and then in pointing out the changes happened in that area within the difference timespan. In all cases, two choices must be made to carry out image registration. The first choice involves the kind of geometric transformation to be considered to find correlations between the given images, while the second one concerns the measure of match (MOM), i.e. the feature on the value of which the goodness of the registration is evaluated. Once made these choices, the MOM can be maximized by using suitable optimization algorithms.

Affine Transformation. The most frequently used transformation model in registration applications is the affine transformation. This model is sufficiently general, since it can handle rotations, translations, scaling and shearing. This transformation can be represented in the most general 3–D case as:

$$\mathbf{x}' = A \cdot \mathbf{x} + \mathbf{b} \tag{1}$$

where A is a 3×3 square matrix accounting for rotations and scalings while $\mathbf{x}$, $\mathbf{x}'$ and $\mathbf{b}$ are three–dimensional arrays representing respectively the original positions, the transformed ones and a translation vector.

Mutual Information. The most widely employed MOM is the Mutual Information (MI) [23,24], which represents the relative entropy of the two images to be registered. In general, given two random variables Y and Z, their MI is:

$$I(Y,Z) = \sum_{y,z} P_{Y,Z}(y,z) \cdot log \frac{P_{Y,Z}(y,z)}{P_Y(y) \cdot P_Z(z)} \tag{2}$$

where $P_Y(y)$ and $P_Z(z)$ are the marginal probability mass functions and $P_{Y,Z}(y,z)$ is the joint probability mass function. MI is related to entropies by:

$$I(Y, Z) = H(Y) + H(Z) - H(Y, Z) \tag{3}$$

with $H(Y, Z)$ being their joint entropy, and $H(Y)$, $H(Z)$ the entropies of Y and Z respectively. The definitions of these entropies are:

$$H(Y) = -\sum_y P_Y(y) \cdot log P_Y(y), \;\; H(Z) = -\sum_z P_Z(z) \cdot log P_Z(z) \tag{4}$$

$$H(Y, Z) = -\sum_{y,z} P_{Y,Z}(y, z) \cdot log P_{Y,Z}(y, z) \tag{5}$$

To employ MI as a similarity measure, the 2–D histogram of an image pair, the joint histogram h, must be utilized. It is defined as a function of two variables Y and Z, the gray–level intensities in the two images. Its value at the coordinate (Y, Z) is the number of corresponding pairs having gray–level Y in the first image and gray–level Z in the second image. The joint probability mass function of an image pair is then obtained by normalizing the joint histogram of the image pair:

$$P_{Y,Z}(y, z) = \frac{h(y,z)}{\sum_{y,z} h(y,z)} \tag{6}$$

From it the two marginal probability mass functions can be obtained as:

$$P_Y(y) = \sum_z P_{Y,Z}(y, z), \;\; P_Z(z) = \sum_y P_{Y,Z}(y, z) \tag{7}$$

The MI registration criterion states that the image pair is geometrically aligned through a geometric transformation $\mathbf{T}$ when $I(Y(\mathbf{x}), Z(\mathbf{T}(\mathbf{x})))$ is maximal. Thus, the aim is to maximize the eq. (3).

3 Differential Evolution

Differential Evolution (DE) is a stochastic, population-based optimization algorithm [17]. Given a maximization problem with m real parameters, DE faces it starting with a randomly initialized population consisting of n individuals each made up by m real values. Then, the population is updated from a generation to the next one by means of many different transformation schemes. In all of these schemes DE basicly generates new individuals by adding to a member a number of weighted difference vectors between couples of population members. We have decided to perturb a random individual by using one difference vector and by applying binomial crossover, so our strategy can be referenced as *DE/rand/1/bin*. In it for the generic i–th individual in the current population three integer numbers r_1, r_2 and r_3 in $[1, n]$ differing one another and different from i are randomly generated. Furthermore, another integer number k in the range $[1, m]$ is randomly chosen. Then, starting from the i–th individual a new trial one i' is generated whose generic j–th component is given by:

$$x_{i',j} = x_{r_3,j} + F \cdot (x_{r_1,j} - x_{r_2,j}) \tag{8}$$

provided that either a random real number ρ in $[0.0, 1.0]$ is lower than a value CR (parameter of the algorithm, in the same range as ρ) or the position j under account is exactly k. If neither is verified then a copy takes place: $x_{i',j} = x_{i,j}$.

Table 1. Problem variable ranges

	a_{11}	a_{12}	a_{21}	a_{22}	b_1	b_2
min	0.500	-0.500	-0.500	0.500	-200.0	-200.0
max	1.500	0.500	0.500	1.500	200.0	200.0

F is a real and constant factor in $[0.0, 1.0]$ which controls the magnitude of the differential variation $(x_{r_1,j} - x_{r_2,j})$, and is a parameter of the algorithm.

This new trial individual i' is compared against the i–th individual in current population and, if fitter, replaces it in the next population, otherwise the old one survives and is copied into the new population. This basic scheme is repeated for a maximum number of generations g.

3.1 DE Applied to Image Registration

Encoding. We have decided to make use of the aforementioned affine transformation model. Since the experiments reported in this paper make reference to couples of two–dimensional images, the eq. (1) reduces to:

$$\begin{cases} x'_1 = a_{11} \cdot x_1 + a_{12} \cdot x_2 + b_1 \\ x'_2 = a_{21} \cdot x_1 + a_{22} \cdot x_2 + b_2 \end{cases} \tag{9}$$

so the whole problem consists in finding the best combination of six real–valued parameters. Therefore, any individual in the DE population is an array with six positions, with the parameters listed as follows: $\mathbf{T} = (a_{11}, a_{12}, a_{21}, a_{22}, b_1, b_2)$ and each parameter can vary within a range of its own.

Fitness. Given two images C and D we take as fitness function their mutual information I, so the aim of the problem becomes to find the best transformation $\mathbf{T}$ for D such that the mutual information of C and $\mathbf{T}(D)$ is maximized.

4 Experiments and Results

We have faced both *Mosaicking* and *Change Discovery* problems typical of remotely sensed image registration. The first, named below as *Mosaic*, accounts for the registration of two images while the second, referred to as *Changes*, looks for the changes in an area by examining two images taken at different times.

In both cases DE parameters have been set as follows: $n = 30$, $g = 200$, $CR = 0.5$ and $F = 0.5$. No preliminary tuning phase has been performed. It is important to remark here that, differently from some papers in literature about use of EAs to solve this task, we have decided to use quite wide ranges for each variable in the $\mathbf{T}$ solution, since we hope that evolution drive the search towards good transformations. The allowed variation ranges are shown in Tab. 1.

For each problem 20 DE runs have been carried out, so as to investigate the dependence of the results on the initial random seed. The best of those runs will be discussed below in terms of image transformation achieved and of evolution.

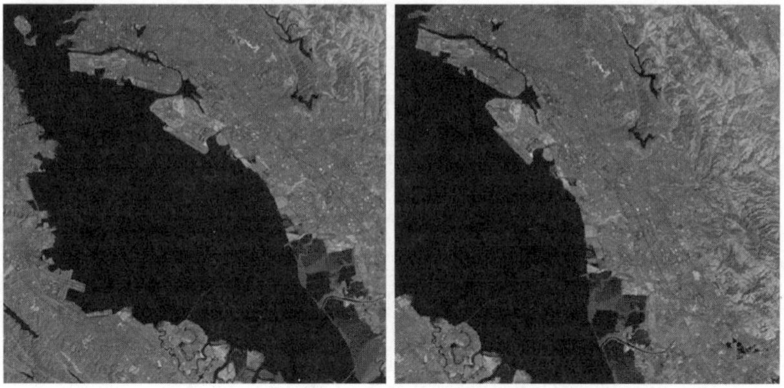

Fig. 1. The two original *Mosaic* images

The *Mosaic* Task. In the first test case we have used two images which are portions of a Landsat Thematic Mapper (TM) digital image recorded on September 7, 1984 over San Francisco bay area (CA, USA) (property of United States Geological Survey [25]). Those images were transformed by us into grey monochannel images, so that each of them is $500 \cdot 500$ pixel large and uses 8 bits to represent each pixel. Figure 1 shows them both. Their I value is 0.1732. Figure 2 (top left) reports the fusion of the two original images. They share a common area, which should be used by the DE algorithm to find their best registration. Namely, the up–left part of the second image overlaps the bottom–right part of the first, and a slight clockwise rotation was applied to the second image with reference to the first one. So, the best affine transformation should contain a slight counterclockwise rotation and two positive shifts for both the coordinates.

In this problem the best value of I obtained in the best execution is 1.1083. The average of the best final values over the 20 runs is 0.9068 and the variance is 0.1378, the worst result being 0.6351. The best affine transformation found is:

$$\begin{cases} x_1' = 0.946 \cdot x_1 - 0.253 \cdot x_2 + 41.858 \\ x_2' = 0.253 \cdot x_1 + 0.946 \cdot x_2 + 49.779 \end{cases} \tag{10}$$

which represents a counterclockwise rotation of about 15 degrees coupled with a translation in both axes. The resulting transformed image is shown in Fig. 2 (top right). Figure 2 (bottom left) depicts the fusion of the first original image with the best transformation found for the second one. The alignment of the two registered images is excellent: any detail in the first image, from the streets to the shoreline to the bridges, is perfectly aligned with the corresponding pixels representing it in the transformed second image. In Fig. 2 (bottom right) we report the behavior of the best run achieved for the *Mosaic* task. In this case, in spite of the very relaxed parameter range allowed, already the initial population achieves an improving solution with respect to the original one. From then on the system proposes many improving affine transformations and both the average and the best fitness values increase over generations until the end of the run.

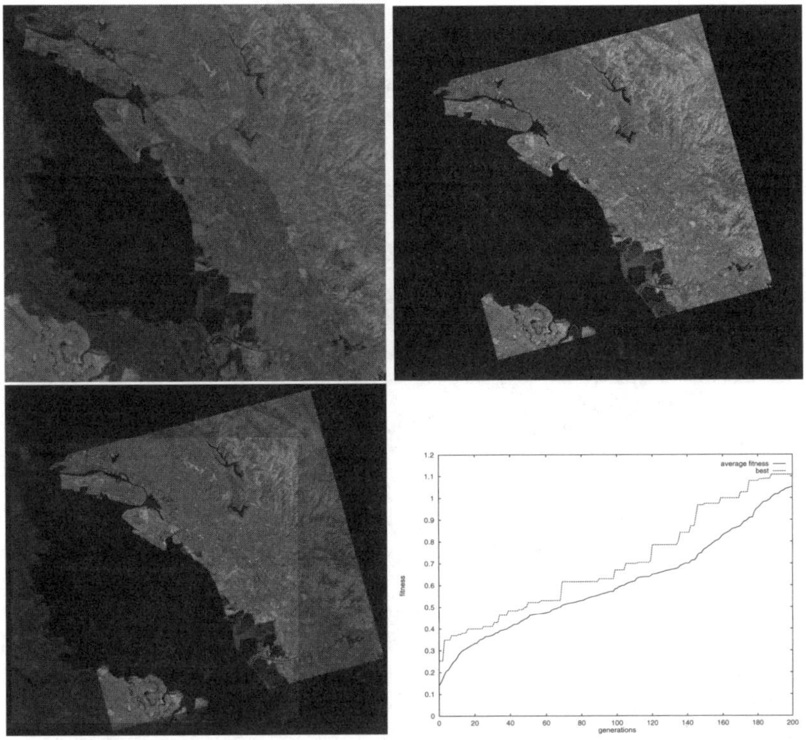

Fig. 2. Top Left: The fusion of the two original images. **Top Right:** The best transformation for the second *Mosaic* image. **Bottom Left:** The first image is fused with the best transformation found for the second one. **Bottom Right:** Behavior of fitness as a function of the number of generations for the best run.

To investigate the goodness of the results achieved, we have compared them against those provided by a publicly available image registration tool, *i.e. ImReg* [21], developed at the Vision Research lab at the University of Santa Barbara, USA [22]. A user needs to upload the two images and to set a parameter about the desired registration quality (fast, normal, quality, extra). *ImReg* is based on automatically retrieving a set of tie points, culling bad tie points and testing the transformation. On this task *ImReg* performs its best registration (in extra quality mode) by the following parameter set $T = (0.984, -0.263, 0.263, 0.984, 53.687, 37.667)$. This results in a value of I equal to 0.3984, much lower than that found by our tool, even in its worst case.

The *Changes* Task. In the second test case we have used two images which refer to about the same area but were taken at different times. Namely, they represent an agricultural area near San Francisco (CA, USA) in 1984 and in 1993 respectively (they too are property of USGS [25]). As before, the original Landsat TM images were transformed by us into grey monochannel images,

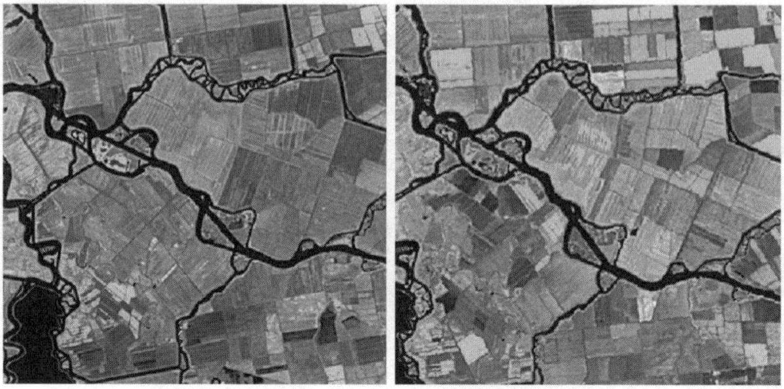

Fig. 3. The two original *Changes* images

so that each of them is $500 \cdot 500$ pixel large with an 8–bit representation for each pixel (see Fig. 3). Their I value is 0.1123. Figure 4 (top left) reports the fusion of the two original images. As it can be observed, they share a common area, which should be used by the DE algorithm to find their best registration. Namely, the right part of the first image overlaps the left part of the second, and a slight clockwise rotation took place when the second image was taken with reference to the first one. So, the best affine transformation should contain a slight counterclockwise rotation and some shifts for both the coordinates.

In this problem the best value of I attained in the best execution is 0.3951. The average of the best final values over the 20 runs is 0.3918 and the variance is 0.0049, the worst result being 0.3803. The best affine transformation found is:

$$\begin{cases} x'_1 = 0.954 \cdot x_1 - 0.083 \cdot x_2 + 16.995 \\ x'_2 = 0.083 \cdot x_1 + 0.953 \cdot x_2 + 20.361 \end{cases} \tag{11}$$

which represents a counterclockwise rotation of about 5 degrees coupled with a translation in both axes. The resulting transformed image is shown in Fig. 4 (top right). Figure 4 (bottom left) shows the fusion of the first original image with the best transformation found for the second one. The alignment of the two registered images is very good: any detail in the first image, from the rivers to the roads to the fields, is well aligned with the corresponding pixels representing it in the transformed second image. Figure 4 (bottom right) presents the behavior of the best run achieved for the *Changes* task. Also in this case, in spite of the very relaxed parameter range allowed, already the initial population achieves an improving solution with respect to the original one and from then on the system proposes many improving affine transformations and both the average and the best fitness values increase over generations until the end of the run.

The computed differences between the first image and the transformed second one are shown in Fig. 5, where only the part in which the two images overlap is

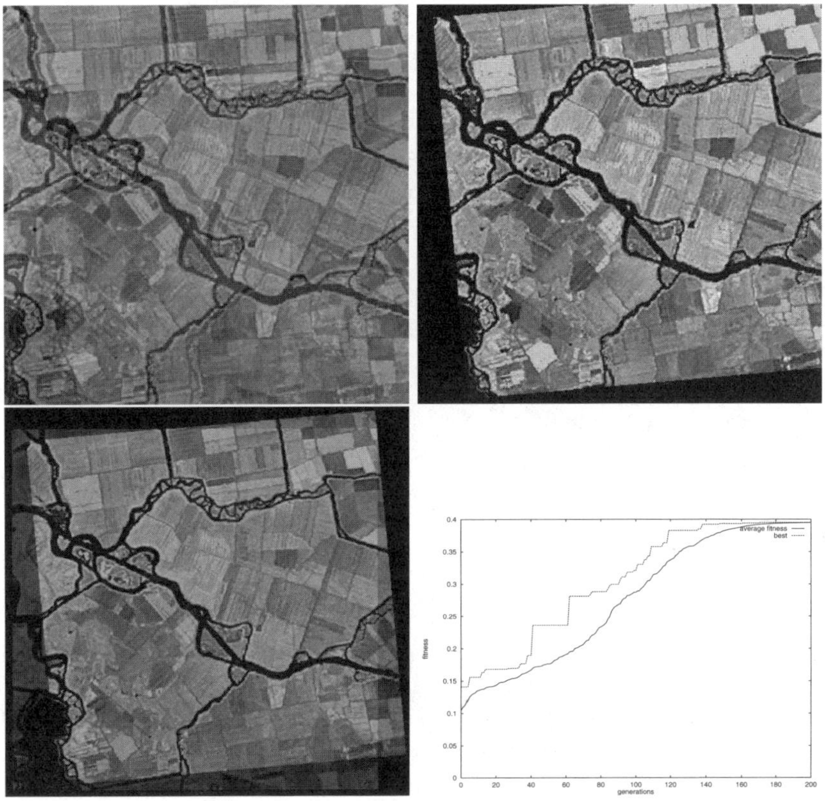

Fig. 4. Top Left: The fusion of the two original images. **Top Right:** The best transformation for the second *Changes* image. **Bottom Left:** The first image is fused with the best transformation found for the second one. **Bottom Right:** Behavior of fitness as a function of the number of generations for the best run.

meaningful. In it the grey color refers to areas where no changes occurred, the black represents areas that were burned in 1984 and recovered by 1993, whereas the white stands for areas more vegetated in 1984 than in 1993 due to differences in the amount of rainfall, or to the density or level of maturity of the vegetation. Light pixels represent areas burned in 1993 or which were natural landscape areas in 1984 that were converted to agricultural lands and recently tilled, and finally dark pixels stand for areas more vegetated in 1993 than in 1984.

Also here we have compared our results against those provided by *ImReg*. In this case this tool in its most performing configuration (extra quality) achieves the following parameter set $T = (1.041, -0.088, 0.088, 1.041, 18.845, 19.437)$, which results in $I = 0.1456$, much lower than that found by our tool, even in its worst case. As a result, for both problems our system achieves better solutions than *ImReg* in terms of higher MI, which proves the quality of our approach.

Fig. 5. *Changes* image in an agricultural area near San Francisco within 1984 and 1993

5 Conclusions and Future Works

In this paper a Differential Evolution strategy has been coupled with affine transformation and Mutual Information maximization to perform registration of remotely sensed images. No control points are needed to accomplish this task. A comparison has been carried out against a publicly available registration tool, *ImReg*, on two classical problems. The results show that our evolutionary system outperforms *ImReg*, and seem to imply that this approach is promising, yet there is plenty of work still to do. Therefore, future works shall aim to shed light on the effectiveness of our system in this field, and on its limitations as well.

Firstly, a wide tuning phase shall be carried out to investigate if some DE parameter settings are, on average, more useful than others. This phase shall take into account lots of image couples taken from different fields.

A comparison must be carried out against the results achieved by other image registration methods, to examine the effectiveness of our proposed approach.

Furthermore we plan to implement a coarse–grained parallel version of the DE algorithm based on the island model, and to run it on a cluster of workstations. Lastly, our final goal is to design and implement a technique which could be useful also for 3–D medical image registration.

References

1. Brown L G (1992) A Survey of Image Registration Techniques. ACM Computing Surveys 24(4): 325–376.
2. Maintz J B A, Viergever M A (1998) A Survey of Medical Image Registration Methods. Medical Image Analysis 2(1):1–37. Oxford University Press.
3. Zitova B, Flusser J (2003) Image Registration Methods: A Survey. Image and Vision Computing, 21:977–1000.

4. Ton J, Jain A K (1989) Registering Landsat Images by Point Matching. IEEE Trans. on Geoscience and Remote Sensing 27 (5):642–651.
5. Fonseca L M G, Manjunath B S (1996) Registration Techniques for Multisensor Remotely Sensed Imagery. Photogrammetric Engineering & Remote Sensing, 62(9): 1049–1056.
6. LeMoigne J (1997) Towards an Intercomparison of Automated Registration Algorithms for Multiple Source Remote Sensing Data. In: Proc. of the Image Registration Workshop, pp. 307–316, November 20–21, NASA GSFC, MD, USA.
7. Lee C, Bethel J (2001) Georegistration of Airborne Hyperspectral Image Data. IEEE Trans. on Geoscience and Remote Sensing 39 (7):1347–1351.
8. Hart G W, Levy S, McLenaghan R (1995) Geometry. In: Zwillinger D (Ed), CRC Standard Mathematical Tables and Formulae. CRC Press, Boca Raton, FL.
9. Goldberg D (1989) Genetic Algorithms in Optimization, Search and Machine Learning. Addison Wesley, New York.
10. Eiben A E, Smith J E (2003) Introduction to Evolutionary Computing. Springer.
11. Jacq J, Roux C (1995) Registration of non-segmented images using a genetic algorithm. Lecture notes in computer science, 905:205-211. Springer–Verlag.
12. Chow C K, Tsui H T, Lee T, Lau T K (2001) Medical image registration and model construction using genetic algorithms. In: Proc. of Int. Workshop on Medical Imaging and Augmented Reality (MIAR '01), pp 174–179, IEEE Computer Society.
13. Fitzpatrick J, Grefenstette J, Gucht D (1984) Image Registration by Genetic Search. In: Proc. of the IEEE SoutheastCon Conf., pp. 460–464.
14. Dasgupta D, McGregor D R (1992) Digital Image Registration using Structured Genetic Algorithms. In: Proceedings of SPIE The Int. Society for Optical Engineering, vol. 1776, pp. 226–234.
15. Chalermwat P, El-Ghazawi T A (1999) Multi-Resolution Image Registration Using Genetics. In: Proc. of the Int. Conf. on Image Processing, vol. 2, pp. 452–456.
16. Kim T, Im Y (2003) Automatic Satellite Image Registration by Combination of Stereo Matching and Random Sample Consensus. IEEE Trans. on Geoscience and Remote Sensing, 41(5):1111–1117.
17. Storn R, Price K (1997) Differential Evolution - a Simple and Efficient Heuristic for Global Optimization over Continuous Spaces. Journal of Global Optimization 11(4):341–359. Kluwer Academic Publishers.
18. Price K, Storn R, Lampinen J (2005), Differential Evolution: A Practical Approach to Global Optimization, Natural Computing Series. Springer-Verlag.
19. Dai X, Khorram S (1999) A Feature-based Image Registration Algorithm using Improved Chain-code Representation Combined with Invariant Moments. IEEE Trans. on Geoscience and Remote Sensing 37(5):2351–2362.
20. Thomas P, Vernon D (1997) Image Registration by Differential Evolution. In: Proc. of the Irish Machine Vision and Image Processing Conf., pp. 221–225, Magee College, University of Ulster, Ireland.
21. http://nayana.ece.ucsb.edu/registration/
22. Fedorov D, Fonseca L M G, Kenney C, Manjunath B S (2002) Automatic Registration and Mosaicking System for Remotely Sensed Imagery. In: Proc of SPIE 9th International Symposium on Remote Sensing, Crete, Greece.
23. Maes F, Collignon A, Vandermeulen D, Marchal G, Suetens P (1997) Multimodality Image Registration by Maximization of Mutual Information. IEEE Trans. on Medical Imaging 16(2):187–198.
24. Pluim J P W, Maintz A J B, Viergever M A (2003) Mutual–information–based Registration of Medical Images: a Survey. IEEE Trans. on Medical Imaging 22:986–1004.
25. http://terraweb.wr.usgs.gov/projects/SFBay/

Geodesic Distance Based Fuzzy Clustering

Balazs Feil and Janos Abonyi

Department of Process Engineering, University of Pannonia
Veszprem P.O. Box 158, H-8201 Hungary
feilb@fmt.uni-pannon.hu,abonyij@fmt.uni-pannon.hu
www.fmt.uni-pannon.hu/softcomp

Summary. Clustering is a widely applied tool of data mining to detect the hidden structure of complex multivariate datasets. Hence, clustering solves two kinds of problems simultaneously, it partitions the datasets into cluster of objects that are similar to each other and describes the clusters by cluster prototypes to provide some information about the distribution of the data. In most of the cases these cluster prototypes describe the clusters as simple geometrical objects, like spheres, ellipsoids, lines, linear subspaces etc., and the cluster prototype defines a special distance function. Unfortunately in most of the cases the user does not have prior knowledge about the number of clusters and not even about the proper shape of prototypes. The real distribution of data is generally much more complex than these simple geometrical objects, and the number of clusters depends much more on how well the chosen cluster prototypes fit the distribution of data than on the real groups within the data. This is especially true when the clusters are used for local linear modeling purposes.

The aim of this paper is not to define a new distance norm based on a problem dependent cluster prototype but to show how the so called geodesic distance that is based on the exploration of the manifold the data lie on, can be used in the clustering instead of the classical Euclidean distance. The paper presents how this distance measure can be integrated within fuzzy clustering and some examples are presented to demonstrate the advantages of the proposed new methods.

1 Introduction

This paper is dealing with clustering of high dimensional data. Various definitions of a cluster can be formulated, depending on the objective of clustering. Generally, one may accept the view that a cluster is a group of objects that are more similar to one another than to members of other clusters. It is important to emphasize that more specific definitions of clusters can hardly be formulated because of the various types of problems and aims. (Besides this problem, another crucial one is the enormous search space.) However, there is a need to cluster the data automatically, and an objective definition should be formulated for the similarity and the quality of the clustering, because if the clustering is embedded by a mathematical model, there is a possibility to solve these problems quickly and effectively.

A. Saad et al. (Eds.): Soft Computing in Industrial Applications, ASC 39, pp. 50–59, 2007.
springerlink.com

The term "similarity" should be understood as mathematical similarity, measured in some well-defined sense. In metric spaces, similarity is often defined by means of a distance norm. Distance can be measured among the data vectors themselves, or as a distance from a data vector to some prototypical object of the cluster. The prototypes are usually not known beforehand, and are sought by the clustering algorithms simultaneously with the partitioning of the data. The prototypes may be vectors of the same dimension as the data objects, but they can also be defined as "higher-level" geometrical objects, such as linear or nonlinear subspaces or functions.

The cluster prototypes chosen depend heavily on the problem and also on the aim of the clustering, and often a priori information should be used to choose the proper one(s). If spherical clusters are to be searched, the classical fuzzy c-means algorithm can be a good choice [7]. If the location of data is more complex, Gustafson–Kessel algorithm can be used that is able to discover ellipsoids with the same size [2]. It uses an adaptive distance norm and the covariance matrices are also searched by the algorithm. A more sophisticated method is Gath–Geva clustering which is able to reveal ellipsoids with different sizes based on an exponential distance norm [3]. If it is known that the data lie on or close to a lower dimensional (linear) subspace of the feature space, fuzzy c-lines or c-regression approaches can be applied. These methods measure the similarity of data or data and cluster prototypes using linear models. Other algorithms use more complex cluster prototypes and/or distance measures to identify locally linear fuzzy models directly [5] or segment a high dimensional time-series using Probabilistic Principal Component models [4].

This paper proposes two approaches to reveal the hidden structure of high dimensional data. Data can form groups and also can lie on a low dimensional (smooth) manifold of the feature space. It is a very common problem if there is a relationship among the variables. It can be the case if a model identification problem has to be solved: the relationship between the output and the input variables has to be revealed. It can happen that even the input variables are correlated e.g. because of the redundant information they contain. In these cases the classical clustering techniques would fail to discover the hidden structure or special cluster prototypes have to be used that can solve the specific problem. The proposed approaches are able to handle data that lie on a low dimensional manifold of the feature space. They are built on clustering methods that use a distance measure called geodesic distance which reflects the true embedded manifold, but various cluster prototypes can be used with this measure, not only special ones. In the graph theory, the distance between two vertices in a weighted graph is the sum of weights of edges in a shortest path connecting them. This is an approximation of the geodesic distance that can be measured on the real manifold the (noiseless) data lie on. The Isomap algorithm (short for isometric feature mapping) [6] seeks to preserve the intrinsic geometry of the data, as captured in the geodesic manifold distances between all pairs of data points. It uses the (approximated) geodesic distances between the data, and it is able to discover nonlinear manifolds and project them into a lower dimensional space.

The first algorithm applies Isomap on the original data and the lower dimensional data projected by Isomap are used for clustering. The second approach applies the geodesic distances between data points directly without previous projection.

This paper organized as follows. Section 2 describe the proposed algorithms in details. Examples can be found in Section 3 to demonstrate the proposed approaches with two often used data sets. Section 4 concludes the paper.

2 Geodesic Distance Based Clustering Algorithms

In this section two algorithms are presented for clustering high-dimensional data in embedded manifolds. Both exploit the geodesic distance between the data because there is no other information about the manifold, and the Euclidean distance measure would fail to discover the hidden structure of data.

The first algorithm (Algorithm I) presented exploits the projection capability of Isomap, and does the clustering on the data projected by Isomap. In other words, this approach leaves the manifold exploration problem wholly to Isomap, and looks for clusters on the explored and projected manifold. After the clustering, the resulting cluster centers can be embedded in the original feature space if this is needed. The main drawback of this method is that if Isomap failed to explore the true structure of the data, the clustering would fail to find clusters in the data. The Isomap projection is the bottleneck of the whole approach.

The second algorithm (Algorithm II) presented can avoid the drawbacks of the former technique. It does the clustering in the original feature space. In order to reveal the hidden structure of data and the embedded manifold they lie on or close to, it uses the geodesic distance to compute the similarity of the cluster prototypes and the measured data. The cluster prototypes are chosen from the data points to guarantee that the clusters lie on the manifold, the proposed method can be considered as a modified version of the fuzzy c-medoid algorithm.

2.1 Algorithm I: Clustering of the Isomap

This approach contains two main steps. In the first step, Isomap is applied to the high dimensional data in the feature space, which tries to find an appropriate projection into the (lower dimensional) real manifold. In the second step, a clustering algorithm is applied to the projected data to find groups in the data set. In the following, these steps will be described briefly because of the well known nature of the applied methods.

The Isomap algorithm [6] builds on classical multidimensional scaling (MDS) (see e.g. in [1]) which tries to find a low dimensional embedding that preserves the interpoint distances. However, Isomap seeks to preserve the intrinsic geometry of the data, as captured in the geodesic manifold distances between all pairs of data points. Because it uses the (approximated) geodesic distances between the data and not the Euclidean distances, it is able to discover nonlinear manifolds of various types and forms unlike multidimensional scaling or principal component analysis. However, it can be applied well only to smooth manifolds like in Figure 2. Several methods have been worked with aims and methods similar to Isomap, see e.g. the Locally Linear Embedding [9, 11, 12].

Isomap works as follows. *In the first step* it is determined which points are neighbors on the manifold, based on the interpoint distances of all pairs. Tenenbaum et. al. proposed two approaches to do that task: connect each point to all points within some radius ϵ (ϵ-Isomap), or to all of its k nearest neighbors (k-Isomap). After the neighborhood has been created, a weighted graph is built by linking all neighboring points and labeling all arcs with the Euclidean distance between the corresponding linked points. *In the second step*, the geodesic distance between two points is approximated by the sum of the arc lengths along the shortest path linking the points. To find the shortest paths in a weighted graph, several well known methods can be used, e.g. Floyd or Dijkstra algorithms. Bernstein et. al. prove that the geodesic distance can be approximated by the graph distance arbitrarily closely, as the density of data points tends to infinity [10]. *In the last step*, classical MDS is applied on the approximated geodesic distance matrix (based on the eigenvector-eigenvalue decomposition of the matrix [1]), to construct an embedding of the data in a low dimensional Euclidean space that best preserves the manifold's estimated intrinsic geometry. It is based on a cost function minimization, and it gives the possibility to measure the appropriateness of the projection and estimate the intrinsic dimensionality of the manifold.

After the embedding, a clustering algorithm can be applied on the projected data. In the following, classical fuzzy c-means will be used [7]. It is based on the minimization of the weighted distances between the cluster prototypes and the data points:

$$J = \sum_{i=1}^{c} \sum_{k=1}^{N} (\mu_{i,k})^m d(\mathbf{x}_k, \mathbf{v}_i)^2 \tag{1}$$

where c and N are the number of clusters and data points, respectively, $\mathbf{v}_i, i = 1, \ldots, c$ are the cluster prototypes (centers), which have to be determined, $m \in \langle 1, \infty)$ is a weighting exponent which determines the fuzziness of the resulting clusters, $d(\mathbf{x}_k, \mathbf{v}_i)$ means the (Euclidean) distance between the kth data point and ith cluster center, and $\mu_{i,k}$ denotes the degree of the membership of the kth observation belongs to the ith cluster. If the data points closest to the cluster centers in the projected space are found after clustering, they can be seen and used as cluster centers in the original feature space. Based on the computed geodesic distances by Isomap, a fuzzy partition of the original data can be calculated, and the goodness of the fuzzy partition can also be determined using the cost function above (1). The number of clusters c can be determined e.g. by cluster validity measures. In the following, the number of clusters is assumed to be known.

The approach described above has one main parameter: the radius in ϵ-Isomap or the number of neighbors in k-Isomap. It is not known how to find the optimal parameter ϵ or k. However, the scale-invariant parameter k is typically easier to set than the neighborhood radius ϵ. The crucial problem is that the number of components (the connected subgraphs) in the graph depends on the number of neighbors chosen in Isomap. If the graph is not connected, the relationship between the components is not known (because the points within are infinitely far from each other). Hence, the independent components can be projected one by one, but the projected data cannot be treated together, therefore the clustering of the whole data set cannot be performed. If the number of neighbors is set larger to get a connected graph, then the exploration of the manifold may be lost because edges large enough can "short circuit" the actual manifold. There is a contradiction in this approach because Isomap is able to project only connected units of the graph but the task of clustering is to explore different groups within the whole data set, so there is a need to know the relationship between groups.

2.2 Algorithm II: Geodesic Distance Based c-Medoid Clustering

To avoid the drawbacks of the Isomap projection, the second approach presented does the clustering in the original feature space. The aim is to explore the hidden structure of data and find groups with similar data points. If the data lie on an (low dimensional) embedded manifold, the classical clustering methods cannot be used mainly because of their distance measure. The crucial question is how to measure the distances between data points to calculate their similarity measures. To reflect the manifold containing the samples, there is a need to measure the distances on the manifold. Hence, geodesic distances have to be used.

The proposed method is built on the fuzzy version of the classical hard k-medoid algorithm (c-medoid method), only the distances is measured by the (approximated) geodesics. The objective function is the same as in fuzzy c-means (1), the difference is that c-medoid accepts measured data points as cluster centers (and not calculated means like in c-means). To find the minimum of the cost function, several methods can be used. The proposed algorithm works well with small data sets as follows.

Step 1. Calculate the (approximated) geodesic distances between all pairs of data points.

Step 2. Use fuzzy c-medoid algorithm:

a) Arbitrarily choose c objects as the initial medoids.
b) Use the calculated geodesic distances to determine how far the data points are from the medoids (the cluster centers).
c) Calculate fuzzy membership degrees as usual by fuzzy partitional clustering methods:

$$\mu_{i,k} = \frac{1}{\sum_{j=1}^{c} \left(d(\mathbf{x}_k, \mathbf{v}_i)/d(\mathbf{x}_k, \mathbf{v}_j)\right)^{2/(m-1)}}, \quad 1 \le i \le c, \ 1 \le k \le N. \tag{2}$$

This expression follows from the minimization of the cost function (1) by Lagrange multiplier method, see e.g. [7].

d) Calculate the objective function terms $\sum_{k=1}^{N}(\mu_{i,k})^m d(\mathbf{x}_k,\mathbf{v}_i)^2$, $\forall i$ with the determined membership degrees, for all $\mathbf{x}_k$ as potential medoids, and choose the data points as new medoids that minimize the objective function.

$$\mathbf{v}_i = \left\{ \mathbf{x}_j | j = \arg\min_j \sum_{k=1}^{N} (\mu_{i,k})^m d(\mathbf{x}_k, \mathbf{v}_j)^2 \right\} \tag{3}$$

e) If there are changes, jump Step 2(b).

This method can only handle small data sets because of Step 2(d). This drawback can be avoided by random sampling of new medoids, or some more sophisticated approach like CLARA (Clustering LARge Applications) [8].

3 Examples

In this section two examples are shown to present the efficiency of the proposed algorithms. The first one is a well known and often used data set in manifold exploration: the S-curve data set. This is a 3 dimensional data set with a 2 dimensional nonlinear embedded manifold as can be seen in Figure 1(b) with 2000 data points. The second one is a 2 dimensional spiral data set with two "arms" as it is shown in Figure 3(a) with 1300 data points. This data set is only two dimensional but relatively more complex than the former one, because there are two manifolds in the feature space that do not "touch" each other.

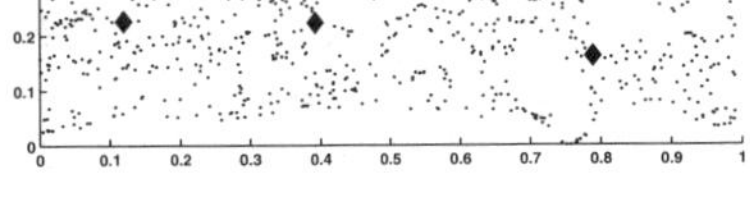

(a) 2 dimensional Isomap projection of the S-curve data set (dots) and the cluster centers by fuzzy c-means on the projected data (diamonds).

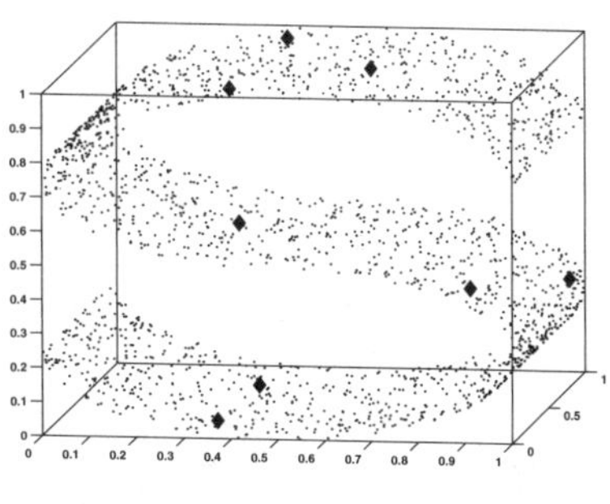

(b) The data points in the feautre space closest to the fuzzy c-means cluster centers.

Fig. 1. Results of fuzzy c-means clustering on the S-curve data set projected by Isomap (Algorithm I)

3.1 S-Curve Data Set

The Isomap algorithm determines the intrinsic dimensionality of the S-curve data set properly even in a wide range of neighbors used. It finds only one component in the neighborhood graph, and the two dimensional projection fits the data set accurately as can be seen in Figure 1(a). The fuzzy c-means is applied on the projected data with 8 clusters. The cluster centers can be seen on the same figure marked with diamonds. The data points closest to the cluster centers can be determined in the projected space, and they can be "re-projected" to the original feature space (Figure 1(b)). These data points can be seen as cluster centers, and as mentioned above in Section 2.1, the cost function value (1) can be calculated using geodesic distances given by Isomap.

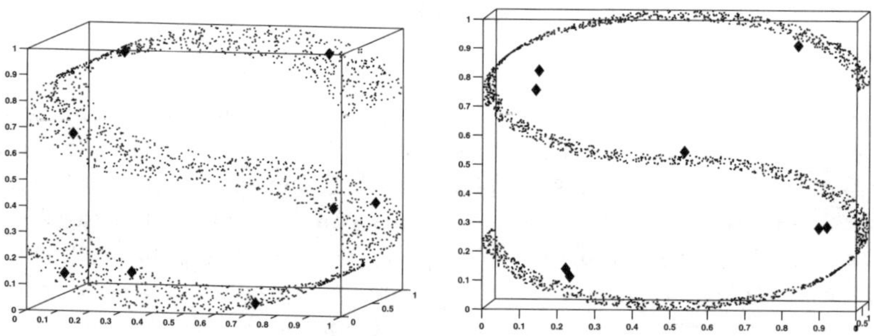

(a) Centers of the geodesic distance based clustering in the feature space (Algorithm II).

(b) Centers of the fuzzy c-means clustering in the feature space.

Fig. 2. Results of clustering on the S-curve data set

The second approach presented in Section 2.2 is also applied to the same data set with the same parameters, i.e. number of neighbors is 10, number of clusters is 8, and the initial cluster centers chosen from the data randomly are the same. The geodesic distance based clustering results can be seen in Figure 2(a). As can be determined, the cluster centers "cover" the whole embedded manifold. If the classical fuzzy c-means is applied with the same parameters initialized from the same centers, the clustering will fail to explore the hidden structure of the data. These results can be seen in Figure 2(b), and can be determined that the centers do not lie on the manifold.

The cost function value of the geodesic distance based clustering can be compared to the one given by the previous approach. The objective function values show that the geodesic distance based clustering is better than the Isomap based clustering in this case.

3.2 Spiral Data Set

The spiral data set is more complex than the previous one because the approximated geodesic distances depend heavily on the number of neighbors. In case of this particular spiral, the graph based on the approximated geodesic distances contains two components (the two arms of the spiral) using up to 4 neighbors. However, if 5 neighbors are used, only one component is given, and it has a huge effect on the clustering. Two clusters are searched in the following.

If two components are given in the geodesic distance based graph, the intrinsic dimensionality of the components is 1 given by the Isomap residuals, but the Isomap will fail to handle the data set as a whole. It is able to handle only the components one by one, and in this case the clustering can be performed only within a component, and it is meaningless from the viewpoint of the whole data set (certainly the components can be seen as clusters or groups of clusters, but they cannot be compared to each other). However, the geodesic distance based clustering is able to cluster the data and explore the two manifolds in the original feature space. The results can be seen in Figure 3(a). Two different markers (circles and dots) denote the points that belong to the two different clusters. The cluster centers are depicted with diamonds. If the classical fuzzy c-means is used, the clustering will totally fail to discover the real structure of data (Figure 3(b)).

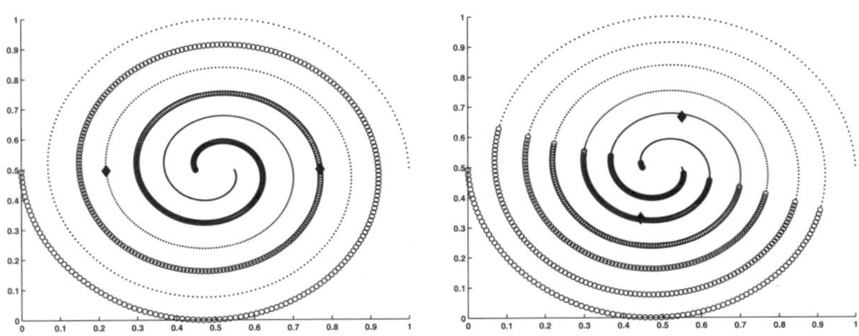

(a) Centers of the geodesic distance based clustering in the feature space (Algorithm II).

(b) Centers of the fuzzy c-means clustering in the feature space.

Fig. 3. Results of clustering on the spiral data set with two components in the geodesic distance based graph

However, if the number of neighbors is chosen equal to or greater than 5, only one component will be given in the geodesic distance based graph (the two arms of the spiral will be connected on the ends, owing to the less density of data). The intrinsic dimensionality of the data set is 2 given by Isomap, and the projected data can be seen in Figure 4(a). Following the steps of the first approach (Section 2.1), the fuzzy c-means is applied on the projected data where the number of clusters is 2. In this case, the method can distinguish the

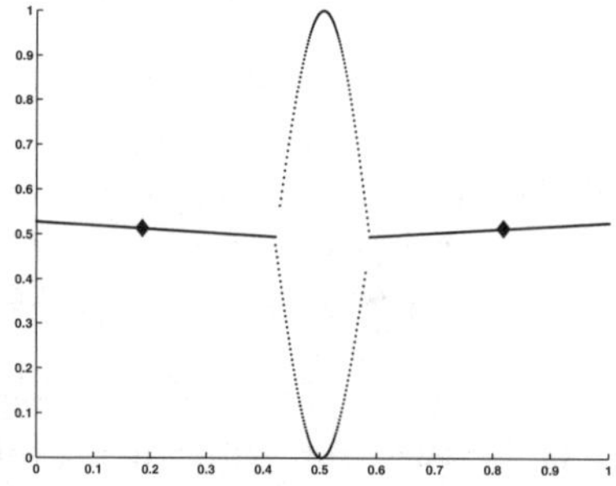

(a) *2 dimensional Isomap projection of the spiral data set (dots) and the cluster centers by fuzzy c-means on the projected data (diamonds).*

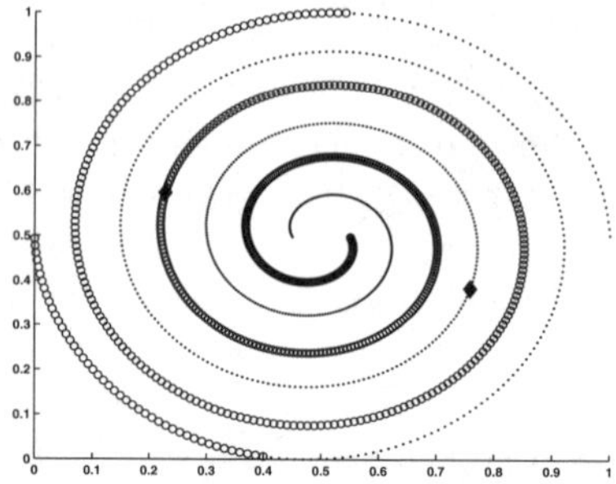

(b) *The data points in the feature space closest to the cluster centers.*

Fig. 4. Results of fuzzy c-means clustering on the spiral data set projected by Isomap with one component in the geodesic distance based graph (Algorithm I)

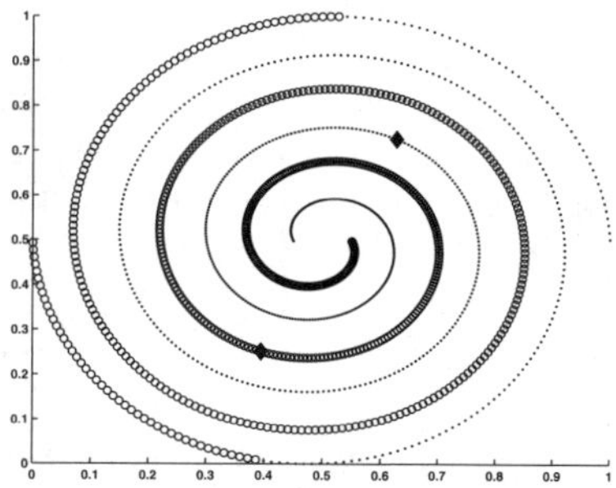

Fig. 5. Centers of the geodesic distance based clustering in the feature space with one component in the geodesic distance based graph (Algorithm II)

main part of the two spirals, only on the ends will fail, where the points from different spirals are directly connected. (The more neighbors are used, the more overlapped the two clusters will be.) The results of the clustering can be seen in Figure 4(b), projected back into the feature space.

If the second proposed approach is used, similar (slightly worse) results are given. These can be seen in Figure 5.

4 Conclusion

This paper proposes two approaches to discover the hidden structure of complex multivariate datasets. The methods are based on clustering of the data, but the classical clustering techniques may fail to explore the (nonlinear) manifolds the data lie on. The real distribution of data is generally much more complex than the simple geometrical objects used for classical cluster prototypes, and the number of clusters depends much more on how well the chosen cluster prototypes fit the distribution of data than the real groups within the data.

The paper presented how the so called geodesic distance that is based on the exploration of the manifold, can be used in the clustering instead of the classical Euclidean distance. Algorithm I is based on the clustering of the Isomap, i.e. the Isomap algorithm is used to explore the hidden (nonlinear) structure of data and the projected data are clustered. Algorithm II is based on the geodesic distances directly, and can be considered as a modification of the fuzzy c-medoid clustering. The examples show the advantages of the proposed methods using benchmark datasets in (manifold) clustering.

References

1. Naud A. *Neural and statistical methods for the visualization of multidimensional data.* PhD thesis, 2001.
2. Gustafson DE and Kessel WC. Fuzzy clustering with fuzzy covariance matrix. *Proceedings of the IEEE CDC*, pages 761–766, 1979.
3. Gath I and Geva AB. Unsupervised optimal fuzzy clustering. *IEEE Transactions on Pattern Analysis and Machine Intelligence*, 7:773–781, 1989.
4. Abonyi J, Feil B, Nemeth S, and Arva P. Modified gath-geva clustering for fuzzy segmentation of multivariate time-series. *Fuzzy Sets and Systems - Fuzzy Sets in Knowledge Discovery*, 149(1):39–56, 2005.
5. Abonyi J, Szeifert F, and Babuska R. Modified gath-geva fuzzy clustering for identification of takagi-sugeno fuzzy models. *IEEE Trans. on Systems, Man and Cybernetics, Part B*, 32(5):612–621, 2002.
6. Tenenbaum JB, Silva V, and Langford JC. A global geometric framework for nonlinear dimensionality reduction. *Science*, 290:2319–2323, 2000.
7. Bezdek JC. *Pattern recognition with fuzzy objective function algorithms.* Plenum Press, 1981.
8. Kaufman L and Rousseeuw PJ. *Finding groups in data: An introduction to cluster analysis.* John Wiley & Sons, 1990.
9. Saul LK and Roweis ST. An introduction to locally linear embedding. Technical report, AT&T Labs - Research, 2001.
10. Bernstein M, Silva V, Langford JC, and Tenenbaum JB. Graph approximations to geodesics on embedded manifolds. Technical report, Department of Psychology, Stanford University, 2000.
11. Souvenir R and Pless R. Manifold clustering. *10th International Conference on Computer Vision*, pages 648–653, 2005.
12. Roweis ST and Saul LK. Nonlinear dimensionality reduction by locally linear embedding. *Science*, 290:2323–2326, 2000.

Part II

Control Systems

Stability Analysis of the Simplest Takagi-Sugeno Fuzzy Control System Using Popov Criterion

Xiaojun Ban[1], X.Z. Gao[2], Xianlin Huang[3], and Hang Yin[4]

[1] Department of Control Theory and Engineering, Harbin Institute of Technology, Harbin, China
xiaojun1978@xinhuanet.com.
[2] Institute of Intelligent Power Electronics, Helsinki University of Technology, Espoo, Finland
gao@cc.hut.fi.
[3] Department of Control Theory and Engineering, Harbin Institute of Technology, Harbin, China
xlinhuang@hit.edu.cn
[4] Department of Control Theory and Engineering, Harbin Institute of Technology, Harbin, China
yinhang@hit.edu.cn

Abstract. In our paper, the properties of the simplest Takagi-Sugeno (T-S) fuzzy controller are first investigated. Next, based on the well-known Popov criterion with graphical interpretation, a sufficient condition in the frequency domain is proposed to guarantee the globally asymptotical stability of the simplest T-S fuzzy control system. Since this sufficient condition is presented in the frequency do-main, it is of great significance in designing the simplest T-S fuzzy controller in the frequency domain.

Keywords: Takagi-Sugeno (T-S) fuzzy controllers, Popov criterion, stability analysis, frequency response methods.

1 Introduction

The Takagi-Sugeno (T-S) fuzzy model [1] is a landmark in the history of fuzzy control theory. Numerous fuzzy control problems, such as stability analysis, systematic design, robustness, and optimality, can be ad-dressed under the framework of this T-S model [2]. Especially, given a T-S fuzzy model, a fuzzy controller design method named Parallel Distributed Compensation (PDC), has been proposed by Sugeno and Kang [3]. The corresponding stability analysis is also discussed in one of their papers [4]. The unique advantage of the PDC technique is that a lot of conventional linear controller design solutions based on both classical and modern control theory, which are actually for linear control systems, can be deployed in designing the nonlinear T-S fuzzy controllers as well.

As we know, frequency response methods have been well-developed, and widely used in industrial applications, which are straightforward and easy to follow by practicing engineers. The negative effect of noise in a control system can be evaluated

A. Saad et al. (Eds.): Soft Computing in Industrial Applications, ASC 39, pp. 63–71, 2007.
springerlink.com

by its frequency response. This advantage is very useful for control system analysis, since unavoidable noise usually deteriorates the overall control performance [5]. Besides, two popular frequency response methods, Bode plots and Nyquist plots, can provide a graphic insight into the control systems under study, and help engineers synthesize the corresponding controllers. Therefore, fusion of the T-S fuzzy model and frequency response methods is emerging in the field of control engineering. It is apparently necessary to analyze the stability of T-S fuzzy control systems in the frequency domain, when the frequency response methods are utilized in designing T-S fuzzy controllers.

The frequency response methods have been employed in both the Mamdani and T-S fuzzy control systems. In [6], the describing function approach is used to analyze the stability of a Mamdani fuzzy control system. Various types of Mamdani fuzzy controllers, e.g., Single-Input-Single-Output (SISO) and Multiple-Input-Multiple-Output (MIMO), are further investigated based on this describing function method in [7]-[10]. [11] and [12] discuss the application of the circle criterion and its graphical interpretation in the stability analysis of the SISO and MIMO Mamdani fuzzy control systems. In [13]-[15], the stability of the Mamdani fuzzy controllers is explored based on the Popov criterion. The describing function method is also used in the T-S fuzzy control systems in [16]. The multi-variable circle criterion is utilized to analyze the stability of the T-S fuzzy controllers in [16] and [17]. In this paper, we investigate the globally asymptotical stability of the simplest T-S fuzzy control system by using the famous Popov criterion with graphical interpretation in the frequency domain.

Our paper is organized as follows: in Section II, the simplest T-S fuzzy control system is first discussed. In Section III, the principles of the Popov criterion are briefly introduced. Next, a sufficient condition is derived to guarantee the globally asymptotically stability of the simplest T-S fuzzy control system in Section IV. A numerical example is presented in Section V to demonstrate how to employ this condition for the stability analysis of the simplest T-S fuzzy controller. Finally, some conclusions are drawn in Section VI.

2 Configuration of the Simplest T-S Fuzzy Control System

The structure of the simplest T-S fuzzy control system to be explored in our paper is shown in Fig. 1, where FLC and *G*(*s*) are the T-S fuzzy controller and plant to be controlled, respectively, *r* is the reference input, *e* is the feedback error, *u* is the controller output, and *y* is the system output.

This simplest T-S fuzzy controller can be described by the following two rules:

$$\text{If } e \text{ is } A, \text{ then } u_1 = k_1 e,$$
$$\text{If } e \text{ is } B, \text{ then } u_2 = k_2 e,$$

where *e* is the input of this T-S fuzzy controller, u_i , i =1,2 are the outputs of the local consequent controllers, which are both proportional controllers here. It should be pointed out that $k_i, i=1,2$, gains of these local controllers, are assumed to be positive in this paper. Both A and B are fuzzy sets, and we use the triangular membership functions to quantify them, as shown in Fig. 2.

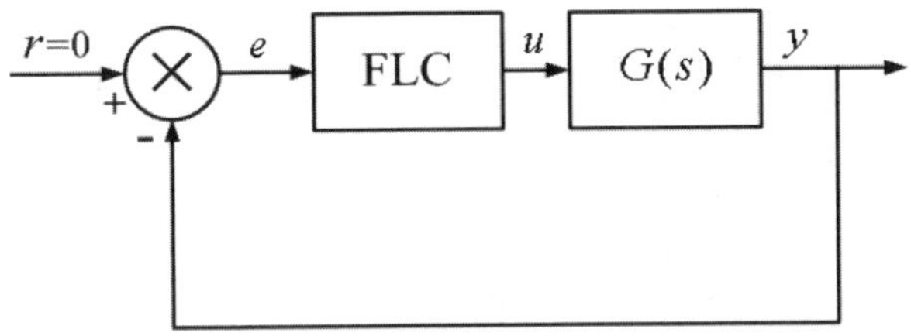

Fig. 1. Structure of the simplest T-S fuzzy control system

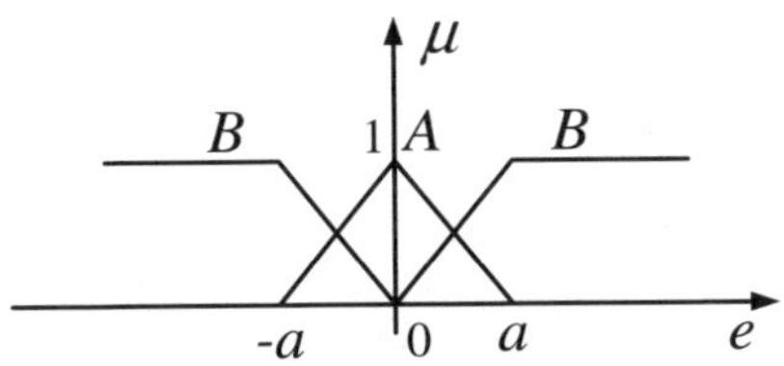

Fig. 2. Membership functions of *A* and *B*

A and *B* can be written as follows:

$$\mu_A(e)=\begin{cases} 0, & e<-a \\ \frac{1}{a}(e+a), & -a\le e<0 \\ -\frac{1}{a}(e-a), & 0\le e<a \\ 0, & e\ge a \end{cases}, \tag{1}$$

$$\mu_B(e)=\begin{cases} 1, & e<-a \\ -\frac{1}{a}e, & -a\le e<0 \\ \frac{1}{a}e, & 0\le e<a \\ 1, & e\ge a \end{cases}. \tag{2}$$

3 Popov Criterion

In this section, the Popov criterion [18] is briefly discussed, which is employed for the stability analysis of our simplest T-S fuzzy control system.

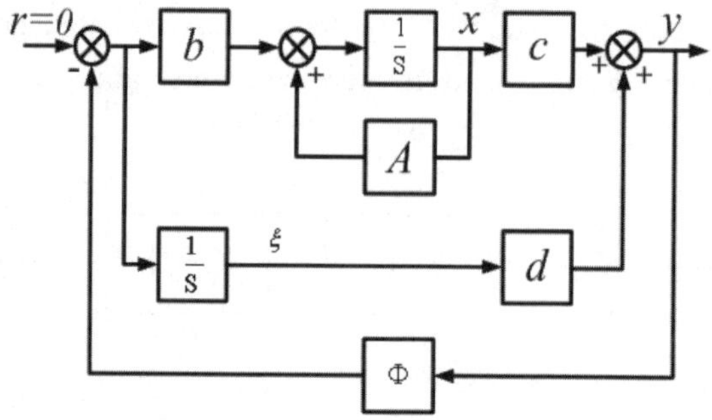

Fig. 3. Structure of nonlinear system for Popov criterion

The structure of the nonlinear system for the Popov criterion is illustrated in Fig. 3, and it can be formulated by the following state equations:

$$\dot{X} = AX + bu , \tag{3}$$

$$\dot{\xi} = u , \tag{4}$$

$$y = cX + d\xi , \tag{5}$$

$$u = -\Phi(y) , \tag{6}$$

where $X \in R^n, \xi, u, y$ are all scalars, and *A, b, c, d* have commensurate dimensions. The nonlinear element $\Phi : R \to R$ is a time-invariant nonlinearity belonging to sector $(0, k)$, where $k>0$ is a finite number.

Here, function $\Phi : R \to R$ belongs to sector $(0, k)$, if both the following two assumptions are met:

(i) $\Phi(0) = 0$, (7)

(ii) $\Phi(y)(ky - \Phi(y)) > 0, \forall y \neq 0$. (8)

According to the state equations of the above nonlinear system, the transfer function of the linear system in the forward path is

$$h(s) = \frac{d}{s} + c(sI - A)^{-1} b \cdot \tag{9}$$

Popov Criterion: Consider the above system, and suppose (i) the matrix *A* is Hurwitz, (ii) the pair (*A, b*) is controllable, (iii) the pair (*c, A*) is observable, (iv) $d>0$, and the nonlinear element Φ belongs to sector $(0, k)$, where $k>0$ is a finite number. Under these conditions, this system is globally asymptotically stable, if there exists a number $r>0$, such that

$$\inf_{\omega \in R} \text{Re}[(1 + j\omega r) h(j\omega)] + \frac{1}{k} > 0 \cdot \tag{10}$$

The graphical interpretation of the Popov criterion can be given as follows: suppose we plot $\omega \, \text{Im}\, h(j\omega)$ *vs.* $\text{Re}\, h(j\omega)$, when ω varies from 0 to ∞, which is known as the Popov plot of *h(s)*, the nonlinear system is globally asymptotically stable if there exists a nonnegative number *r*, such that the Popov plot of *h(s)* lies to the right of a straight line passing through the point (-1/*k*,0) with a slope of 1/*r*.

Proof. Refer to [18] for the proof of this theorem.

4 Analysis of the Simplest Takagi-Sugeno Fuzzy Control System

In this section, the Popov criterion is employed to analyze the stability of the above simplest T-S fuzzy control system. Several theorems and lemmas are proved to demonstrate that if certain hypotheses are satisfied, the stability of the nonlinear system illustrated in Fig. 1 can be analyzed by using the Popov criterion.

Theorem 1. Let $\Phi(e)$ represent the functional mapping achieved by the simplest T-S fuzzy control system, the following equation holds:

$$\Phi(-e) = -\Phi(e)\,. \tag{11}$$

Proof. The proof is omitted for the simplicity of our presentation.

In fact, the theorem states that the functional mapping achieved by this simplest T-S fuzzy controller is symmetric about the origin.

Based on Theorem 1, the nonlinear system described in Fig. 1 can be recast into the system represented in Fig. 4, in which the minus in front of variable *e* is taken from before the FLC to after the module. In other words, the two systems are equivalent based on Theorem 1. It is observed from Fig. 4, if the functional mapping achieved by the simplest fuzzy T-S controller belongs to some sector, and *G*(*s*) satisfies the assumptions in the Popov criterion, this theorem can be directly applied to the simplest T-S fuzzy control system.

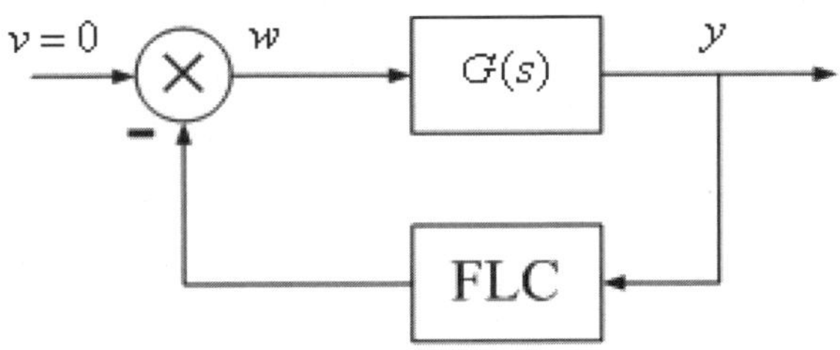

Fig. 4. Equivalent diagram of the fuzzy system in Fig. 1

Lemma 1. The following two statements are equivalent:

(i) $\Phi(y)(ky - \Phi(y)) > 0, \forall y \neq 0, and, \Phi(0) = 0$, (12)

(ii) $0 < y\Phi(y) < ky^2, \forall y \neq 0, and, \Phi(0) = 0$, (13)

where *k* is a positive number.

Proof. The proof is omitted here for convenience.

Theorem 2. Let $\Phi(e)$ denote the functional mapping of the T-S fuzzy controller in Figs. 4 or 1, $\Phi(e)$ belongs to sector $(0, k_2 + \varepsilon)$, where ε is a sufficiently small positive number, i.e., the following inequality holds:

(i) $\Phi(0) = 0$, (14)

(ii) $\Phi(e)[(k_2 + \varepsilon)e - \Phi(e)] > 0, \forall e \neq 0$. (15)

Proof

The output of the simplest T-S fuzzy controller can be represented as:

$$\Phi(e) = \frac{\mu_A(e)u_1 + \mu_B(e)u_2}{\mu_A(e) + \mu_B(e)} = (\frac{\mu_A(e)k_1 + \mu_B(e)k_2}{\mu_A(e) + \mu_B(e)})e \cdot \tag{16}$$

Note that $\mu_A(e) \geq 0$ and $\mu_B(e) \geq 0$.

It is obvious that $\Phi(e) = 0$, if and only is e=0. Furthermore, there are

$$k_1 \leq \frac{\mu_A(e)k_1 + \mu_B(e)k_2}{\mu_A(e) + \mu_B(e)}) \leq k_2, \tag{17}$$

$$k_1 e^2 \leq \frac{\mu_A(e)k_1 + \mu_B(e)k_2}{\mu_A(e) + \mu_B(e)})e^2 \leq k_2 e^2, \tag{18}$$

$$k_1 e^2 \leq \Phi(e)e \leq k_2 e^2. \tag{19}$$

Obviously, there exists a sufficiently small positive number ε, such that $0 < \Phi(e)e < (k_2 + \varepsilon)e^2$, when e is not equal to zero. In view of Lemma 1, $\Phi(e)$ belongs to sector $(0, k_2 + \varepsilon)$.

Now, we can apply the Popov criterion to the stability analysis of our simplest T-S fuzzy control systems, as stated by the following theorem.

Theorem 3. The fuzzy control system shown in Fig. 1 is globally asymptotically stable, if the following set of conditions hold:

(i) $G(s)$ can be represented by the state equations from (3) to (6),
(ii) the matrix A is Hurwitz,
(iii) the pair (A, b) is controllable, and the pair (c, A) is observable,
(iv) $d > 0$,
(v) there exists a number $r > 0$, such that

$$\inf_{\omega \in R} \mathrm{Re}[(1 + j\omega r)h(j\omega)] + \frac{1}{k_2 + \varepsilon} > 0, \tag{20}$$

where ε is a sufficiently small positive number.

Similar to the graphical interpretation of the Popov criterion, the graphical interpretation of the above theorem is that suppose we plot $\mathrm{Re}\, h(j\omega)$ vs. $\omega \mathrm{Im} h(j\omega)$, as ω varies from 0 to ∞, the equilibrium of our simplest T-S fuzzy system is globally asymptotically stable, if there exists a nonnegative number r, such that the Popov plot of $h(s)$ lies to the right of a straight line passing through the point $\left(\frac{-1}{k_2 + \varepsilon}, 0\right)$ with a slope of $1/r$.

As a matter of fact, if there exists a nonnegative number r, such that the Popov plot of $h(s)$ lies to the right of a straight line passing through point $\left(\frac{-1}{k_2},0\right)$, it is guaranteed that there is a line passing through point $\left(\frac{-1}{k_2+\varepsilon},0\right)$, such that the Popov plot of $h(s)$ lies to the right of this straight line. The discussion can be formally stated as the following corollary.

Corollary 1. The fuzzy control system shown in Fig. 1 is globally asymptotically stable, if the following set of conditions hold:

(i) - (iv) are the same with those in Theorem 3.
(v) there exists a number r>0, such that

$$\inf_{\omega\in R} \mathrm{Re}[(1+j\omega r)h(j\omega)]+\frac{1}{k_2}>0\cdot \tag{21}$$

In the next section, a numerical example is presented to demonstrate how to employ Theorem 3 or Corollary 1 in analyzing the stability of our T-S fuzzy control system.

5 Simulations

Example. In this example, a stable plant to be controlled is:

$$G(s)=\frac{1}{s(s+1)^2}\cdot \tag{22}$$

Two suitable proportional gains, $k_1=0.2, k_2=0.5$, are obtained based on the Bode plot of $G(s)$. A simplest T-S fuzzy controller with the following two rules is constructed:

If e is A, then $u_1=0.2e$,

If e is B, then $u_2=0.5e$.

'a', a characteristics parameter of the input membership functions (refer to (1) and (2)), is $\pi/8$. Note, $\pi/8$ is chosen here only for convenience. In fact, a has no effect on the stability of the closed-loop fuzzy control system. In this example, the proportional compensators are designed only to show how to utilize Theorem 3 in the stability analysis of our fuzzy control system, which does not require the control performance of the system be perfect. Both the Popov plot of $G(s)$ and a straight line passing through point $(\frac{1}{k_2},0)$ with the slope of 0.5 are shown in Fig. 5. It is argued in Theorem 3 that the whole nonlinear system is globally asymptotical stable, if there exists a nonnegative number r, such that the Popov plot of $h(s)$ lies to the right of a straight line passing through point $\left(\frac{-1}{k_2},0\right)$ with a slope of $1/r$. Hence, the T-S fuzzy control system in this example is globally asymptotical stable, since we can easily find such a straight line passing through point $\left(\frac{-1}{k_2},0\right)$, provided that k_2 is less than 2.

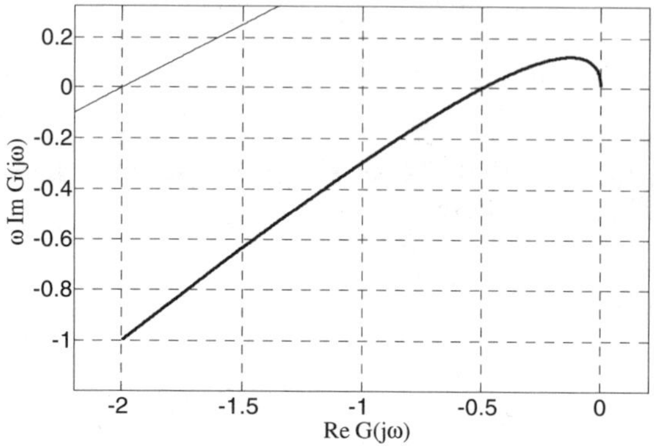

Fig. 5. Popov plot of G(s) and a straight line (solid line represents Popov plot of G(s), and thin line represents straight line passing through point $(\frac{1}{k_2},0)$ with a slope of 0.5)

6 Conclusions

In our paper, the properties of the simplest T-S fuzzy controller are first investigated. A sufficient condition is next proposed to guarantee the globally asymptotical stability of the equilibrium point of this simplest T-S fuzzy control system by using the famous Popov criterion. The theorem derived based on the Popov theorem has a good graphical interpretation in the frequency domain. Thus, it can be employed in designing a T-S fuzzy controller in the frequency domain. Additionally, it can be observed that *a*, which is a characteristics parameter of the input membership functions, has no effect on the stability of the fuzzy control system. However, *a* does affect the dynamical control performance. With the decrease and growth of *a,* the simplest T-S fuzzy controller converts to $u = k_2 e$ and $u = k_1 e$, respectively. Therefore, how to choose appropriate *a* for achieving the optimal control performance is a challenging topic. We also emphasize that although only two fuzzy rules are examined here, the proposed stability analysis method is still applicable to the single-input T-S fuzzy controllers with multiple rules. Although the theorems derived in this paper have a graphical interpretation, their conditions imposed on the controlled plant are stringent, and the T-S fuzzy controller that has been analyzed is simple. In the future research, we are going to explore new stability theorems with graphical interpretation in the frequency domain for a wider class of plants as well as more complex T-S fuzzy controllers.

Acknowledgment. X. Z. Gao's research work was funded by the Academy of Finland under Grant 201353.

References

1. Takagi, T., Sugeno, M.: Fuzzy Identification of Systems and Its Applications to Modeling and Control. IEEE Transactions on Systems, Man, and Cybernetics. Vol. 15. (1985) 116-132
2. Tanaka, K., Wang, H. O.: Fuzzy Control Systems Design and Analysis: A Linear Matrix Inequality Approach. Wiley-Interscience Publication: New York (2001)
3. Sugeno, M., Kang, G. T.: Fuzzy Modeling and Control of Multilayer Incinerator. Fuzzy Sets and Systems. Vol. 18. (1986) 329-346
4. Tanaka, K., Sugeno, M.: Stability Analysis and Design of Fuzzy Control Systems. Fuzzy Sets and Systems. Vol. 45. (1992) 135-156
5. D'azzo, J. J., Houpis, C. H.: Linear Control System Analysis and Design. Fourth Edition, McGraw-Hill Companies Inc (1995)
6. Kickert, W. J. M., Mamdani, E. H.: Analysis of a Fuzzy Logic Controller. Fuzzy Sets and Systems. Vol. 1. (1978) 29-44
7. Abdelnour, G., Cheung, J. Y., Chang, C-H., Tinetti, G.: Application of Describing Functions in the Transient Response Analysis of a Three-Term Fuzzy Controller. IEEE Transactions on Systems, Man, Cybernetics. Vol. 23. (1993) 603-606
8. Abdelnour, G., Cheung, J. Y., Chang, C-H., Tinetti, G.: Steady-state Analysis of a Three-Term Fuzzy Controller. IEEE Transactions on Systems, Man, Cybernetics. Vol. 23. (1993) 607-610
9. Gordillo, F., Aracil, J., Alamo, T.: Determining Limit Cycles in Fuzzy Control Systems. Proceedings of the IEEE International Conference on Fuzzy Systems, Barcelona, Spain (1997) 193-198
10. Kim, E., Lee, H., Park, M.: Limit-cycle Prediction of a Fuzzy Control System Based on Describing Function Method. IEEE Transactions on Fuzzy Systems. Vol. 8. (2000) 11-22
11. Ray, K. S., Majumder, D. D.: Application of Circle Criteria for Stability Analysis of Linear SISO and MIMO Systems Associated with Fuzzy Logic Controller. IEEE Transactions on Systems, Man, and Cybernetics. Vol. 14. (1984) 345-349
12. Guerra, R. E., Schmitt-Braess, G., Haber, R. H., Alique, A., Alique, J. R.: Using Circle Criteria for Verifying Asymptotic Stability in Pl-like Fuzzy Control Systems: Application to the Milling Process. IEE Proceedings-Control Theory and Applications. Vol. 150. (2003) 619-627
13. Furutani, E., Saeki, M., Araki, M.: Shifted Popov Criterion and Stability Analysis of Fuzzy Control Systems. Proceedings of the 31st IEEE Conference on Decision and Control, Tucson, AZ (1992) 2790-2795
14. Kandel, A., Luo, Y., Zhang, Y. Q.: Stability Analysis of Fuzzy Control Systems. Fuzzy Sets and Systems. Vol. 105. (1999) 33-48
15. Fuh, C. C., Tung, P. C.: Robust Stability Analysis of Fuzzy Control Systems. Fuzzy Sets and Systems.Vol. 88. (1997) 289-298
16. Cuesta, F., Gordillo, F., Aracil, J., Ollero, A.: Stability Analysis of Nonlinear Multivariable Takagi-Sugeno Fuzzy Control Systems. IEEE Transactions on Fuzzy Systems. Vol. 7. (1999) 508-520
17. Cuesta, F., Ollero, A.: Fuzzy Control of Reactive Navigation with Stability Analysis Based on Conicity and Lyapunov Theory. Control Engineering Practice. Vol. 12. (2004) 625-638
18. Vidyasagar, M.: Nonlinear Systems Analysis. Prentice-Hall Inc.: Englewood Cliffs, NJ 1993

Identification of an Experimental Process by B-Spline Neural Network Using Improved Differential Evolution Training

Leandro dos Santos Coelho[1] and Fabio A. Guerra[2]

[1] Production and Systems Engineering Graduate Program, PPGEPS
Pontifical Catholic University of Parana, PUCPR
Imaculada Conceição, 1155, Zip code 80215-901, Curitiba, Parana, Brazil
[2] Institute of Technology for Development, LACTEC
Low Voltage Technology Unit, UTBT
Centro Politécnico UFPR, Zip code 81531-980, Curitiba, Parana, Brazil

Abstract. B-spline neural network (BSNN), a type of basis function neural network, is trained by gradient-based methods, which may fall into local minimum during the learning procedure. To overcome the problems encountered by the conventional learning methods, differential evolution (DE) — an evolutionary computation methodology — can provide a stochastic search to adjust the control points of a BSNN are proposed. DE incorporates an efficient way of self-adapting mutation using small populations. The potentialities of DE are its simple structure, easy use, convergence property, quality of solution and robustness. In this paper, we propose a modified DE using chaotic sequence based on logistic map to train a BSNN. The numerical results presented here indicate that the chaotic DE is effective in building a good BSNN model for nonlinear identification of an experimental nonlinear yo-yo motion control system.

Keywords: B-spline neural network, system identification, differential evolution.

1 Introduction

The use of neural networks to nonlinear identification problems has attracted some attention in recent years [1]-[3]. A relevant approach is to find the best approximation with respect to certain class of basis functions for neural networks representation. In this case, there are many possible choices of basis functions, such as radial basis function, associate memory networks, wavelets, and B-spline function.

The main advantage of the B-spline functions over other radial functions e.g., the Bezier curve, is the local control of the curve shape, as the curve only changes in the vicinity of a few control points that have been changed [4]. A B-spline neural network (BSNN) consists of the piecewise polynomials with a set of local basis functions to model an unknown function for which a finite set of input-output samples are available. The performance of the identification depends largely of an optimization algorithm for the training procedure of BSNN in order to avoid any possible local minima.

A. Saad et al. (Eds.): Soft Computing in Industrial Applications, ASC 39, pp. 72–81, 2007.
springerlink.com

In this context, the development of training methods and improvements for BSNN is an emergent research area. Several heuristics have been developed in recent years to improve the performance and set up the parameters of the BSSN design and also fuzzy systems approaches [5]-[11].

The differential evolution (DE) algorithm [12], [13] is an evolutionary algorithm (EA) which uses a rather greedy and less stochastic approach to problem solving than do classical evolutionary algorithms such as genetic algorithms, evolutionary programming, and evolution strategies. DE also incorporates an efficient way of self-adapting mutation using small populations.

In this paper, we propose a modified DE using chaotic sequence to train a BSNN. Chaos describes the complex behavior of a nonlinear deterministic system [14]. The application of chaotic sequences instead of random sequences in DE is a powerful strategy to diversify the DE population and improve the DE's performance in preventing premature convergence to local minima. Numerical results for identification of the nonlinear dynamics of an experimental yo-yo motion system show the feasibility and effectiveness of the proposed approach.

2 B-Spline Neural Networks

BSNN is introduced as a class of one-hidden-layer feedforward neural networks composed of B-spline functions. Each basis function is composed of q polynomial segments. There exists a simple and stable recursive relationship for evaluating the membership of a B-spline basis function of order k,

$$N_q^j(x) = \left(\frac{x-\lambda_{j-q}}{\lambda_{j-1}-\lambda_{j-q}}\right) N_{q-1}^{j-1}(x) + \left(\frac{\lambda_j - x}{\lambda_j - \lambda_{j-q+1}}\right) N_{q-1}^{j}(x) \tag{1}$$

$$N_1^j(x) = \begin{cases} 1 & if\,(x \in I_j) \\ 0 & otherwise \end{cases} \tag{2}$$

where $N_q^j(\cdot)$ is defined as the j-th univariate basis function of order q and λ_j the j-th knot and I_j is the j-th interval. The output of neural network is

$$\hat{o}_k = f(x_k) = \sum_{j=1}^{p} w_j N_q^j(x_k) \tag{3}$$

where x_k and $\hat{o}_k$ are the inputs and output of network, respectively, w_j is the weight attached to the j-th basis function and $N_q^j(\cdot)$ is given by the recursive form (2). The index j is associative with the region of local support $\lambda_{(j-q)} \le x \le \lambda_{(j)}$, whereas the index q indicates the order of the basis functions [13].

The quality of approximation depends on the placement of knots of B-spline functions. The objective of optimization of BSNNs by DE is determination of the knots of each B-spline basis functions. In particular, the number of basis functions in this work is choice of user.

3 Differential Evolution (DE) for BSNN Training

The variant implemented here of DE was the DE/*rand*/1/*bin*, which involved the following steps and procedures:

Step 1: *Parameter setup*

The user chooses the parameters of population size, the boundary constraints of optimization variables, the mutation factor (f_m), the crossover rate (CR), and the stopping criterion of maximum number of iterations (generations), G_{max}.

Step 2: *Initialization of an individual population*

Set generation t=0. Initialize a population of i=1,..,M individuals (real-valued n-dimensional solution vectors) with random values generated according to a uniform probability distribution in the n dimensional problem space. These initial individual values are chosen at random from within user-defined bounds (boundary constraints).

Step 3: *Evaluation of the individual population*

Evaluate the fitness value of each individual.

Step 4: *Mutation operation (or differential operation)*

Mutation is an operation that adds a vector differential to a population vector of individuals according to the following equation:

$$z_i(t+1) = x_{i,r_1}(t) + f_m\,[x_{i,r_2}(t) - x_{i,r_3}(t)] \tag{4}$$

where i =1,2,...,M is the individual's index of population; j=1,2,..., n is the position in n dimensional individual; t is the time (generation); $x_i(t) = \left[x_{i_1}(t), x_{i_2}(t), \ldots, x_{i_n}(t)\right]^{\mathrm{T}}$ stands for the position of the i-th individual of population of N real-valued n-dimensional vectors; $z_i(t) = \left[z_{i_1}(t), z_{i_2}(t), \ldots, z_{i_n}(t)\right]^{\mathrm{T}}$ stands for the position of the i-th individual of a *mutant vector*; r_1, r_2 and r_3 are mutually different integers and also different from the running index, i, randomly selected with uniform distribution from the set $\{1, 2, \cdots, i-1, i+1, \cdots, N\}$; $f_m > 0$ is a real parameter called *mutation factor*, which controls the amplification of the difference between two individuals so as to avoid search stagnation and is usually taken from the range [0.1, 1].

Step 5: *Recombination operation*

Following the mutation operation, recombination is applied to the population. Recombination is employed to generate a trial vector by replacing certain parameters of the target vector with the corresponding parameters of a randomly generated donor vector.

For each vector, $z_i(t+1)$, an index $rnbr(i) \in \{1, 2, \cdots, n\}$ is randomly chosen using uniform distribution, and a *trial vector*, $u_i(t+1) = \left[u_{i_1}(t+1), u_{i_2}(t+1), \ldots, u_{i_n}(t+1)\right]^T$, is generated with

$$u_{i_j}(t+1)=\begin{cases} z_{i_j}(t+1), \text{ if } (randb(j)\leq CR) \text{ or } (j=rnbr(i)),\\ x_{i_j}(t), \text{ if } (randb(j)> CR) \text{ or } (j\neq rnbr(i)) \end{cases} \quad (5)$$

In the above equations, *randb(j)* is the *j*-th evaluation of a uniform random number generation with [0, 1] and *CR* is a *crossover* or *recombination rate* in the range [0, 1]. The performance of a DE algorithm usually depends on three variables: the population size *N*, the mutation factor f_m, and the recombination rate *CR*.

Step 6: *Selection operation*

Selection is the procedure of producing better offspring. To decide whether or not the vector $u_i(t + 1)$ should be a member of the population comprising the next generation, it is compared with the corresponding vector $x_i(t)$. Thus, if *f* denotes the objective function under minimization, then

$$x_i(t+1)=\begin{cases} u_i(t+1), \text{ if } \ f(u(t+1))<f(x_i(t)),\\ x_i(t), \text{ otherwise} \end{cases} \quad (6)$$

In this case, the cost of each trial vector $u_i(t+1)$ is compared with that of its parent target vector $x_i(t)$. If the cost, *f*, of the target vector $x_i(t)$ is lower than that of the trial vector, the target is allowed to advance to the next generation. Otherwise, the target vector is replaced by the trial vector in the next generation [26].

Step 7: *Verification of stop criterion*

Set the generation number for $t = t + 1$. Proceed to Step 3 until a stopping criterion is met, usually G_{max}. The stopping criterion depends on the type of problem.

4 DE Using Chaotic Sequences for BSNN Training

Optimization algorithms based on chaos theory [14], [15] are stochastic search methodologies that differ from any of the existing EAs. EAs are optimization approaches with concepts bio-inspired on genetics and natural evolution. In contrast, chaotic optimization approaches are based on ergodicity, stochastic properties and irregularity [16]. These approaches are unlike some stochastic optimization algorithms, which escape from local minima by accepting bad solutions according to a certain probability [17]. Chaotic optimization approaches can escape from local minima more easily than other stochastic optimization algorithms can [18].

In the DE context, the concepts of chaotic optimization methods can be useful. The parameters *M*, *CR* and f_m of DE are generally the key factors affecting the DE's convergence. However, the f_m and *CR* parameters cannot ensure the optimization's ergodicity completely in the search phase because they are constant factors in traditional DE. Therefore, this paper offers three new approaches that introduce chaotic mapping with ergodicity, irregularity and the stochastic property into DE to improve its global convergence. The utilization of chaotic sequences in EAs can be useful to escape more easily from local minima than with the traditional EAs [16].

One of the simplest dynamic systems evidencing chaotic behavior is the iterator called the logistic map [14], [15], whose equation is given by:

$$y(t) = \mu \cdot y(t-1) \cdot [1 - y(t-1)] \tag{7}$$

where t is the sample, and μ is a control parameter, $0 \leq \mu \leq 4$. The behavior of the system of equation (7) is greatly changed with the variation of parameter μ. The value of μ determines whether y stabilizes at a constant size, oscillates between a limited sequence of sizes, or behaves chaotically in an unpredictable pattern. A very small difference in the initial value of y causes substantial differences in its long-time behavior [17]. Equation (7) is deterministic, displaying chaotic dynamics when $\mu = 4$ and $y(1) \notin \{0, 0.25, 0.50, 0.75, 1\}$. In this case, $y(t)$ is distributed in the range (0,1) provided the initial $y(1) \in (0,1)$ and $y(1)=0.48$, as was adopted here.

The design of methods to improve the convergence of DE is a challenging task in EA design. This work proposes three new DE approaches which, in combination with chaotic sequences (DEC) based on logistic map, are described as follows:

Approach 1 – DEC(1): The parameter f_m of equation (4) is modified by the formula (8) through the following equation:

$$z_i(t+1) = x_{i,r_1}(t) + f_1(t)[x_{i,r_2}(t) - x_{i,r_3}(t)] \tag{8}$$

$$f_1(t) = \mu \cdot f_1(t-1) \cdot [1 - f_1(t-1)] \tag{9}$$

$$f_1(t) \in (0,1) \tag{10}$$

where t is the current generation, $f_1(t)$ is the new mutation factor based on logistic map, and $\mu = 4$.

Approach 2 – DEC(2): The parameter f_m of equation (4) is incremented with the evolution of generations. The value of f_m is modified by the formula (11) using the following equation:

$$z_i(t+1) = x_{i,r_1}(t) + f_2(t)[x_{i,r_2}(t) - x_{i,r_3}(t)] \tag{11}$$

$$f_1(t) = \mu \cdot f_1(t-1) \cdot [1 - f_1(t-1)] \tag{12}$$

$$f_2(t) = \left[(f_{2f} - f_{2i}) \frac{G}{G_{\max}} + f_{2i} \right] f_1(t) \tag{13}$$

where $f_2(t)$ is a new chaotic mutation factor based on evolution of generations; f_{2i} and f_{2f} are constants (adopted $f_{2i} = 1.5$ and $f_{2f} = 0.5$), and G (the value of G is equal to t) is the current generation number.

Approach 3 – DEC(3): The value of parameter f_m of equation (4) is reduced as the generations evolve. The value of f_m is modified by same equations as those of DEC(2). However, here, f_{2i} and f_{2f} are constants (adopted f_{2i}=0.5 and f_{2f}=1.5).

5 Description of the yo-yo Motion System

Yo-yo playing is considered a representative example of open-loop unstable control problems that involve intermittent dynamic environments. Stable control of yo-yo playing relies on a proper phase relationship between the controller's action and the motion of the yo-yo [18].

The development of automatic control systems that efficiently control a yo-yo represents a significant challenge for the development of electromechanical designs [19], [20]. One of the main difficulties is the lack of sensors to obtain the motion measure of the toys. Another difficulty is the lack of mathematical models of this measurement device type, which justifies the use of the BSNNs to identify the dynamic behavior of a yo-yo motion in a real system.

The control system prototype employed in this work uses a yo-yo, and a direct current (DC) motor for its motion presents nonlinearity and complex behavior. A block diagram of the described system and a photograph of the system are presented in Figure 1 [21].

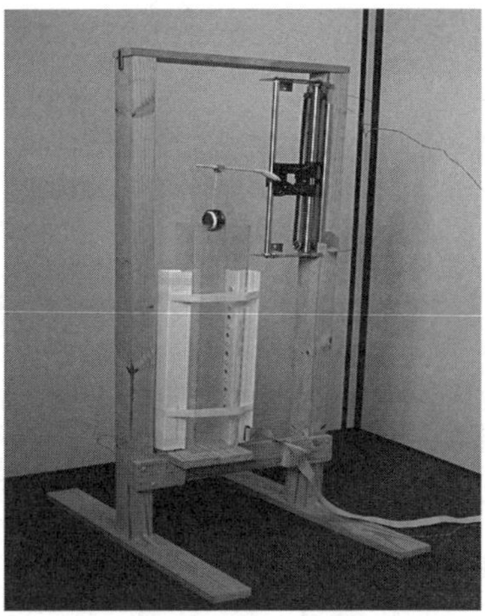

Fig. 1. Photograph of prototype of yo-yo motion system

The prototype modules are composed of hardware and firmware and are connected to the same printed circuit board, called the control board. The control board contains two hardware modules and communicates with a personal computer through the RS-232 I/O interface. All the components used for the yo-yo system are off the shelf items to keep the cost minimal.

6 Simulation Results of Nonlinear Identification

The mathematical model employed in this work to represent the yo-yo motion system is a NARX (Nonlinear AutoRegressive with eXogenous inputs). In this case, the

NARX model with series-parallel conception is used for one-step-ahead prediction of the BSNN system.

A computer with a data acquisition board for generating the control signal (identification in closed-loop using a proportional controller design) and position value of the yo-yo was used to obtain system measurements. In the identification procedure based on the BSNN model, 290 samples of input (tension applied to the DC motor) and output (position of yo-yo) were collected with a time sampling of 40ms (see Fig. 2). The tension value corresponds to the maximum value configuration of the driver in PWM (Pulse Width Modulation) control of a DC motor.

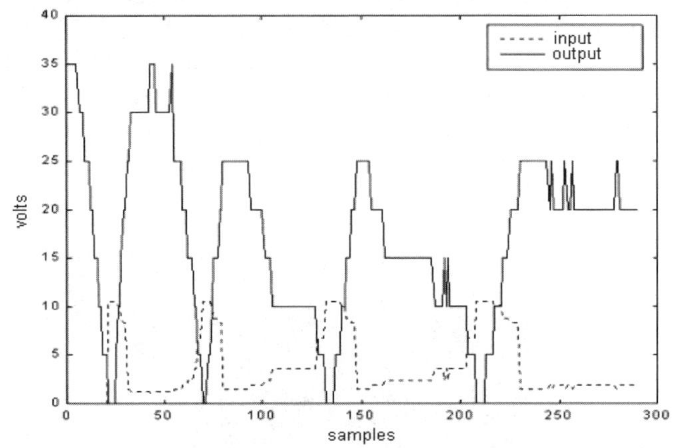

Fig. 2. Yo-yo motion system input and output data

Experiments for the estimation phase of the mathematical model of the yo-yo motion system are carried out using samples 1 to 150. For the validation phase, the BSNN model uses the input and output signals of samples 151 to 290. The system identification by BSNN model based on DE optimization is appropriate if a performance index is in permissible values for the user's needs. In this case, the fitness function for maximization proposes using DE and is given by the harmonic mean of multiple correlation indices of estimation and validation phases. The fitness function (to be maximized) is calculated using the expression of R^2_{est} given by:

$$R^2_{est} = 1 - \frac{\sum_{t=1}^{150} [y(t) - \hat{y}(t)]^2}{\sum_{t=1}^{150} [y(t) - \bar{y}]^2} \tag{14}$$

where R^2_{est} is the multiple correlation index of the estimation phase, $y(t)$ is the output of the real system, $\hat{y}(t)$ is the output estimated by the BSNN, and $\bar{y}$ is the mean value of the system's output. For the validation phase (verification of generalization capability) of optimized BSSN is employed the R^2_{val} index give by

$$R_{val}^2 = 1 - \frac{\sum_{t=151}^{290} [y(t) - \hat{y}(t)]^2}{\sum_{t=151}^{290} [y(t) - \bar{y}]^2} \tag{15}$$

where R_{val}^2 is the multiple correlation index of the validation phase. When the value R^2 =1.0 (estimation or validation phases), it indicates an accurate approach of the model for the measured data of the system. A R^2 value between 0.9 and 1.0 is considered sufficient for applications in designs of identification and model-based controller [22].

All the computational programs were run on a 3.2 GHz Pentium IV processor with 3 MB of RAM. In each case study, 30 independent runs were made for each of the optimization methods involving 30 different initial trial solutions for each optimization method. The setup of classical DE approaches used was the following:

- DE(1): classical DE using a constant mutation factor given by f_m = 0.4 and a crossover rate of CR = 0.8;
- DE(2): DE using a linear reduction of f_m with initial and final values of 1.5 and 0.3, respectively;
- DE(3): DE using a linear increase of f_m with initial and final values of 0.3 and 1.5, respectively.

In these case studies, the population size N was 20 and the stopping criterion G_{max} was 100 generations for the DE and DEC approaches. The three chosen vectors of BSNN's input were [$u(t$-1); $y(t$-2); $y(t$-1)]. The space searches for knots of each B-spline basis functions are [-1.0; 1.0]. Simulation tests using 3 to 7 knots in each input of BSNN were realized. However, the best performance of DE and DEC approaches was using 5 knots in each input of BSNN.

Table 1. presents the simulation results (best of 30 experiments with 100 generations for each run) for DE and DEC for optimization of BSNN using 5 knots. As indicated in Table 1, the results of the optimized BSNN present precision and provide an appropriate experimental mathematical model for the yo-yo motion system.

Table 1. Results obtained in estimation and validation phases by the maximization of R_{est}^2 using DE and DEC approaches (it is adopted the solution vector with best R_{est}^2 after the accomplishment of 30 runs)

DE type	R_{est}^2 maximum	R_{est}^2 mean	R_{est}^2 minimum	R_{est}^2 standard deviation	R_{val}^2 (of best R_{est}^2)
classical DE	0.95363	0.95410	**0.95449**	0.00035	0.8785
DEC(1)	0.95427	0.95371	0.95304	0.00055	0.8845
DEC(2)	0.95482	0.95440	0.95405	**0.00028**	0.8880
DEC(3)	**0.95486**	**0.95448**	0.95404	0.00031	0.8936

For the case study of the BSNN optimization, there is a consistent performance pattern across tested approaches with similar results in terms of R_{est}^2. The DE(3) approach presents better results in relation to the mean and maximum fitness than does the classical DE, DE(1) and DE(2). The best results shown in Figure 3 represent the BSNN (using DE(3)) with 5 knots for each network input.

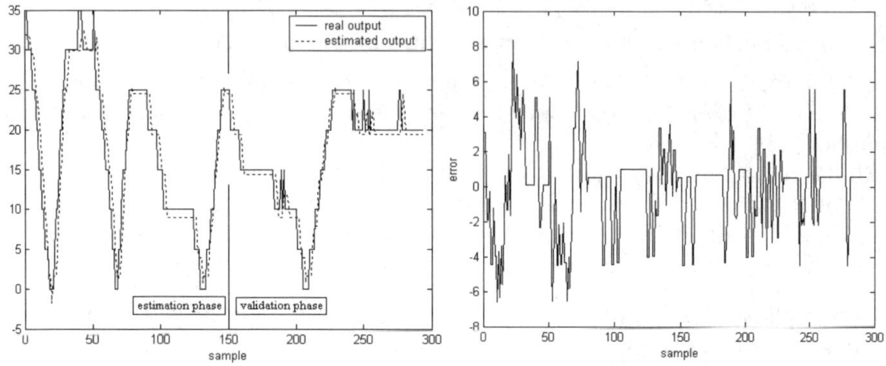

Fig. 3. Best result of R_{est}^2 using BSNN with 5 knots for each input and optimization based on DEC(3) (see Table 1)

7 Conclusion and Future Research

In this work, new DE approaches using chaotic sequence based on logistic map to adjust the control points of a BSNN were proposed. Simulation results show the potential of DE and DEC approaches to the BSNNs optimization in systems identification. In this context, more works need be done to test the DEC approaches on benchmark optimization problems.

References

[1] M. Kuczmann and A. Iványi, "Neural network model of magnetic hysteresis," *COMPEL: The Int. J. Computation and Math. in Electrical and Electronic Eng.*, vol. 21, no. 3, pp. 364-376, 2002.

[2] Y. Tan, X. Dang, F. Liang and S. -Y. Su, "Dynamic wavelet neural network for nonlinear dynamic system identification," *Proceedings of the IEEE Int. Conf. on Control Applications*, Anchorage, AL, USA, p. 214-219, 2000.

[3] S. Mcloone, M. D. Brown, G. Irwin and A. Lightbody, "A hybrid linear/nonlinear training algorithm for feedforward neural networks," *IEEE Trans. on Neural Networks*, vol. 9, no. 4, pp. 669-684, 1998.

[4] W. M. Newmann and R. F. Sproull, *Principles of Interactive Computer Graphics*, Mc-Graw-Hill, Inc, 1979.

[5] G. Starrenburg, W. T. C. van Luenen, W. Oelen, and J. van Amerongen, "Learning feedforward controller for a mobile robot vehicle," *Control Eng.Practice*, vol. 4, no. 9, pp. 1221-1230, 1996.

[6] J. Zhang and A. Knoll, "Designing fuzzy controllers by rapid learning," *Fuzzy Sets and Systems*, vol. 101, pp. 287-301, 1999.
[7] K. F. C. Yiu, S. Wang, K. L. Teo, and A. C. Tsoi, "Nonlinear system modeling via knot-optimizing B-spline networks," *IEEE Transactions on Neural Networks*, vol. 12, no. 4, pp. 1013-1022, 2001.
[8] A. Saranli and B. Baykal, "Complexity reduction in radial basis function (RBF) networks by using radial B-spline functions," *Neurocomputing*, vol. 18, pp. 183-194, 1998.
[9] K. Shimojima, T. Fukuda, and F. Arai, "Self-tuning fuzzy inference based on spline function," in *Proceedings of IEEE Int. Conference on Fuzzy Systems*, Orlando, FL, USA, pp. 690-695, 1994.
[10] V. K. Chu and M. Tomizuka, "Rule generation for fuzzy systems based on B-splines," in *Proceedings of IEEE Int. Conference on Neural Networks*, Perth, Australia, pp. 6098-611, 1995.
[11] S. Haykin, *Neural Networks*, 2nd edition, Prentice-Hall, Upper Saddle River, New Jersey, 1996.
[12] R. Storn and K. Price, "Differential evolution: a simple and efficient adaptive scheme for global optimization over continuous spaces," *TR-95-012*, Int. Computer Science Inst., Berkeley, 1995.
[13] R. Storn, "Differential evolution — a simple and efficient heuristic for global optimization over continuous spaces," *Journal of Global Optimization*, vol. 11, no. 4, pp. 341-359, 1997.
[14] T. S. Parker and L. O . Chua, *Practical numerical algorithms for chaotic system*, Springer-Verlag, Berlim, Germany, 1989.
[15] K. T. Alligood, T. D. Sauer and J. A. Yorke, *Chaos: an Introduction to Dynamical Systems*, Springer, 1996.
[16] L. S. Coelho and V. C. Mariani, "Combining of chaotic differential evolution and quadratic programming for economic dispatch optimization with valve-point effect," *IEEE Transactions on Power Systems*, vol. 21, no. 2, pp. 989-996, 2006.
[17] Shengsong, L., Min, W. and H. Zhijian, "Hybrid algorithm of chaos optimisation and SLP for optimal power flow problems with multimodal characteristic," *IEE Proceedings in Generation, Transmission, and Distribution*, vol. 150, no. 5, pp. 543-547, 2003.
[18] Li, B. and W. Jiang, "Optimizing complex functions by chaos search," *Cybernetics and Systems*, vol. 29, no. 4, pp. 409-419, 1998.
[19] Yan, X. F., Chen, D. Z. and S. X. Hu, "Chaos-genetic algorithm for optimizing the operating conditions based on RBF-PLS model," *Computers and Chemical Eng.*, vol. 27, pp. 1393-1404.
[20] H. -L. Jin and M. Zacksenhouse, "Oscillatory neural networks for robotic yo-yo control," *IEEE Transactions on Neural Networks*, vol. 14, no. 2, pp. 317-325, 2003.
[21] L. Zlajpah and B. Nemec, "Control strategy for robotic yo-yo," *Proceedings of the IEEE/RSJ Int. Conference on Intelligent Robots and Systems*, Las Vegas, Nevada, USA, pp. 767-772, 2003.
[22] K. Hashimoto and N. Toshiro, "Modeling and control of robotic yoyo with visual feedback," *IEEE International Conference on Robotics and Automation*, Minneapolis, Minnesota, USA, vol. 3, pp. 2650-2655, 1996.
[23] B. M. Herrera, L. V. Ribas, and L. S. Coelho, "Nonlinear identification method of a yo-yo system using fuzzy model and fast particle swarm optimization," *9th World Conference on Soft Computing in Industrial Applications*, [Online], Available: http://www.cs.nmt.edu/~wsc9/ , 2005.
[24] B. Schaible, H. Xie, and Y. C. Lee, "Fuzzy logic models for ranking process effects," *IEEE Transactions on Fuzzy Systems*, vol. 5, no. 4, pp. 545-556, 1997.

Applying Particle Swarm Optimization to Adaptive Controller

Leandro dos Santos Coelho[1] and Fabio A. Guerra[2]

[1] Production and Systems Engineering Graduate Program, PPGEPS
Pontifical Catholic University of Parana, PUCPR
Imaculada Conceição, 1155, Zip code 80215-901, Curitiba, Parana, Brazil
[2] Institute of Technology for Development, LACTEC
Low Voltage Technology Unit, UTBT
Centro Politécnico UFPR, Zip code 81531-980, Curitiba, Parana, Brazil

Abstract. A design for a model-free learning adaptive control (MFLAC) based on pseudo-gradient concepts and optimization procedure by particle swarm optimization (PSO) is presented in this paper. PSO is a method for optimizing hard numerical functions on metaphor of social behavior of flocks of birds and schools of fish. A swarm consists of individuals, called particles, which change their positions over time. Each particle represents a potential solution to the problem. In a PSO system, particles fly around in a multi-dimensional search space. During its flight each particle adjusts its position according to its own experience and the experience of its neighboring particles, making use of the best position encountered by itself and its neighbors. The performance of each particle is measured according to a pre-defined fitness function, which is related to the problem being solved. The PSO has been found to be robust and fast in solving non-linear, non-differentiable, multi-modal problems. Motivation for application of PSO approach is to overcome the limitation of the conventional MFLAC design, which cannot guarantee satisfactory control performance when the plant has different gains for the operational range when designed by trial-and-error by user. Numerical results of the MFLAC with particle swarm optimization for a nonlinear control valve are showed.

Keywords: particle swarm optimization, adaptive control, model-free adaptive control.

1 Introduction

Model-based control techniques are usually implemented under the assumption of good understanding of process dynamics and their operational environment. These techniques, however, cannot provide satisfactory results when applied to poorly modeled processes, which can operate in ill-defined environments. This is often the case when dealing with complex dynamic systems for which the physical processes are either highly nonlinear or are not fully understood [1].

The conventional proportional-integral-derivative (PID) algorithm is still widely used in process industries because its simplicity and robustness. PID controllers are the most common controllers in industry. In fact, 95% of control loops use PID and the majority is PI control [2]. However, its performance is not adequate in many

A. Saad et al. (Eds.): Soft Computing in Industrial Applications, ASC 39, pp. 82–91, 2007.
springerlink.com

chemical processes. A change in the signal and the directionality of the process gain is a complex practical situation and, so, still becoming complex the design of a control system [3].

In addition, several approaches have been proposed in the literature for controlling nonlinear processes, such as model predictive control, neural control, fuzzy control, robust control, sliding mode control, and adaptive control.

The aim of this paper is to merge for nonlinear systems, the model-free learning adaptive control structure [4], [5] with the controller design optimization based on particle swarm optimization (PSO) [6].

PSO methods explore the search space using a population of particles, each with a particle or agent, starting from a random location and velocity vector. Each particle in the swarm represents a candidate solution (treated as a point) in an n-dimensional space for the optimization problem, which adjusts its own "flying" according to other particles. Several heuristics have been developed in recent years to improve the performance and set up the parameters of the PSO algorithm [7]-[11].

2 Model-Free Learning Adaptive Control

In this paper, the direct adaptive control of the following general discrete SISO (Single-Input and Single-Output) nonlinear system is considered

$$y(k+1)=f\left(y(k),\cdots,y(k-n_a),u(k),\cdots,u(k-n_b)\right) \tag{1}$$

where n_a and n_b are the orders of system output, $y(k)$, and input, $u(k)$, respectively, and $f(\cdot)$ is a general nonlinear function. The plant (equation 1) can be rewritten as follows:

$$y(k+1)=f\left(Y(k),u(k),U(k-1)\right) \tag{2}$$

where $Y(k)$ and $U(k\text{-}1)$ are the sets of system outputs and inputs up to sampling instant k and k-1.

The following assumptions are considered about the controlled plant: (A1) the system (1) and (2) is observable and controllable; (A2) the partial derivative of $f(\cdot)$ with respect to control input $u(k)$ is continuous; and (A3) the system (1) is generalized Lipschitz.

For a nonlinear system (2), satisfying assumptions (A1-A3), then there must exist $\phi(k)$, called pseudo-gradient vector, when control change $\Delta u(k)\neq 0$, and

$$\Delta y(k+1)=\phi^T(k)\Delta u(k) \tag{3}$$

where the control change $\Delta u(k) = u(k) - u(k\text{-}1)$; $\|\phi(k)\|\leq L$, and L is a constant.

Details of the theoretical basis and the mathematical proof of the MFLAC are given in [4] and [5]. In this proof, the equation $y(k+1)=f\left(Y(k),u(k),U(k-1)\right)$ gives

$$\Delta y(k+1)=f\left(Y(k),u(k),U(k-1)\right)-f\left(Y(k-1),u(k-1),U(k-2)\right) \tag{4}$$

or

$$\begin{aligned}\Delta y(k+1)=&f\left(Y(k),u(k),U(k-1)\right)-f\left(Y(k),u(k-1),U(k-1)\right)+\\&f\left(Y(k),u(k),U(k-1)\right)-f\left(Y(k-1),u(k-1),U(k-2)\right)\end{aligned} \tag{5}$$

Using assumption (A2) and the mean value theorem, equation (5) gives

$$\Delta y(k+1) = \frac{\partial f^-}{\partial u(k)} \Delta u(k) + \xi(k) \tag{6}$$

where $\frac{\partial f^-}{\partial u(k)}$ denotes the value of gradient vector of $f(Y(k), u(k), U(k-1))$ with respect to u at some point between $u(k-1)$ and $u(k)$, and $\xi(k)$ given by

$$\xi(k) = f(Y(k), u(k-1), U(k-1)) - f(Y(k-1), u(k-1), U(k-2)) \tag{7}$$

Considering the following equation

$$\xi(k) = \eta^T(k) \Delta u(k) \tag{8}$$

where $\eta(k)$ is a variable. Since condition $\Delta u(k) \neq 0$, equation (8) must have solution $\eta(k)$.

$$\text{Let } \phi(k) = \frac{\partial f^-}{\partial u(k)} + \eta(k) \tag{9}$$

From (8) and (9), then (7) can be rewritten as $\Delta y(k+1) = \phi^T(k) \Delta u(k)$. This is the same as (3). In this case, by using (3) and assumption (A3), and $\Delta u(k) \neq 0$, we have

$$\left| \phi^T(k) \Delta u(k) \right| \le L \|\Delta u(k)\| \tag{10}$$

Hence $\|\phi(k)\| \le L$.

For the learning control law algorithm, a weighted one-step-ahead control input cost function is adopted, and given by

$$J(u(k)) = [y(k+1) - y_r(k+1)]^2 + \lambda \|\Delta u(k)\|^2 \tag{11}$$

For the control design, where $y_r(k+1)$ is the expected system output signal (true output of the controlled plant), and λ is a positive weighted constant. The equation (3) can be rewrite as follows

$$y(k+1) = y(k) + \phi^T(k) \Delta u(k) \tag{12}$$

Substituting (12) into (11), differentiating (11) with respect to $u(k)$, solving the equation $\partial J(u(k)) / \partial u(k) = 0$, and using the matrix-inversion-lemma gives the control law as follows:

$$u(k) = u(k-1) + \frac{\rho_k \phi(k)}{\lambda + \|\phi(k)\|^2} [y_r(k+1) - y(k)] \tag{13}$$

The control law (13) is a kind of control that has no relationship with any structural information (mathematical model, order, structure, etc.) of the controlled plant. It is designed only using I/O data of the plant.

The cost function proposed by Hou *et al.* [5] for parameter estimation is used in this paper as

$$J(\phi(k)) = \left[y(k) - y(k-1) - \phi^T \Delta u(k-1)\right]^2 + \mu\left\|\phi(k) - \hat{\phi}(k-1)\right\|^2 \quad (14)$$

Using the similar procedure of control law equations, we can obtain the parameter estimation algorithm as follows:

$$\hat{\phi}(k) = \hat{\phi}(k-1) + \frac{\eta \Delta u(k-1)}{\mu + \left\|\Delta u(k)\right\|^2}\left[\Delta y(k) - \hat{\phi}^T(k-1)\Delta u(k-1)\right] \quad (15)$$

Summarizing, the MFLAC scheme is

$$\hat{\phi}(k) = \hat{\phi}(k-1) + \frac{\eta \Delta u(k-1)}{\mu + \left\|\Delta u(k)\right\|^2}\left[\Delta y(k) - \hat{\phi}^T(k-1)\Delta u(k-1)\right] \quad (16)$$

$\hat{\phi}(k) = \hat{\phi}(1)$ if

$$sign(\phi(1)) \neq sign(\hat{\phi}(k)) \quad (17)$$

$\hat{\phi}(k) = \hat{\phi}(1)$ if

$$\left\|\hat{\phi}(k)\right\| \geq M, \text{or} \left\|\hat{\phi}(k)\right\| \leq \varepsilon \quad (18)$$

$$u(k) = u(k-1) + \frac{\rho_k \phi(k)}{\lambda + \left\|\phi(k)\right\|^2}\left[y_r(k+1) - y(k)\right] \quad (19)$$

where step-size series ρ and η, and the weighted constants λ and μ are design parameters optimized by differential evolution in this paper. The parameter ε is a small positive constant (adopted 0.00001), M is adopted with value 10, and $\hat{\phi}(k) = \hat{\phi}(1)$ is the initial estimation value of $\phi(k)$.

3 Optimization Using PSO

The proposal of PSO algorithm was put forward by several scientists who developed computational simulations of the movement of organisms such as flocks of birds and schools of fish. Such simulations were heavily based on manipulating the distances between individuals, i.e., the synchrony of the behavior of the swarm was seen as an effort to keep an optimal distance between them. Sociobiologist Edward Osbourne Wilson outlined a link of these simulations for optimization problems [6].

PSO, originally developed by Kennedy and Eberhart in 1995, is a population-based swarm algorithm [12], [13]. In the PSO computational algorithm, population dynamics simulates bio-inspired behavior, i.e., a “bird flock's” behavior which involves social sharing of information and allows particles to to take profit from the discoveries and previous experience of all the other particles during the search for food.

Each particle in PSO has a randomized velocity associated to it, which moves through the problem space.

Each particle in PSO keeps track of its coordinates in the problem space, which are associated with the best solution (fitness) it has achieved so far. This value is called *pbest* (personal best). Another "*best*" value that is tracked by the *global* version of the particle swarm optimizer is the overall best value. Its location, called *gbest* (global best), is obtained by any particle in the population. The past best position and the entire best overall position of the group are employed to minimize (or maximize) the solution.

The PSO concept consists, in each time step, of changing the velocity (acceleration) of each particle flying toward its *pbest* and *gbest* locations (global version of PSO). Acceleration is weighted by random terms, with separate random numbers being generated for acceleration toward *pbest* and *gbest* locations, respectively. The procedure for implementing the global version of PSO is given by the following steps:

Step 1: *Initialization random swarm positions and velocities*: Initialize a population (array) of particles with random positions and velocities in the n dimensional problem space using uniform probability distribution function.

Step 2: *Evaluation of particle's fitness*: Evaluate each particle's fitness value.

Step 3: *Comparison to pbest (personal best)*: Compare each particle's fitness with the particle's *pbest*. If the current value is better than *pbest*, then set the *pbest* value equal to the current value and the *pbest* location equal to the current location in n-dimensional space.

Step 3: *Comparison to gbest (global best)*: Compare the fitness with the population's overall previous best. If the current value is better than *gbest*, then reset *gbest* to the current particle's array index and value.

Step 4: *Updating of a particle's velocity and position*: Change the velocity, v_i, and position of the particle, x_i, according to equations (20) and (21):

$$v_i(t+1) = w \cdot v_i(t) + c_1 \cdot ud_i(t) \cdot [p_i(t) - x_i(t)] + c_2 \cdot Ud_i(t) \cdot [p_g(t) - x_i(t)] \quad (20)$$

$$x_i(t+1) = x_i(t) + \Delta t \cdot v_i(t+1) \quad (21)$$

where $i=1,2,\ldots,N$ indicates the number of particles of population (swarm); $t=1,2,\ldots t_{max}$, indicates the iterations, w is a parameter called the inertial weight; $v_i = [v_{i1}, v_{i2}, \ldots, v_{in}]^T$ stands for the velocity of the i-th particle, $x_i = [x_{i1}, x_{i2}, \ldots, x_{in}]^T$ stands for the position of the i-th particle of population, and $p_i = [p_{i1}, p_{i2}, \ldots, p_{in}]^T$ represents the best previous position of the i-th particle. Positive constants c_1 and c_2 are the cognitive and social components, respectively, which are the acceleration constants responsible for varying the particle speed towards *pbest* and *gbest*, respectively. Index g represents the index of the best particle among all the particles in the swarm. Variables $ud_i(t)$ and $Ud_i(t)$ are two random functions in the range [0,1]. Equation (1) represents the position update, according to its previous position and its velocity, considering $\Delta t = 1$.

Step 5: *Repeating the evolutionary cycle*: Return to step (ii) until a stop criterion is met, usually a sufficiently good fitness or a maximum number of iterations (generations).

In this work, a time-varying modification of c_1 and c_2 was used that can be represented as follows [14]:

$$c_1 = (c_{1f} - c_{1i}) \cdot \frac{t}{t_{max}} + c_{1i} \tag{22}$$

$$c_2 = (c_{2f} - c_{2i}) \cdot \frac{t}{t_{max}} + c_{2i} \tag{23}$$

where c_{1i}, c_{1f}, c_{2i} and c_{2f} are constants. In this work, an improved solution based on preliminary tests was observed when changing c_1 from 2.05 to 0.4 and changing c_2 from 0.4 to 2.5, i.e., the values c_{1i} = 2.05, c_{1f} = 0.4, c_{2i} = 0.4 and c_{2f} = 2.05 were adopted in the simulations done here.

The inertial weight w represents the degree of the momentum of the particles. The use of the variable w, inertial weight, is responsible for dynamically adjusting the speed of the particles.

The velocity of *i-th* particles in each dimension is associated with a maximum velocity V_{max}. If the sum of accelerations causes the velocity in that dimension to exceed V_{max}, which is a parameter specified by the user, then the velocity in that dimension is limited to V_{max}. The parameter, V_{max}, is used to determine the resolution with which the regions around the current solutions are searched. If V_{max} is too high, the PSO facilitates global search, and particles may fly past good solutions; if it is too small, the PSO facilitates local search, and the particles may not explore sufficiently beyond locally good regions.

The choice of the PSO approach for optimization of MFLAC design is based on its useful features such as [11]: (i) it is a stochastic search algorithm that is originally motivated by the mechanisms of swarm intelligence, (ii) it is less likely become trapped in a local optimum because it searches for the global optimal solution by manipulating a population of candidate solutions, and (iii) it is very effective for solving the optimization problems with nonsmooth objective functions as it does not require the derivative information.

In this paper, a PSO-based optimization technique is adopted to obtain $\phi(1), \rho$, η, λ and μ for the MFLAC design. The setup of PSO used in this work was the following:

- number of particles (swarm population size): 30;
- inertial weight using a linear reduction equation with initial and final values of 0.7 and 0.4, respectively;
- stop criterion: 20 generations.

The objective of the PSO in the MFLAC optimization is to maximize the fitness equation given by

$$f = \frac{\xi}{1 + \left\{ \sum_{i=1}^{t} \left| y(k) - y_r(k) \right| + 0.001 [u(k) - u(k-1)]^2 \right\}} \quad (24)$$

where $u(k)$ is the control signal, $y(k)$ is the process output, and $y_r(k)$ is the reference (setpoint), and ξ is a scale factor (adopted $\xi = 0.3$).

4 Simulation Results

The control valve system is an opening with adjustable area. Normally it consists of an actuator, a valve body and a valve plug. The actuator is a device that transforms the control signal to movement of the stem and valve plug. Wigren [15] describes the plant where the control valve dynamic is described by a Wiener model (the nonlinear element follows linear block) and it is given by

$$x(k) = 1{,}5714x(k-1) + 0{,}6873x(k-2) + 0{,}0616u(k\text{-}1) + 0{,}0543u(k\text{-}2) \quad (25)$$

$$y(k) = f_n[x(k)] = \frac{x(k)}{\sqrt{0{,}10 + 0{,}90[x(k)]^2}} \quad (26)$$

where $u(k)$ is the control pressure, $x(k)$ is the stem position, and $y(k)$ is the flow through the valve which is the controlled variable. The input to the process, $u(k)$, is constrained between [0; 1.2]. The nonlinear behavior of the control valve described by equation (26) is shown in Figure 1.

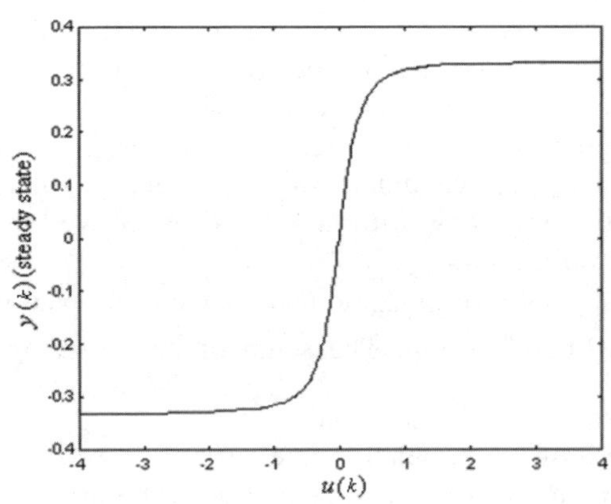

Fig. 1. Static characteristic of a control valve

The space search adopted in PSO setup is: $0.01 \le \phi(1) \le 0.50$, $0.10 \le \rho \le 5.00$, $-1.00 \le \eta \le 1.00$, $0.01 \le \lambda \le 1.00$, and $1.00 \le \mu \le 5.00$.

For the MFLAC design, the optimization procedure by PSO obtains $\phi(1) = 0.366618$, $\rho = 0.499131$, $\eta = 3.662461$, $\lambda = 1.375923$, $\mu = 0.391867$ and fitness $f = 0.8273$ (best results in 30 runs).

Simulation results for servo and regulatory responses of MFLAC are shown in Figures 2 and 3, respectively. Regulatory behavior analysis of the MFLAC was based on parametric changes in the plant output when: (i) sample 60: $y(k) = y(k) + 0.2$; (ii) sample 160: $y(k) = y(k) - 0.2$; (iii) sample 260: $y(k) = y(k) - 0.4$; (iv) sample 360: $y(k) = y(k) + 0.4$; and (v) sample 460: $y(k) = y(k) + 0.4$.

Numerical results presented in Figures 2 and 3 show that the MFLAC using PSO approach have precise control performance. In Table 1, a summary of simulation results and performance of the MFLAC design based on PSO is presented.

Table 1. Indices for the best MFLAC design using PSO

MFLAC	servo behavior	regulatory behavior
mean of *u*	0.5474	0.5535
variance of *u*	0.1227	0.1260
Mean of error	0.0160	0.0123
variance of error	0.0015	0.0025

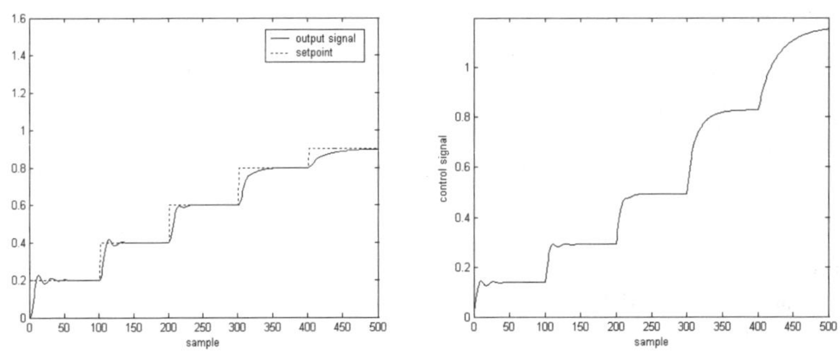

Fig. 2. Input and output signals for the MFLAC (servo behavior)

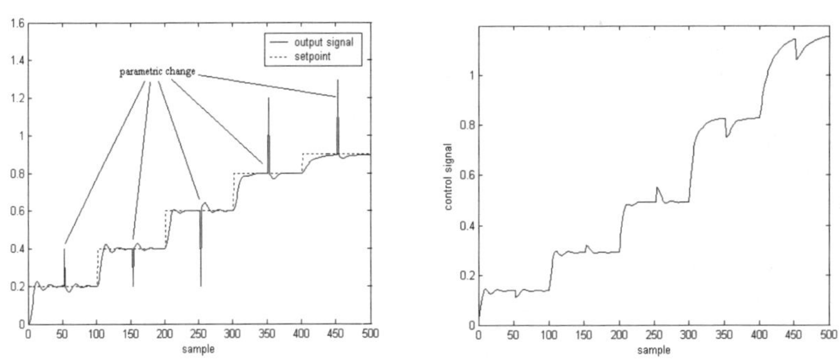

Fig. 3. Input and output signals for the MFLAC (regulatory behavior)

5 Conclusion and Future Research

Numerical results for controlling a control valve have shown the efficiency of the proposed MFLAC that guaranteed the convergence of the tracking error for servo and regulatory responses. However, it still has a distance to industrial applications and more practical issues must be done. A further investigation can be directed to analyze the PSO for model-free adaptive control methods [16] in essential control issues such as control performance, robustness and stability.

References

[1] F. Karray, W. Gueaieb, and S. Al-Sharhan, "The hierarchical expert tuning of pid controllers using tools of soft computing," *IEEE Transactions on Systems, Man, and Cybernetics —Part B: Cybernetics*, vol. 32, no. 1, pp. 77-90, 2002.
[2] K. J. Åström and T. Hägglund, *PID controllers: theory, design, and tuning*. Instrument Society of America, ISA, 1995.
[3] B. H. Bisowarno, Y. -C. Tian, and M. O. Tade, "Model gain scheduling control of an ethyl *tert*-butyl ether reactive distillation column," *Ind. Eng. Chem. Res.*, vol. 42, pp. 3584-3391, 2003.
[4] Z. Hou and W. Huang, "The model-free learning adaptive control of a class of siso nonlinear systems," *Proceedings of the American Control Conference*, Albuquerque, NM, pp. 343-344, 1997.
[5] Z. Hou, C. Han, and W. Huang, "The model-free learning adaptive control of a class of MISO nonlinear discrete-time systems," *IFAC Low Cost Automation*, Shenyang, P.R. China, pp. 227-232, 1998.
[6] J. F. Kennedy, R. C. Eberhart and R. C. Shi, *Swarm intelligence*, Morgan Kaufmann Pub, San Francisco, USA, 2001.
[7] Y. Shi and R. C. Eberhart, "Parameter selection in PSO optimization," *Proceedings of the 7th Annual Conf. Evolutionary Programming*, San Diego, CA, USA, pp. 25-27, 1998.
[8] K. Yasuda, A. Ide, and N. Iwasaki, "Adaptive particle swarm optimization," *Proceedings of IEEE Int. Conf. on Systems, Man and Cybernetics*, Washington, DC, USA, vol. 2, pp. 1554-1559, 2003.
[9] D. Devicharan and C. K. Mohan, "Particle swarm optimization with adaptive linkage learning," *Proceedings of the IEEE Congress on Evol. Computation*, Portland, OR, USA, 530-535, 2004.
[10] R. Mendes and J. F. Kennedy, "The fully informed particle swarm: simper, maybe better," *IEEE Transactions on Evolutionary Computation*, vol. 8, no. 3, pp. 204-210, 2004.
[11] R. A. Krohling, F. Hoffmann, and L. S. Coelho, "Co-evolutionary particle swarm optimization for min-max problems using Gaussian distribution," Proceedings of Congress on Evolutionary Computation, Portland, USA, 959-964, 2004.
[12] J. F. Kennedy and R. C. Eberhart, "Particle swarm optimization," *Proceedings of IEEE International Conference on Neural Networks*, Perth, Australia, pp. 1942-1948, 1995.
[13] R. C. Eberhart and J. F. Kennedy, "A new optimizer using particle swarm theory," *Proceedings of International Symposium on Micro Machine and Human Science*, Japan, pp. 39-43, 1995.

[14] A. Ratnaweera, S. K. Halgamuge, and H. C. Watson, "Self-organizing hierarchical particle swarm optimizer with time-varying acceleration coefficients," *IEEE Transactions on Evolutionary Computation*, vol. 8, no. 3, pp. 240-255, 2004.
[15] T. Wigren, "Recursive prediction error identification using the nonlinear Wiener model," *Automatica*, vol. 29, no. 4, pp. 1011-1025, 1993.
[16] J. C. Spall and J. A. Cristion, "Model-free control of nonlinear systems with discrete time measurements," *IEEE Transactions on Automatic Control*, vol. 43, pp. 1198-1210, 1998.

B-Spline Neural Network Using an Artificial Immune Network Applied to Identification of a Ball-and-Tube Prototype

Leandro dos Santos Coelho and Rodrigo Assunção

Production and Systems Engineering Graduate Program, PPGEPS
Pontifical Catholic University of Parana, PUCPR
Imaculada Conceição, 1155, Zip code 80215-901, Curitiba, Parana, Brazil

Abstract. B-spline neural network (BSNN), a type of basis function neural network, is trained by gradient-based methods that may fall into local minima during the learning procedure. When using feed-forward BSNNs, the quality of approximation depends on the control points (knots) placement of spline functions. This paper describes the application of an artificial immune network inspired optimization method – the opt-aiNet – to provide a stochastic search to adjust the control points of a BSNN. The numerical results presented here indicate that artificial immune network optimization methods useful for building a good BSNN model for the nonlinear identification of an experimental nonlinear ball-and-tube system.

Keywords: B-spline neural network, artificial immune system, nonlinear identification.

1 Introduction

The use of neural networks to model chaotic systems and nonlinear identification problems has attracted considerable attention in recent years [1]-[4]. A relevant approach is to find the best approximation with respect to a certain class of basis functions for the representation neural of networks. In this case, there are many possible choices of basis functions, such as radial basis function [5], associated memory networks [6], wavelets [7], and B-spline functions [8].

The main advantage of the B-spline functions over other radial functions are the local controls of the curve shape, since the curve only changes in the vicinity of a few control points that have been changed [9]. A B-spline neural network (BSNN) consists of the piecewise polynomials with a set of local basis functions to model an unknown function for which a finite set of input-output samples are available. The identification performance depends largely on an optimization algorithm for the training procedure of the BSNN in order to avoid any possible local minima.

In this context, the development of training methods and improvements for BSNN is an emerging research area. Several heuristics have been developed in recent years to improve the performance and set up the parameters of the BSSN design and the fuzzy system approaches [10]-[17]. Recently, as an alternative to the conventional mathematical approaches based on gradient information [18], modern heuristic

A. Saad et al. (Eds.): Soft Computing in Industrial Applications, ASC 39, pp. 92–101, 2007.
springerlink.com

optimization techniques such as evolutionary algorithms [19] and swarm intelligence [20] have received much attention by many researchers due to their ability to find an almost global optimal solution.

Artificial immune systems (AIS) are learning and optimization methods that can be used for the solution of many different types of optimization problems [21], [22], [23]. A meta-heuristic optimization approach employing artificial immune networks called opt-aiNET algorithm to solve the knots of BSNN is proposed in this paper. The aiNET algorithm is a discrete immune network algorithm based on the artificial immune systems paradigm that was developed for data compression and clustering [24], and was also extended slightly and applied to optimization to create the algorithm opt-aiNET [25]. Opt-aiNET, proposed in [25], evolves a population, which consists of a network of antibodies (considered as candidate solutions to the function being optimized). These undergo a process of evaluation against the objective function, clonal expansion, mutation, selection and interaction between themselves.

In this paper, we propose modified opt-aiNET approach to train a BSNN. Numerical results for identification of the nonlinear dynamics of an experimental ball-and-tube system confirm the feasibility and effectiveness of the proposed approach.

2 B-Spline Neural Network

BSNN is introduced as a class of one-hidden-layer feedforward neural networks composed of B-spline functions. BSNN is an example of associate memory networks. The input space is defined over an n-dimensional lattice with basis functions defined for each cell. Each basis function is composed of q polynomial segments. A simple, stable recursive relationship exists to evaluate the membership of a B-spline basis function of order k,

$$N_q^j(x) = \left(\frac{x - \lambda_{j-q}}{\lambda_{j-1} - \lambda_{j-q}} \right) N_{q-1}^{j-1}(x) + \left(\frac{\lambda_j - x}{\lambda_j - \lambda_{j-q+1}} \right) N_{q-1}^j(x) \tag{1}$$

$$N_1^j(x) = \begin{cases} 1 & if\,(x \in I_j) \\ 0 & otherwise \end{cases} \tag{2}$$

where $N_q^j(\cdot)$ is defined as the j-th univariate basis function of order q and λ_j the j-th knot and I_j is the j-th interval. The output of neural network is

$$\hat{o}_k = f(x_k) = \sum_{j=1}^{p} w_j N_q^j(x_k) \tag{3}$$

where x_k and $\hat{o}_k$ are, respectively, the inputs and output of the network, w_j is the weight attached to the j-th basis function, and $N_q^j(\cdot)$ is given by the recursive form (2).

The index j is associative with the region of local support $\lambda_{(j-q)} \leq x \leq \lambda_{(j)}$, whereas the index q indicates the order of the basis functions [20].

The main advantage of B-spline functions over other radial functions e. g., the Bezier curve, is the local control of the shape of the curve, as the curve only changes in the vicinity of a few control points that been changed [9].

The quality of approximation depends on the placement of knots of B-spline functions. The purpose of optimizing a BSNN by opt-aiNET is to determine the knots of each B-spline basis function. In particular, the number of basis functions here depends on the user's choice.

3 Opt-aiNET

Opt-aiNET is capable of performing local and global search, as well as to adjust dynamically the size of population [27]. Opt-aiNET creates a memory set of antibodies (points in the search space) that represent (over time) the best candidate solutions to the objective function. Opt-aiNET is capable of either unimodal or multimodal optimization and can be characterized by five main features [26]: (i) the population size is dynamically adjustable; (ii) it demonstrates exploitation and exploration of the search space; (iii) it determines the locations of multiple optima; (iv) it has the capability of maintaining many optima solutions; and (v) it has defined stopping criteria. The steps of opt-aiNET are summarized as follows:

A. Initialization of the parameter setup

The user must choose the key parameters that control the opt-aiNET, i.e., population size (M), suppression threshold (σ_s), number of clones generated for each cell (N_c), percentage of random new cells each iteration (d), scale of affinity proportion selection (β), and maximum number of iterations allowed (stop criterion), N_{gen}.

B. Initialization of cell populations

Set iteration t=0. Initialize a population of i=1,..,M cells (real-valued n-dimensional solution vectors) with random values generated according to a uniform probability distribution in the n dimensional problem space. Initialize the entire solution vector population in the given upper and lower limits of the search space.

C. Evaluation of each network cell

Evaluate the fitness value of each cell (in this work, the objective of the fitness function is to maximize the cost function).

D. Generation of clones

Generate a number N_c of clones for each network cell. The clones are offspring cells that are identical copies of their parent cell [25].

E. Mutation operation

Mutation is an operation that changes each clone proportionally to the fitness of the parent cells, but keeps the parent cell. Clones of each cell are mutated according to the affinity (Euclidean distance between two cells) of the parent cell. The affinity proportional mutation is performed according to equations (4) and (5), given by:

$$c' = c + \alpha \cdot N(0,1) \tag{4}$$

$$\alpha = \beta^{-1} e^{-f^*} \tag{5}$$

where c' is a mutated cell c, $N(0,1)$ is a Gaussian random variable of zero mean and unitary standard deviation, β is a parameter that controls the decay of the inverse exponential function, and f^* is the fitness of an individual normalized in the interval [0,1].

F. Evaluation the fitness of all network cells

Evaluate the fitness value of all network cells of the population including new clones and mutated clones.

G. Selection of fittest clones

For each clone select the most fit and remove the others.

H. Determination of affinity of all network cells

Determine the affinity network cells and perform network suppression.

I. Generate randomly d network cells

Introduce a percentage d of randomly generated cells. Set the generation number for $t = t + 1$. Proceed to step C until a stopping criterion is met, usually a maximum number of iterations, t_{max}. The stopping criterion depends on the type of problem.

4 Case Study: Identification of Ball-and-Tube Prototype

The prototype of the ball-and-tube process consists in a plastic tube, a DC motor and a polystyrene sphere. The aim of this project is to control the height of the sphere (ball) by applying a flux of air through the tube base, considering that this air flux is generated by the application of tension in the DC motor.

A photograph and a general view of the ball-and-tube prototype design are shown in figures 1 and 2, respectively. The project contains three main modules: (i) ball-and-tube structure, (ii) hardware, and (iii) software.

The hardware module of ball-and-tube prototype involves the distance sensor, DC motor, interface circuit, and firmware. The interface circuit is composed by the serial communication circuit, the distance sensor multiplexer circuit, and the PWM (Pulse Width Modulator) circuit, as shown in figure 3. The Circuit Maker Student Version software was chosen to build up the circuit diagrams.

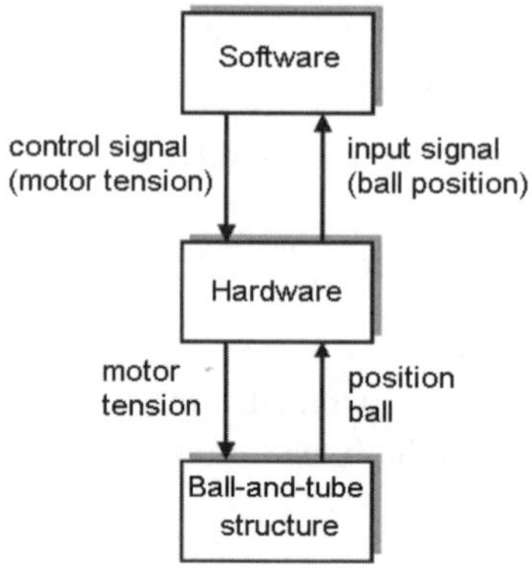

Fig. 1. General view of ball-and-tube prototype

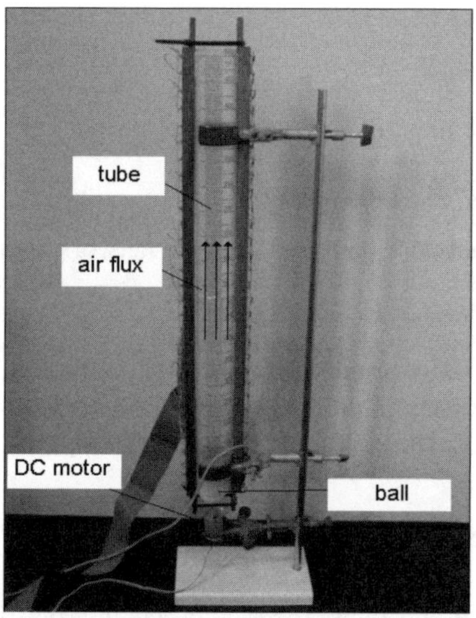

Fig. 2. Photograph of ball-and-tube system, a low cost prototype

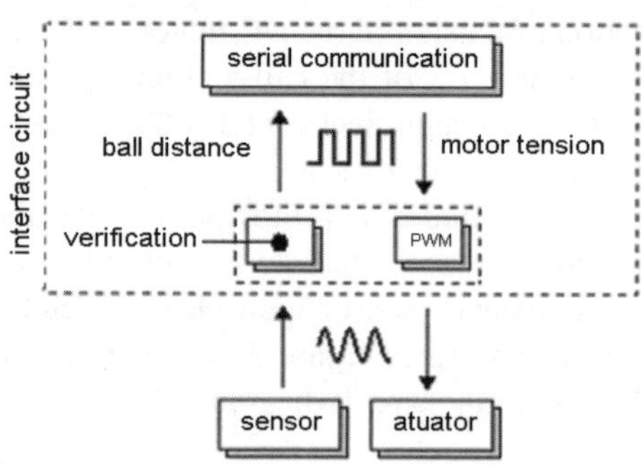

Fig. 3. Hardware module of ball-and-tube prototype

A personal computer with an acquisition data board to transmit the control sign and to register the value of the height of the sphere in the tube (by the use of a group of sensors based on phototransistors) is utilized. Through the acquisition data board, a tension is converted into an output tension to be acquired by the acquisition data board. In figure 4 it is presented the ball-and-tube prototype with 55 cm of height and 4.5 cm of diameter. The chosen DC motor (fan generation) allows change the input tension from 0 to 20 V, those converted to values between 0 and 5 V (TTL level) to facilitate the construction of the power electronics circuit.

The conception of this low cost ball-and-tube prototype aims at validating nonlinear identification methods and the configuration of adaptive and fuzzy control algorithms.

5 Identification Results Using BSNN with Opt-aiNET

System identification is a procedure to identify a model of an unknown process, for purposes of forecasting and/or understanding the dynamic behavior of a dynamic system. In practice, system identification is an iterative procedure. The lack of *a priori* information regarding the process model will require that each step initially be examined superficially. The mathematical model employed in this work to represent the yo-yo motion system is a NARX (Nonlinear AutoRegressive with eXogenous inputs). In this case, the NARX model with series-parallel conception is used for one-step ahead forecasting of the BSNN model (see figure 4).

A computer with a data acquisition board for generating the input signal was used to obtain system measurements. In the identification procedure based on the BSNN model, 330 samples of input (tension applied to the DC motor) and output (ball height) were collected with a time sampling of 200ms. The tension value corresponds to the maximum value configuration of the driver in PWM control of a DC motor. The operation range used in experiments was between 9 cm (0.094 V) and 53 cm (4.707 V) of full height of tube (see details of data in figure 5).

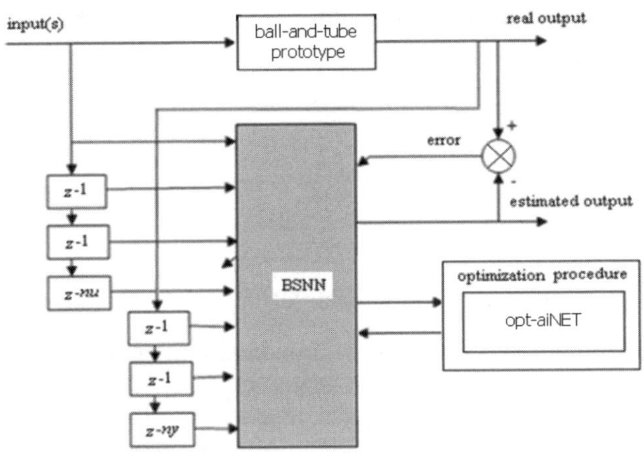

Fig. 4. Structure of NARX model used in this work

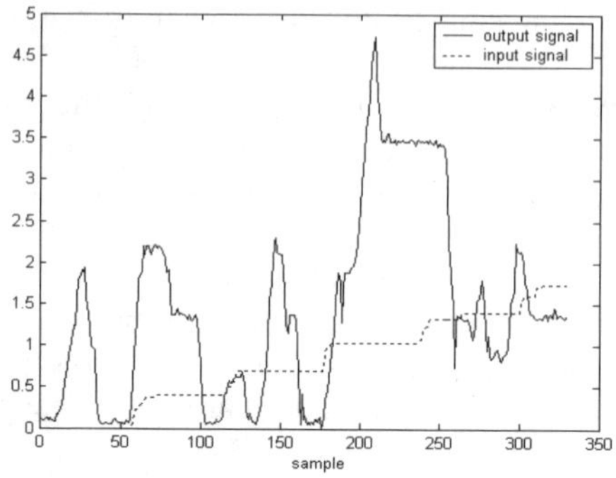

Fig. 5. Input and output signals applied to ball-and-tube prototype

Experiments for the estimation phase of the mathematical model of the ball-and-tube system are carried out using samples 1 to 170. For the validation phase, the BSNN model uses the input and output signals of samples 171 to 330. The system identification by BSNN model based on opt-aiNET is appropriate if a performance index is in values permissible for the user's needs. In this case, the fitness function for maximization proposes using opt-aiNET and is given by the harmonic mean of multiple correlation indices of estimation and validation phases. The fitness function (to be maximized) is calculated using the expression of R_{est}^2 given by:

$$R_{est}^2 = 1 - \frac{\sum_{t=1}^{170} [y(t) - \hat{y}(t)]^2}{\sum_{t=1}^{170} [y(t) - \bar{y}]^2} \tag{6}$$

where R_{est}^2 is the multiple correlation index of the estimation phase, $y(t)$ is the output of the real system, $\hat{y}(t)$ is the output estimated by the BSNN, and $\bar{y}$ is the mean value of the system's output. For the validation phase (verification of generalization capability) of optimized BSNN, we have employed the R_{val}^2 index give by

$$R_{val}^2 = 1 - \frac{\sum_{t=171}^{330} [y(t) - \hat{y}(t)]^2}{\sum_{t=171}^{330} [y(t) - \bar{y}]^2} \tag{7}$$

where R_{val}^2 is the multiple correlation index of the validation phase. When the value R^2 =1.0 (estimation or validation phases), it indicates the model's accurate approach to the system's measured data. A R^2 value between 0.9 and 1.0 is considered sufficient for applications in identification and model-based controller designs [28].

All the computational programs were run on a 3.2 GHz Pentium IV processor with 3 MB of RAM. In each case study, 30 independent runs were made for each of the optimization methods involving 30 different initial trial solutions for each optimization method. The setup of opt-aiNET algorithm used was: suppression threshold = 5, percentage of newcomers: d=40%, scale of the affinity proportional selection using a linear reduction of β with initial and final values of 10 and 100, respectively, and the number of clones generated for each cell is N_c=10.

In the above case studies, the population size N was 20 and the stopping criterion, t_{max}, was 200 generations for the opt-aiNET algorithm. The three chosen vectors of the BSNN's input were [$u(t-1)$; $y(t-2)$; and $y(t-1)$]. The space searches for the knots of each B-spline basis function are [-1.0; 1.0]. Simulation tests were conducted using 3 to 5 knots in each input of the BSNN.

Table 1. presents the simulation results (best of 30 experiments) using opt-aiNET for optimization of the BSNN using 4 knots. As indicated in Table 1, the results of the optimized BSNN are precise, providing an appropriate experimental mathematical model for the ball-and-tube prototype. The best result shown in Figure 6 represents the BSNN using opt-aiNET with 4 knots for each network input.

Table 1. Results obtained by the maximization of R^2_{est} using opt-aiNET (adopted here is the solution with the best R^2_{est} after completing 30 runs)

knots for each BSNN's input	R^2_{est}				R^2_{val}
	maximum (best)	mean	minimum (worst)	standard deviation	(using R^2_{est} best result)
3	0.9331	0.9246	0.9178	0.0078	0.8109
4	**0.9472**	0.9411	0.9359	0.0056	0.8398
5	0.9470	0.9361	0.9143	0.0189	0.8211

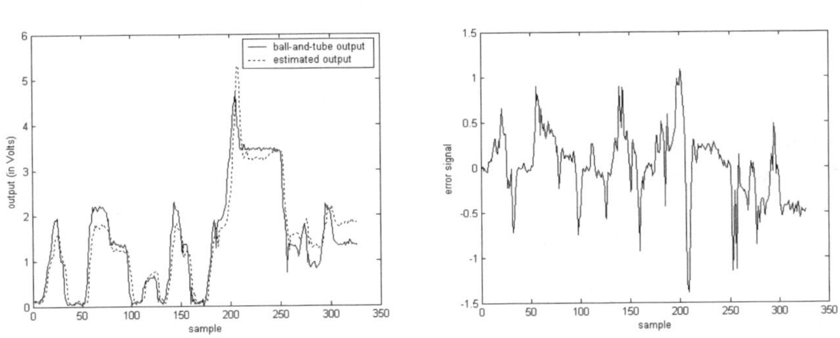

Fig. 6. Best result of R^2_{est} for BSNN using opt-aiNET (see Table 1)

6 Conclusion and Future Research

There is a natural parallel between the immune system and optimization. In this context, one well-known immune inspired algorithm for function optimization is the opt-aiNET. The op-aiNET is inspired by the idiotypic network theory for explaining the immune system dynamics.

This paper describes the application of opt-aiNET algorithm to adjust the control points of a BSNN. Our simulation results confirmed the potential of opt-aiNET algorithm for BSNN optimization for the nonlinear identification of an experimental nonlinear ball-and-tube system. Further studies are needed to test the opt-aiNET algorithm on benchmark optimization problems in system identification and power systems.

References

[1] Narendra, K. S.; Parthasarathy, K. (1990). Identification and control of dynamical systems using neural networks, *IEEE Transactions on Neural Networks*, Vol. 1, No. 1, pp. 4-27.

[2] Karayiannis, N. B.; Randolph-Gips, M. M. (2003). On the construction and training of reformulated radial basis function neural networks, *IEEE Transactions on Neural Networks*, Vol. 14, No. 14, pp. 835-846.

[3] Mcloone, S.; Brown, M. D.; Irwin, G.; Lightbody, A. (1998). A hybrid linear/nonlinear training algorithm for feedforward neural networks, *IEEE Transactions on Neural Networks*, Vol. 9, No. 4, pp. 669-684.

[4] Sierakowski, C. A..; Guerra, F. A.; Coelho, L. S. (2005). Particle swarm optimization approach for multi-step-ahead prediction using radial basis function neural network, *Proceedings of 16th IFAC World Congress*, Prague, Czech Republic.

[5] Yingwei, L.; Sundararajsn, N.;. Saratchandran, P. (1997). Identification of time-varying nonlinear systems using minimal radial basis function neural networks, *IEE Proc.-Control Theory Applications*, Vol. 144, No. 2, pp. 202-208.

[6] Harris, C. J.; Moore, C. G.; Brown, M. (1993). *Intelligent control: aspects of fuzzy logic and neural nets*, Word Scientific.

[7] Zhang, Q.; Benveniste, A. (1992). Wavelet networks, *IEEE Transactions on Neural Networks*, Vol. 3, No. 6, pp. 889-898.

[8] Lightbody, G.; O'Reilly, P.; Irwin, G. W.; Kelly, K.; McCormick, J. (1997). Neural modelling of chemical plant using MLP and B-spline networks, *Control Engineering Practice*, Vol. 5, No. 11, pp. 1501-1515.

[9] Newmann, W. M.; Sproull, R. F. (1979). *Principles of interactive computer graphics*, McGraw-Hill, Inc.

[10] Wang, K; Lei, B. (2001). Using B-splines neural network to extract fuzzy rules for a centrifugal pump monitoring, Journal of Intelligent Manufacturing, Vol. 12, No. 1, pp. 5-11.

[11] Wang, H.; Oyebande, B. (1995). On the application of neural network modeling to a wet end chemical process in paper making, *Proceedings of IEEE Conference on Control Applications*, Albany, NY, pp. 657-662.

[12] Starrenburg, G.; van Luenen, W. T. C.; Oelen, W.; van Amerongen, J. (1996). Learning feedforward controller for a mobile robot vehicle, *Control Engineering Practice*, Vol. 4, No. 9, pp. 1221-1230.

[13] Zhang, J.; Knoll, A. (1999). Designing fuzzy controllers by rapid learning, *Fuzzy Sets and Systems*, Vol. 101, pp. 287-301.

[14] Yiu, K. F. C.; Wang, S.; Teo, K. L.; Tsoi, A. C. (2001). Nonlinear system modeling via knot-optimizing B-spline networks, *IEEE Trans. on Neural Networks*, Vol. 12, No. 4, pp. 1013-1022.

[15] Saranli, A.; Baykal, B. (1998). Complexity reduction in radial basis function (RBF) networks by using radial B-spline functions, *Neurocomputing*, Vol. 18, No. 1-3, pp. 183-194.

[16] Shimojima, K.; Fukuda, T.; Arai, F. (1994).Self-tuning fuzzy inference based on spline function, *Proceedings of IEEE International Conference on Fuzzy Systems*, Orlando, FL, USA, pp. 690-695.

[17] Chua, V. K.; Tomizuka, M. (1995). Rule generation for fuzzy systems based on B-splines, *Proc. IEEE Int. Conference on Neural Networks*, Perth, Australia, pp. 6098-611.

[18] Chan, C. W.; Jin, H.; Cheung, K. C. ; Zhang, H. Y. (1998). An adaptive learning rate for the training of B-spline networks, *Proc. UKACC Int. Conf. Control*, Swansea, UK, pp. 342-347.

[19] Wang, W. -Y.; Lee, T. -T.; Hsu, C. -C.; Li, Y. -H. (2002). GA-based learning of BMF fuzzy-neural network, *Proc. Int. Conference on Fuzzy Systems*, Honolulu, HI, USA, pp. 1234-1238.

[20] Coelho, L. S.; Krohling, R. A. (2006). Nonlinear system identification based on B-spline neural network and modified particle swarm optimization, *Proc.IEEE World Conference on Computational Intelligence* (WCCI), Int. Joint-Conf. on Neural Networks, Vancouver, Canada.

[21] Ishida, Y.; Hirayama, H.; Fujita, H.; Ishiguro, Mori, K. (eds.) (1998). *Immunity-based systems – intelligent systems by artificial immune systems*, Corona Pub. Co. Japan (in Japonese).

[22] Dasrupta, D. (ed.) (1999). *Artificial immune systems and their applications*, Springer-Verlag.

[23] De Castro, L. N.; Timmis, J. I. (2003). Artificial immune systems as a novel soft computing paradigm, *Soft Computing Journal*, Vol. 7, No. 7, pp. 526-544

[24] De Castro, L. N.; Von Zuben, F. (2001). AINET: an artificial immune network for data analysis, *Data Mining: a heuristic approach*, Abbas, H.; Sarker, R.; Newton, C. (eds.), Idea Group Publishing.

[25] De Castro, L. N.; Timmis, J. (2002) An artificial immune network for multimodal function optimization, *Proc.IEEE Congress on Evolutionary Computation*, Hawaii, USA, pp. 699-674.

[26] Timmis, J.; Edmonds, C. (2004). A comment on opt-AiNET: an immune network algorithm for optimisation, D. Kalyanmoy et al. (eds.), *Genetic and Evolutionary Computation*, vol. 3102, Lecture Notes in Computer Science, Springer, pp. 308-317.

[27] Campelo, F.; Guimarães, F. G.; Igarashi, H.; Ramírez, J. A.; Noguchi, S. (2006). A modified immune network algorithm for multimodal electromagnetic problems, *IEEE Transactions on Magnetics*, Vol. 42, No. 4, pp. 1111-1114.

[28] Schaible, B.; Xie, H.; Lee, Y. C. (1997). Fuzzy logic models for ranking process effects, *IEEE Transactions on Fuzzy Systems*, Vol. 5, No. 4, pp. 545-556.

Part III

Pattern Recognition

Pattern Recognition for Industrial Security Using the Fuzzy Sugeno Integral and Modular Neural Networks

Patricia Melin, Alejandra Mancilla, Miguel Lopez, Daniel Solano, Miguel Soto, and Oscar Castillo

Department of Computer Science
Tijuana Institute of Technology
Tijuana, Mexico
pmelin@tectijuana.mx

Abstract. We describe in this paper the evolution of modular neural networks using hierarchical genetic algorithms for pattern recognition. Modular Neural Networks (MNN) have shown significant learning improvement over single Neural Networks (NN). For this reason, the use of MNN for pattern recognition is well justified. However, network topology design of MNN is at least an order of magnitude more difficult than for classical NNs. We describe in this paper the use of a Hierarchical Genetic Algorithm (HGA) for optimizing the topology of each of the neural network modules of the MNN. The HGA is clearly needed due to the fact that topology optimization requires that we are able to manage both the layer and node information for each of the MNN modules. Simulation results prove the feasibility and advantages of the proposed approach.

Keywords: Evolution, Neural Networks, Pattern Recognition, Biometrics.

1 Introduction

Today, a variety of methods and techniques are available to determine unique identity, the most common being fingerprint, voice, face, and iris recognition. Of these, fingerprint and iris offer a very high level of certainty as to a person's identity, while the others are less exact. A large number of other techniques are currently being examined for suitability as identity determinants. These include (but are not limited to) retina, gait (walking style), typing style, body odour, signature, hand geometry, and DNA. Some wildly esoteric methods are also under development, such as ear structure, thermal imaging of the face and other parts of the body, subcutaneous vein patterns, blood chemistry, anti-body signatures, and heart rhythm, to name a few.

The four primary methods of biometric authentication in widespread use today are face, voice, fingerprint, and iris recognition. All of these are supported in our approach, some more abundantly than others. Generally, face and voice are considered to be a lower level of security than fingerprint and iris, but on the other hand, they have a lower cost of entry. We describe briefly in this section some of these biometric methods.

A. Saad et al. (Eds.): Soft Computing in Industrial Applications, ASC 39, pp. 105–114, 2007.
springerlink.com

Face Recognition. Facial recognition has advanced considerably in the last 10 to 15 years. Early systems, based entirely on simple geometry of key facial reference points, have given way to more advanced mathematically-based analyses such as Local Feature Analysis and Eigenface evaluation. These have been extended though the addition of "learning" systems, particularly neural networks.

Face recognition systems are particularly susceptible to changes in lighting systems. For example, strong illumination from the side will present a vastly different image to a camera than neutral, evenly-positioned fluorescent lighting. Beyond this, however, these systems are relatively immune to changes such as weight gain, spectacles, beards and moustaches, and so on. Most manufacturers of face recognition systems claim false accept and false reject rates of 1% or better.

Voice Recognition. Software systems are rapidly becoming adept at recognising and converting free-flowing speech to its written form. The underlying difficulty in doing this is to flatten out any differences between speakers and understand everyone universally. Alternatively, when the goal is to specifically identify one person in a large group by their voice alone, these very same differences need to be identified and enhanced.

As a means of authentication, voice recognition usually takes the form of speaking a previously-enrolled phrase into a computer microphone and allowing the computer to analyse and compare the two sound samples. Methods of performing this analysis vary widely between vendors. None is willing to offer more than cursory descriptions of their algorithms--principally because, apart from LAN authentication, the largest market for speaker authentication is in verification of persons over the telephone.

Fingerprint Recognition. The process of authenticating people based on their fingerprints can be divided into three distinct tasks. First, you must collect an image of a fingerprint; second, you must determine the key elements of the fingerprint for confirmation of identity; and third, the set of identified features must be compared with a previously-enrolled set for authentication. The system should never expect to see a complete 1:1 match between these two sets of data. In general, you could expect to couple any collection device with any algorithm, although in practice most vendors offer proprietary, linked solutions.

A number of fingerprint image collection techniques have been developed. The earliest method developed was optical: using a camera-like device to collect a high-resolution image of a fingerprint. Later developments turned to silicon-based sensors to collect an impression by a number of methods, including surface capacitance, thermal imaging, pseudo-optical on silicon, and electronic field imaging.

As discussed, a variety of fingerprint detection and analysis methods exist, each with their own strengths and weaknesses. Consequently, researchers vary widely on their claimed (and achieved) false accept and false reject rates. The poorest systems offer a false accept rate of around 1:1,000, while the best are approaching 1:1,000,000. False reject rates for the same vendors are around 1:100 to 1:1000.

Iris Recognition. Iris recognition is based entirely on a concept originated by Drs. Leonard Flom and Aran Safir, and a software process developed by Dr. John Daugman, all of Cambridge University, England. US Patent 5,291,560 issued in the name of Daugman has been assigned to Iridian Corp., one of the world's principal companies of iris-based systems. Extensive research has determined that the human iris is essentially unchanged in structure and appearance from the eighth month of gestation until a few minutes after death. Although a neonatal eye can be darkly coloured until a few months after birth, this darkness is not an influence in the infrared wavelengths normally used to collect an iris image.

In the history of iris recognition, there has *never* been a false acceptance. In fact, the equal error rate is 1:1,200,000, with a typical false accept rate of 1:100,000,000 and false reject rate of 1:200,000. Note that these are theoretical values based on strong analysis of limited data (only 5 to 10 million iris scans have ever been performed); they also do not take into account the perceived level of difficulty in using the system. Overall, iris scanning is the system to use if you are concerned about strongly authenticating users.

2 Proposed Approach for Recognition

Our proposed approach for human recognition consists in integrating the information of the three main biometric parts of the person: the voice, the face, and the fingerprint. Basically, we have an independent system for recognizing a person from each of its biometric information (voice, face, and fingerprint), and at the end we have an integration unit to make a final decision based on the results from each of the modules. In Figure 1 we show the general architecture of our approach in which it is clearly seen that we have one module for voice, one module for face recognition, and one module for fingerprint recognition. At the top, we have the decision unit integrating the results from the three modules.

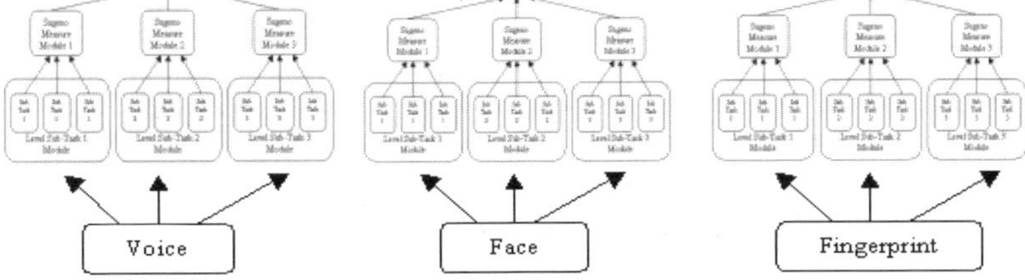

Fig. 1. Architecture of the proposed approach

3 Modular Neural Networks

This section describes a particular class of "modular neural networks", which have a hierarchical organization comprising multiple neural networks; the architecture basically consists of two principal components: local experts and an integration unit, as illustrated in Figure 2. In general, the basic concept resides in the idea that combined (or averaged) estimators may be able to exceed the limitation of a single estimator. The idea also shares conceptual links with the "divide and conquer" methodology. Divide and conquer algorithms attack a complex problem by dividing it into simpler problems whose solutions can be combined to yield a solution to the complex problem. When using a modular network, a given task is split up among several local experts NNs. The average load on each NN is reduced in comparison with a single NN that must learn the entire original task, and thus the combined model may be able to surpass the limitation of a single NN. The outputs of a certain number of local experts (O_i) are mediated by an integration unit. The integrating unit puts the outputs together using estimated combination weights (g_i). The overall output Y is

$$Y_i = \Sigma g_i \, O_I \tag{1}$$

Nowlan, Jacobs, Hinton, and Jordan [7] described modular networks from a competitive mixture perspective. That is, in the gating network, they used the "softmax" function, which was introduced by McCullagh and Nelder [9]. More precisely, the gating network uses a softmax activation g_i of ith output unit given by

$$G_i = \exp(ku_i) / \Sigma_j \exp(ku_j) \tag{2}$$

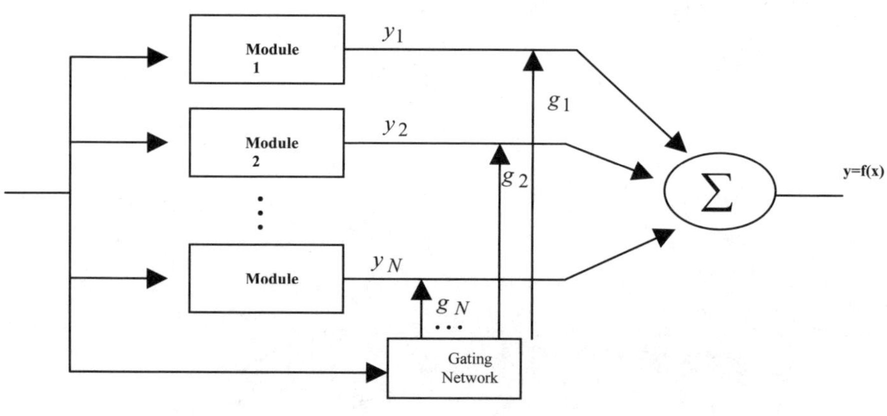

Fig. 2. Architecture of a modular neural network

Where u_i is the weighted sum of the inputs flowing to the ith output neuron of the gating network. Use of the softmax activation function in modular networks provides a sort of "competitive" mixing perspective because the ith local expert's output O_i with a minor activation u_i does not have a great impact on the overall output Y_i.

4 Genetic Algorithm for Neural Network Optimization

The bottleneck problem for NN application lies within the optimization procedures that are used to obtain an optimal NN topology. Hence, the formulation of the Hierarchical Genetic Algorithm (HGA) is applied for this purpose [10]. The HGA differs from the standard GA with a hierarchy structure in that each chromosome consists of multilevel genes. Each chromosome consists of two types of genes, i.e. control genes and connection genes. The control genes in the form of bits, are the genes for layers and neurons for activation. The connection genes, a real value representation, are the genes for connection weightings and neuron bias.

With such a specific treatment, a structural chromosome incorporates both active and inactive genes. It should be noted that the inactive genes remain in the chromosome structure and can be carried forward for further generations. Such an inherent genetic variation in the chromosome avoids any trapping at local optima, which has the potential to cause premature convergence. Thus it maintains a balance between exploiting its accumulated knowledge and exploring the new areas of the search space. This structure also allows larger genetic variations in chromosome while maintaining high viability by permitting multiple simultaneous genetic changes. As a result, a single change in high level genes will cause multiple changes (activation or deactivation in the whole level) in lower level genes. In the case of the traditional GA, this is only possible when a sequence of many random changes takes place. Hence the computational power is greatly improved.

5 The Problem of Speech Recognition

Speaker recognition, which can be classified into identification and verification, is the process of automatically recognizing who is speaking on the basis of individual information included in speech waves. This technique makes it possible to use the speaker's voice to verify their identity and control access to services such as voice dialing, banking by telephone, telephone shopping, database access services, information services, voice mail, security control for confidential information areas, and remote access to computers.

Speaker identification is the process of determining which registered speaker provides a given utterance. Speaker verification, on the other hand, is the process of accepting or rejecting the identity claim of a speaker. Most applications in which a voice is used as the key to confirm the identity of a speaker are classified as speaker verification. Speaker recognition methods can also be divided into text-dependent and text-independent methods. The former require the speaker to say key words or sentences having the same text for both training and recognition trials, whereas the latter do not rely on a specific text being spoken. Both text-dependent and independent methods share a problem however. These systems can be easily deceived because someone who plays back the recorded voice of a registered speaker saying the key words or sentences can be accepted as the registered speaker. To cope with this problem, there are methods in which a small set of words, such as digits, are used as key words and each user is prompted to utter a given sequence of key words that is randomly chosen every time the system is used. Yet even this method is not completely reliable, since it can be deceived with advanced electronic recording equipment that can reproduce key words in a requested order. Therefore, a text-prompted speaker recognition method has recently been proposed by [11, 12].

6 Simulation Results for Voice Recognition

We describe below some simulation results of our approach for speaker recognition. First, in Figure 3 we have the signal of the word "example" in Spanish with noise. Next, in Figure 4 we have the identification of the word "example" without noise. We also show in Figure 5 the word "layer" in Spanish with noise. In Figure 6, we show the identification of the correct word "layer" without noise.

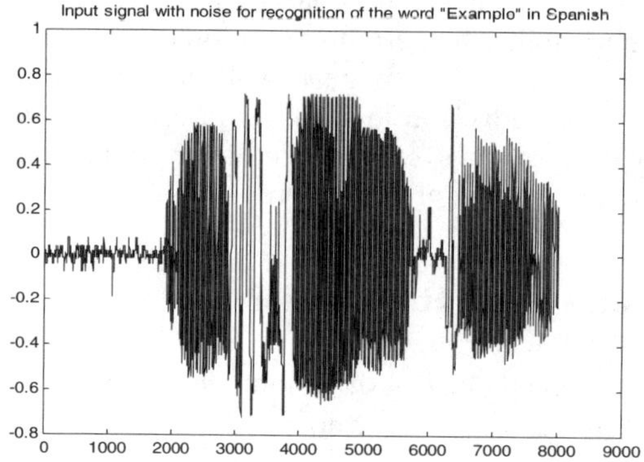

Fig. 3. Input signal of the word "example" in Spanish with noise

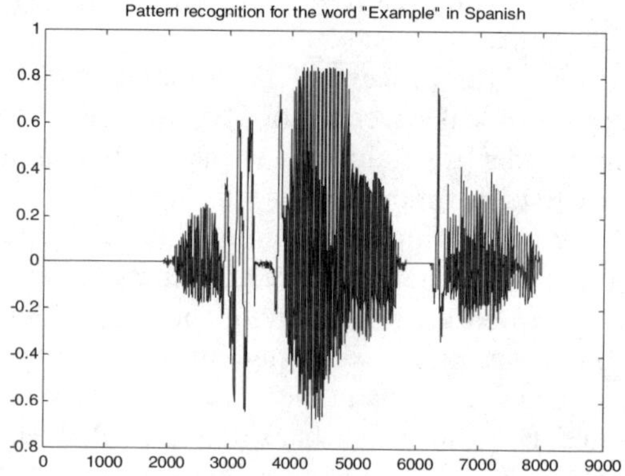

Fig. 4. Indentification of the word "example"

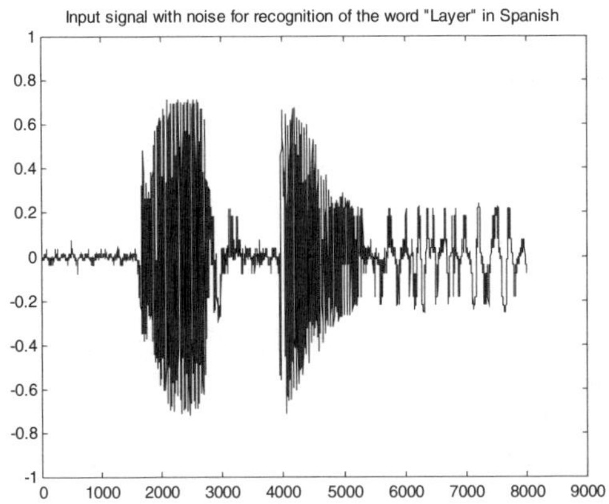

Fig. 5. Input signal of the word "layer" in Spanish with noise added

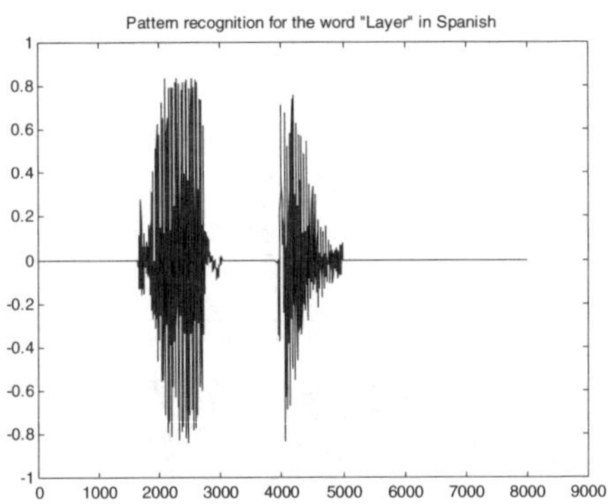

Fig. 6. Identification of the word "layer"

The idea behind this data partition is that learning will be easier in each of the modules, i.e. a simple NN can learn more easily the behavior of the function in one of the regions. We used three-layer feed-forward NNs for each of the modules with the Levenberg-Marquardt training algorithm.

The fitness function used in this work combines the information the error objective and also the information about the number of nodes as a second objective. This is shown in the following equation.

$$f(z)=\left(\frac{1}{\alpha * Ranking(ObjV1)+\beta * ObjV2}\right)*10 \tag{3}$$

The first objective is basically the average sum of squared of errors as calculated by the predicted outputs of the MNN compared with real values of the function. This is given by the following equation.

$$f_1=\frac{1}{N}\sum_{i=1}^{N}(Y_i-y_i) \tag{4}$$

The parameters of the genetic algorithm are as follows:

Type of crossover operator: Two-point crossover
Crossover rate: 0.8
Type of mutation operator: Binary mutation
Mutation rate: 0.05
Population size per generation: 10
Total number of generations: 100

We show in Figure 7 the topology of the final evolved modules of the neural network for the problem of function identification. As we can appreciate, from this Figure, module 2 is the smallest one and module 3 is the largest one. The result of MNN evolution is a particular architecture with different size of the modules (neural networks).

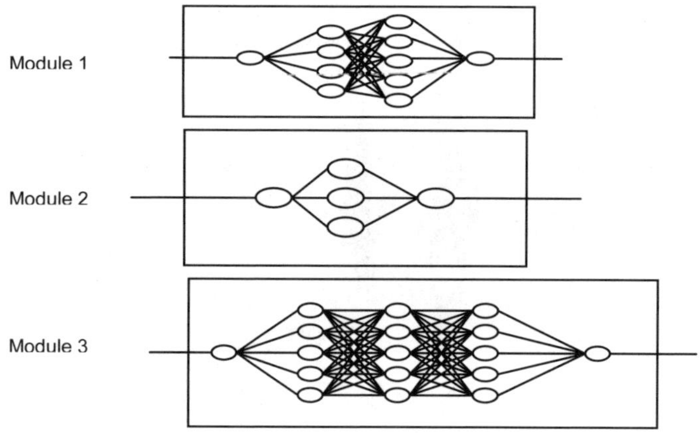

Fig. 7. Topology of the final evolved MNN for pattern recognition

The MNN architecture shown in Figure 7 is the best one for the specific problem of system identification. It is worthwhile noting that this network topology is difficult to design manually, for this reason the HGA approach is a good choice for neural network design and optimization.

7 Integration of Results for Person Recognition

We also have developed previously methods for face and fingerprint recognition with modular neural networks [15] and now we need to integrate the results of the three biometric aspects of a person, to make a decision on the identity of a person. We show in Figure 8 the architecture of the fuzzy system used for this decision process.

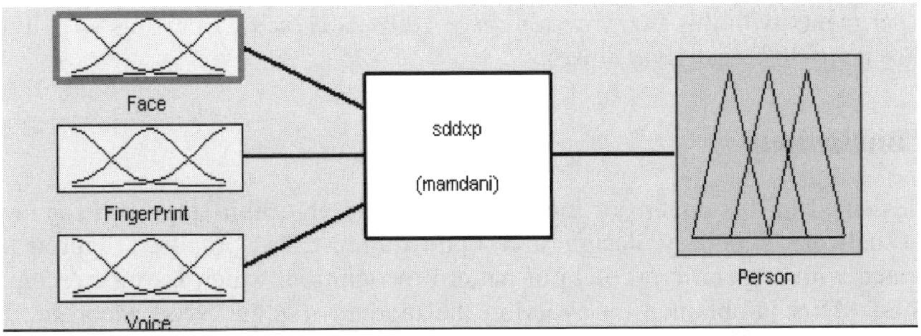

Fig. 8. Architecture of the fuzzy system for person recognition

We show in Table 1 the complete set of fuzzy rules used in the fuzzy system for person recognition.

Table 1. Fuzzy rule base for person recognition

Rules	Face	Fingerprint	Voice	Decision
1	A	A	A	A
2	A	A	B	A
3	A	A	C	A
4	A	B	A	A
5	A	B	B	B
6	A	B	C	B
7	A	C	A	A
8	A	C	B	C
9	A	C	C	C
10	B	A	A	A
11	B	A	B	B
12	B	A	C	A
13	B	B	A	B
14	B	B	B	B
15	B	B	C	B
16	B	C	A	C
17	B	C	B	B
18	B	C	C	C
19	C	A	A	A
20	C	A	B	A
21	C	A	C	C
22	C	B	A	B
23	C	B	B	B
24	C	B	C	C
25	C	C	A	C
26	C	C	B	C
27	C	C	C	C

Experiments with this fuzzy system give 100% accuracy 27 persons with a degree of noise up to 80% (gaussian noise).

8 Conclusions

We described in this paper our hierarchical genetic algorithm approach for modular neural network topology design and optimization. The proposed approach was illustrated with a specific problem of pattern recognition, which is voice recognition. The best MNN is obtained by evolving the modules (single NNs) according to the error of identification and also the complexity of the modules. The results for the problem of voice recognition are very good and show the feasibility of the HGA approach for MNN topology optimization. A fuzzy system was used to integrate the results of voice, face and fingerprints, to make a decision on the identity of a person.

References

[1] G. Barna and K. Kaski, "Choosing optimal network structure", Proceedings of the International Neural Network Conference (INNC90), pp. 890-893, 1990.

[2] O. Castillo and P. Melin,"Hybrid Intelligent Systems for Time Series Prediction using Neural Networks, Fuzzy Logic and Fractal Theory", IEEE Transactions on Neural Networks, Vol. 13, no. 6, pp. 1395-1408, 2002.

[3] D. J. Chalmers, "The Evolution of Learning: An Experiment in Genetic Connectionism", Proceedings of the 1990 Connectionist Models Summer School, Morgan Kauffman, 1990.

[4] J. Feldman, "Neural representation of conceptual knowledge", in Neural connections, mental computation (Nadel and et al., eds.), MIT Press, 1989.

[5] F. Fogelman-Soulie, "Multi-modular neural network-hybrid architectures: a review", Proceedings of 1993 International Joint Conference on Neural Networks, 1993.

[6] B. Happel and J. Murre, "Design and evolution of modular neural network architectures", Neural Networks, vol. 7, pp. 985-1004, 1994.

[7] R. A. Jacobs, M. I. Jordan, S. J. Nowlan and G. E. Hinton, "Adaptive Mixtures of Local Experts", Neural Computation, vol. 3, pp. 79-87, 1991.

[8] R. Jenkins and B. Yuhas, "A simplified neural network solution through problem decomposition: The case of the truck backer-upper", IEEE Transactions on Neural Networks, vol. 4, no. 4, pp. 718-722, 1993.

[9] M. I. Jordan and R. A. Jacobs, "Hierarchical Mixtures of Experts and the EM Algorithm", Neural Computation, vol. 6, pp. 181-214, 1994.

[10] K. F. Man, K. S. Tang and S. Kwong, "Genetic Algorithms: Concepts and Design", Springer-Verlag, 1999.

[11] M. Mitchell, "An Introduction to Genetic Algorithms", MIT Press, 1996.

[12] C. Monrocq, "A probabilistic approach which provides and adaptive neural network architecture for discrimination", Proceedings of the International Conference on Artificial Neural Networks, vol. 372, pp. 252-256, 1993.

[13] D. J. Montana and L. D. Davis, "Training Feedforward Networks using Genetic Algorithms", Proceedings of the International Joint Conference on Artificial Intelligence, Morgan Kauffman, 1989.

[14] H. Simon, The sciences of the artificial, MIT Press, 1981.

[15] P. Melin and O. Castillo, "Hybrid Intelligent Systems for Pattern Recognition", 2005.

Application of a GA/Bayesian Filter-Wrapper Feature Selection Method to Classification of Clinical Depression from Speech Data

Juan Torres[1], Ashraf Saad[2], and Elliot Moore[1]

[1] School of Electrical and Computer Engineering, Georgia Institute of Technology, Savannah, GA 31407, USA
juan.torres@gatech.edu, emoore@gtsav.gatech.edu

[2] Computer Science Department, School of Computing, Armstrong Atlantic State University, Savannah, GA 31419, USA
ashraf@cs.armstrong.edu

Summary. This paper builds on previous work in which a feature selection method based on Genetic Programming (GP) was applied to a database containing a very large set of features that were extracted from the speech of clinically depressed patients and control subjects, with the goal of finding a small set of highly discriminating features. Here, we report improved results that were obtained by applying a technique that constructs clusters of correlated features and a Genetic Algorithm (GA) search that seeks to find the set of clusters that maximizes classification accuracy. While the final feature sets are considerably larger than those previously obtained using the GP approach, the classification performance is much improved in terms of both sensitivity and specificity. The introduction of a modified fitness function that slightly favors smaller feature sets resulted in further reduction of the feature set size without any loss in classification performance.

1 Introduction

In previous work [1], we addressed the problem of selecting discriminating features for the detection of clinical depression from a very large feature set (300–1200 features) that was extracted from a speech database of clinically depressed patients and non-depressed control subjects. The approach taken in [1] was based on Genetic Programming (GP) and consisted of measuring the selection frequency of each feature in the final classifier trees over a large number of GP training episodes. While it was found that a handful of features were selected more frequently than most, the average classification accuracy of the final GP classifiers was not very high (77.4%). However, a naïve Gaussian Mixture Model (GMM) based classifier using the 16 most frequently selected features from the GP algorithm yielded significant improvements (88.8% accuracy, on average), while a simpler naïve Bayesian classifier with a Gaussian assumption performed almost as well as the GMM (88.7% average accuracy). One of the limitations of the previous feature selection approach is that features were ranked individually.

A. Saad et al. (Eds.): Soft Computing in Industrial Applications, ASC 39, pp. 115–121, 2007.
springerlink.com

That is, there was no regard for dependency or redundancy between groups of features. In this paper, classification performance is further improved by applying a two-stage Genetic Algorithm (GA) based feature selection approach that takes the correlation between features into account.

2 Speech Database

The features used in this paper were obtained from a database populated with the speech data of 18 patients (9 male, 9 female) with no history of mental disorders, and 15 patients (6 male, 9 female) who were undergoing treatment for a depressive disorder at the time of the study [2]. The speech corpus for each speaker consisted of a single recording session of the speaker reading a short story. The 65 sentences contained in the corpus were stored individually. Male and female speech samples were analyzed separately, meaning that the database is intended for training gender-specific, binary decision classifiers.

A set of raw features were extracted from each voiced speech frame (approximately 25–30 ms in duration). For each gender, two separate observation groupings were considered. The first grouping (G1) divided the corpus into 13 observations of 5 sentences each while the second grouping (G2) divided the corpus into 5 observations of 13 sentences each. A set of statistical measures was computed for each feature across each sentence. The resulting sentence-level statistics were then subjected to the same set of statistical measures, this time across all sentences in an observation. The procedure resulted in a large vector of *observation feature statistics* (OFS). The reader is referred to [1, 2, 3] for a detailed description of the features contained in the database and their corresponding extraction algorithms.

In this paper, we work with the second observational grouping in the database (G2), which contains 75 and 90 observations for the male and female experiments, respectively. The male and female feature sets contain, respectively, 298 and 857 features.

3 Feature Clustering

The feature selection method under consideration was first proposed in [4] and is designed to exploit correlation in the feature space. It consists of an initial filter in which correlated features are grouped into clusters, followed by two stages of a GA-based wrapper scheme. The initial filtering stage produces clusters that contain groups of features which are highly correlated. The correlation matrices for male and female speech (Fig. 1) show that there is significant correlation between pairs of features in both cases. Correlated feature groups are identified through hierarchical agglomerative clustering with complete linkage [5], which proceeds as follows: The distance between pairs of features is given as $1 - c_{ij}$, where c_{ij} is the correlation coefficient between features i and j. Initially every feature is assigned to its own cluster, and the two closest clusters are merged at each stage of the algorithm. In the complete linkage rule, which tends to produce compact clusters, the distance between two clusters D_x and D_y is given as

$$d(D_x, D_y) = \max_{\substack{i \in D_x \\ j \in D_y}} (1 - c_{ij}) . \tag{1}$$

That is, the distance between two clusters equals the distance between its two most distant members.

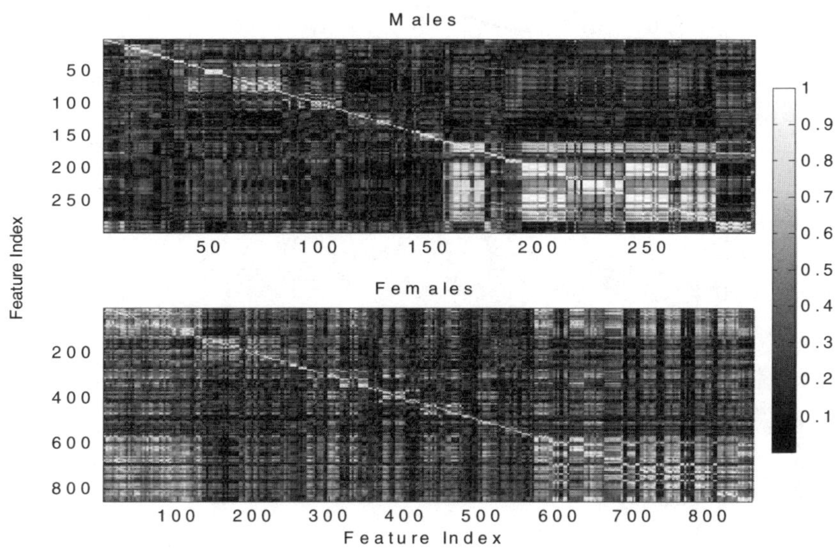

Fig. 1. Correlation Matrices

Clusters continue to be merged until the maximum distance between members of a cluster reaches a particular threshold. Therefore, this threshold controls the minimum correlation that is allowed between features in the same cluster. The number of clusters versus the value of this threshold is shown in Fig. 2. Because each cluster will be represented by a single feature in the first stage of GA (Sect. 4), it is necessary for all features in a cluster to be similar. Following the recommendation in [4], the threshold was set to $c_{ij} = 0.8$, which yielded 127 clusters for males and 282 for females. Figure 3 shows the size of each cluster obtained from the male speech data. Once the clusters are obtained, the feature closest to each cluster center (i.e., mean) is chosen as its representative feature. It should be noted that the final result will contain many clusters of size 1, which represent features having a correlation coefficient of less than 0.8 with respect to every other feature in the dataset.

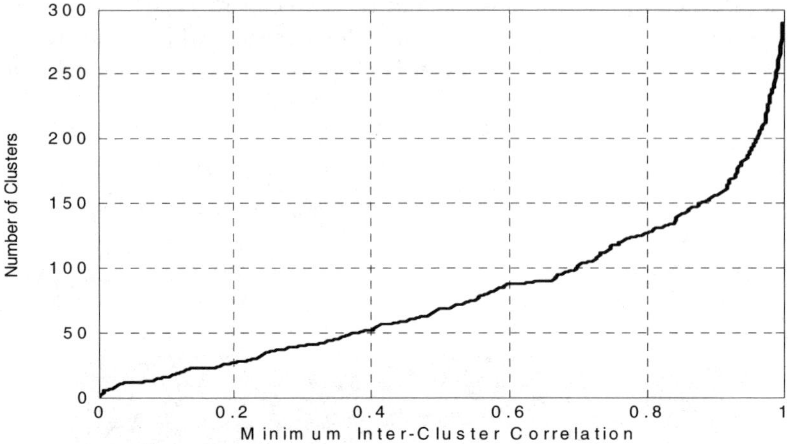

Fig. 2. Threshold Value vs. Number of Clusters (Males)

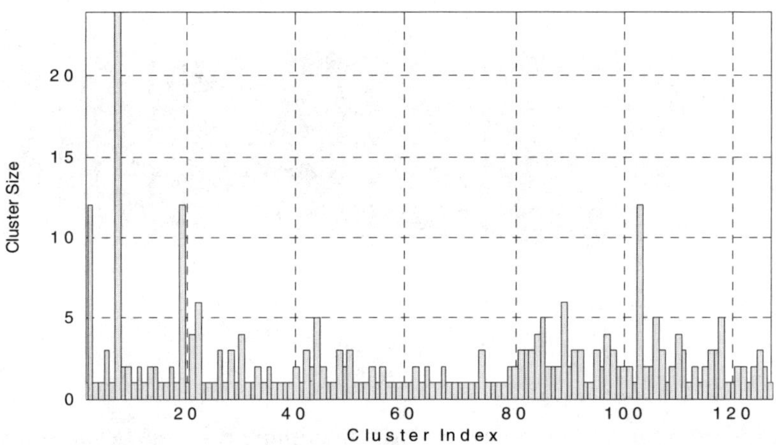

Fig. 3. Cluster Sizes (Males)

4 Two-Stage Feature Selection

Upon completion of the initial filtering step, we then proceed with the first stage of GA, in which the feature subspace to be searched consists of a single representative feature for each cluster. The goal of this stage is to find the cluster subset that maximizes classification accuracy. In this stage, the GA chromosome of an individual consists of a binary vector equal in length to the number of feature clusters. This vector forms a bit-mask that determines which feature clusters are to be used for classification. The fitness function suggested in [4] consists of the classification accuracy as determined by the classifier of choice. In this paper, we propose a fitness function that slightly penalizes larger feature sets:

$$f = Accuracy - 0.01(\frac{n}{N}) , \quad (2)$$

where n is the dimensionality (i.e., the number of clusters) of the feature subset under consideration, N is the dimensionality of the entire feature space, and *Accuracy* is defined in the interval [0, 1]. Thus, the penalty given to larger subsets is equivalent to at most 1% classification accuracy. Because the penalty is so small, the algorithm is still focused on maximizing classification accuracy, but when presented with several feature subsets that yield nearly identical classification accuracy, preference will be given to the smallest subset.

Once an optimal set of feature clusters is found, we proceed into the second GA stage, in which these clusters are opened and a search is performed to find the optimal representative feature for each cluster. In this stage, the GA chromosome consists of an integer vector whose size is equal to the number of clusters selected in stage 1. Each entry in the vector is a feature index into a particular cluster.

Both stages share the same evolutionary rules. The initial population is randomly chosen from a uniform distribution on the entire chromosome space. During reproduction, elitism is used to ensure that the best candidate from each generation is copied to the next. Then, two-point crossover is performed on a given fraction of the population. Parents for the crossover operation are chosen by tournament selection. Population members not chosen by elitism or for crossover are mutated uniformly with a fixed probability. Evolution stops after a fixed number of generations. The complete list of parameters that were used is given in Table 1.

Because of the small number of observations in the database, classification performance is evaluated using the leave-one-out cross-validation technique. Initially, the intention was to use a GMM for classification because of its ability to approximate arbitrary likelihood functions (given enough mixtures). However, GMM training proved to be too computationally intensive to be used within the wrapper scheme, since an entire cross-validation run must be performed for each individual in the GA population in every generation. In addition, the limited amount of available training data restricted the number of mixtures we could use. A naïve Bayesian classifier with a Gaussian assumption on the likelihood functions was shown in [1] to perform comparably to GMM on our speech database, and was chosen here for its computational simplicity. Moreover, because the initial clustering stage essentially filters-out the most correlated features, the naïve Gaussian classifier turns out to be a good match to the GA search stages. Although uncorrelated features from arbitrary probability distributions are not necessarily independent, the equivalence between correlation and dependence does hold for Gaussian distributions. Thus, while the Gaussian assumption limits the classifiers ability to model features that are not normally distributed, the initial filtering step reduces the penalty associated with the naïve assumption when the Gaussian assumption is correct.

Table 1. GA Parameters

Parameter	Value
Crossover fraction	0.80
Mutation fraction	0.20
Mutation probability	0.1
Population Size	1000
Tournament Size	10
Number of Generations	10

Table 2. Classification Performance

	Naïve Bayes Using GPFS Features		Two-Stage GA Original Fitness Function		Two-Stage GA New Fitness Function	
Gender	M	F	M	F	M	F
Accuracy	89.3	91.1	97.3	95.6	97.3	95.6
Sensitivity	93.3	95.6	100.0	97.8	100.0	97.8
Specificity	86.7	86.7	95.6	93.3	95.6	93.3
Selected features	16	16	43	122	34	89
% of feature space	5.4	1.9	14.4	14.2	11.4	10.4

5 Results

Performance was evaluated in terms of classification accuracy, sensitivity (true positive rate), and specificity (false negative rate). We performed 10 runs of GA for each experiment. The results of the best run, given in Table 2, show a significant improvement over the previous GP-based feature selection approach, especially for the male subjects. Furthermore, the use of (2) as a fitness function resulted in a significant reduction of feature set size without any loss in classification performance. Another interesting result is that the second GA stage did not find better solutions than those found during the first stage. This suggests that it may be advantageous to use a lower threshold during clustering in order to reduce the number of clusters at the expense of larger feature variability within clusters. The reduction in the number of available clusters for the first GA stage may in turn lead to further reduction in the number of features in the final solution.

6 Discussion and Future Work

The work described in the previous sections has resulted in a large improvement in classification accuracy relative to the work in [1]. However, the number of selected features is rather high in relation to the number of observations in the dataset. While decreasing the clustering threshold in the initial filtering step could possibly reduce the number of features in the final GA solutions, it would also result in the clustering of features that are not as significantly correlated, which would prevent some potentially effective combinations of features from being considered during the 2- stage GA procedure. One way to get around this issue would be to integrate the correlation filter into the GA via modified reproduction operators. In addition, correlation is a rather limited measure of redundancy between features, since it only considers pairs of features and does not necessarily imply independence between them. Therefore, it may be beneficial to use instead a more general measure of redundancy between features, such as Joint Mutual Information [6]. Finally, work in [7] has shown that it is possible to train a Self-Organizing Map (SOM) on the GA chromosomes to obtain an approximate 2-D representation of the feature space. The map can then be used to guide the GA search into unexplored regions, thereby increasing the chance of convergence to a global minimum.

We have succeeded in applying a hybrid GA and Bayesian approach to the feature selection problem in a difficult domain. By judicious choice of a fitness function for the first GA stage, we were able to improve classification performance while simultaneously reducing the size of the feature subset. The resulting feature subsets, although larger than those obtained using the GP-based approach reported in [1], provide excellent classification performance. The current GA search and Bayesian classifier implementation serve as a solid baseline for future investigation into finding useful features to diagnose clinical depression from speech.

References

1. Torres, J., Saad, A., Moore, E.: Evaluation of objective features for classification of clinical depression in speech by genetic programming. In: WSC11: 11th Online World Conference on Soft Computing in Industrial Applications. (September 2006)
2. Moore, E., Clements, M., Peifer, J., Weisser, L.: Comparing objective feature statistics of speech for classifying clinical depression. In: Proc., 26th IEEE Conf. Eng. in Medicine and Biology. Volume 1., San Francisco, CA (2004) 17–20
3. Moore, E., Clements, M., Peifer, J., Weisser, L.: Analysis of prosodic variation in speech for clinical depression. In: Proc., 25th IEEE Conf. Eng. in Medicine and Biology. Volume 3., Cancún, México (2003) 2925–2928
4. Van Dijck, G., Van Hulle, M., Wevers, M.: Genetic algorithm for feature subset selection with exploitation of feature correlations from continuous wavelet transform: a real-case application. In: International Conference on Computational Intelligence, Istanbul, Turkey (2004) 34–38
5. Duda, R., Hart, P., Stork, D.: Pattern Classification. Wiley, New York, NY (2001)
6. Tourassi, G., Frederick, E., Markey, M., Floyd, C.: Application of the mutual information criterion for feature selection in computer-aided diagnosis. Medical Physics **28**(12) (2001) 2394–402
7. Amor, H.B., Rettinger, A.: Intelligent exploration for genetic algorithms: Using self-organizing maps in evolutionary computation. In: GECCO Genetic and Evolutionary Computation Conference, Washington, DC, USA (June 2005) 1531–1538

Comparison of PSO-Based Optimized Feature Computation for Automated Configuration of Multi-sensor Systems

Kuncup Iswandy and Andreas Koenig

Institute of Integrated Sensor Systems, University of Kaiserslautern,
67663 Kaiserslautern, Germany
{kuncup@rhrk, koenig@eit}.uni-kl.de

Abstract. The design of intelligent sensor systems requires sophisticated methods from conventional signal processing and computational intelligence. Currently, a significant part of the overall system architecture still has to be manually elaborated in a tedious and time consuming process by an experienced designer. Clearly, an automatic method for auto-configuration of sensor systems would be salient. In this paper, we contribute to the optimization of the feature computation step in the overall system design, investigating multi-level thresholding (MLT) and Gaussian windowing. Our goals are to compare these two feature computation methods and two evolutionary optimization techniques, i.e., genetic algorithm (GA) and particle swarm optimization (PSO). To compare with previous research work gas sensor benchmark data is used. In the comparison of GA and PSO the latter method provided superior results of 100% recognition in generalization for thresholding, which proved to be more powerful method.

1 Introduction

Intelligent sensor systems find more and more widespread application, for instance, as the most recent example the fields of ambient intelligence and sensor networks can be mentioned. This remarkable increase of application is also due to the growing spectrum of available sensor principles and implementations. These require, however, also a larger variety sensor electronics and sensor signal processing techniques to efficiently employ them in systems. The design of intelligent sensor systems is still predominantly conducted manually, where the design process goes through the principal steps of sensor selection and scene optimization, choice of signal and feature processing, dimensionality reduction, and classification (see Fig. 1). The processing steps of dimensionality reduction and classification are more and more subject of automation efforts employing learning and optimization techniques. However, the decisive task of heuristic signal processing and feature computation method selection, combination, and parameter setting is currently left to the human designer as a tedious, time and labor consuming task with potentially suboptimal outcome. In particular, the strong diversity of available methods and tools from conventional signal processing to computational intelligence techniques imposes severe challenges on the experiences and qualifications of the designer.

A. Saad et al. (Eds.): Soft Computing in Industrial Applications, ASC 39, pp. 122–131, 2007.
springerlink.com

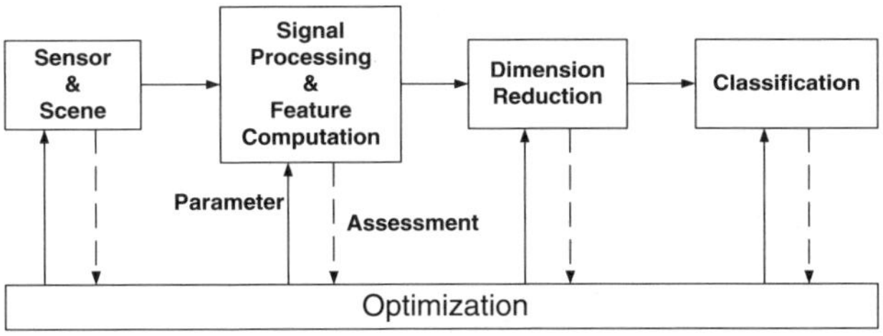

Fig. 1. General architecture of intelligent sensor systems

It is our overall research goal to contribute to the design automation activities for intelligent multi-sensor systems, in this paper we focus on the optimization of feature computation regarding two standard methods, multilevel thresholding [1] and Gaussian windowing [2], which will be optimized by evolutionary techniques, genetic algorithms (GA) [3] and particle swarm optimization (PSO) [4] according to benchmark data from gas sensor data classification.

In the next section, the two feature computation methods will be described, which are the first two instances of a feature computation method library we are currently establishing in our work. In the third section, the employed GA and PSO techniques are explained along with the method parameter settings. In the fourth section we will present experiment and results for the benchmark data. Concluding, we will give an outlook of our envisioned next steps toward automated design of intelligent sensor systems.

2 Feature Computation Methods

The roles of feature computation techniques are to extract the meaningful information of raw data of sensor response patterns and to reduce the dimension size of feature vector of a pattern, where it can increase the speed of computation and the accuracy of pattern classification. In particular, with regard to the application of gas sensor systems two feature computation techniques have been proposed, i.e., multi-level thresholding (MLT) [1] and Gaussian windowing [2].

In the applying of MLT, the first derivative of the conductance or slope curves as a further processing is used. The MLT techniques compute the features with similar to histogram and amplitude distribution computation. There are two ways to compute features using multi-level thresholding techniques, i.e., differential and cumulative modes, which count number of samples of signal resposes lying in range between two thresholds. Figure 2 illustrates cumulative and differential feature computation. The features of MLT differential mode can be computed as

$$z_i = \sum_{s=1}^{Nr} \delta(y_s, T_p, T_q), \tag{1}$$

$$\delta(y_s, T_p, T_q) = \begin{cases} 1 & T_p \leq y_s \leq T_q \\ 0 & \text{otherwise,} \end{cases} \tag{2}$$

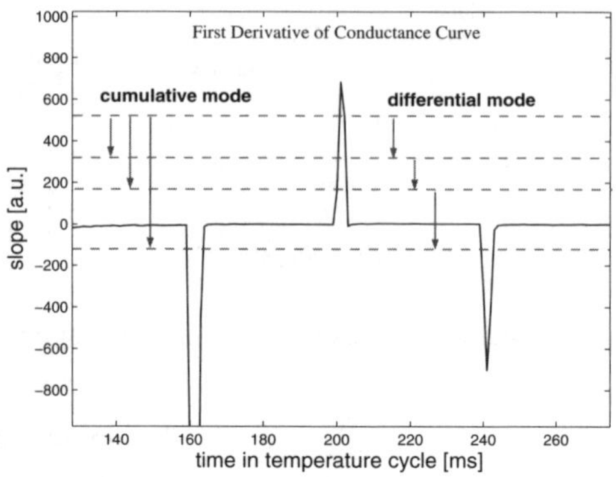

Fig. 2. Multi-level threshold for extracted features from gas sensor data [1]

where y_s is magnitude value of sensor signal with $s = 1,2,...,Nr$ and Nr is total samples of a pattern; i is the number of features ($i = T - 1$) and T is number of thresholds used; and T_p and T_q are level-values with $q = 2,3,..T$ and $p = q - 1$. On the other hand, the MLT cumulative mode is computed in the similar way, except that $q = T$ and $p = 1,2,...,T-1$.

The Gaussian windows or kernels extract the features directly from conductance curves or transient responses. Each kernel is a Gaussian exponential function with a given mean and standard deviation. Each of these kernels is multiplied by sensor response and integrated with respect to time (see Fig. 3). The number of features used is same with the number of kernels. The features of Gaussian windowing can be computed as

$$z_i = \sum_{s=1}^{Nr} y_s.G(s,\mu_i,\sigma_i), \tag{3}$$

$$G(s,\mu_i,\sigma_i) = \exp^{-\frac{1}{2}(\frac{s-\mu_i}{\sigma_i})^2}. \tag{4}$$

3 Optimization Methods

3.1 Genetic Algorithms

Genetic algorithms (GA) are search algorithms based on the mechanics of natural selection and natural genetics. Genetic algorithms consider a population of chromosomes (individuals) encoding potential solutions to a given problem. The main characteristics of GA are the intensive use of randomness and genetic-inspired operations, i.e., selection, recombination (crossover) and mutation to evolve a set of candidate solutions.

We adopt the main steps of GA applied in our previous work [1]. Briefly, the main steps of the GA adapted to the requirements of automated sensor system design are initialization (generate an initial population), selection for recombination (Roulette Wheel Selection), recombination (one point crossover), mutation, reproduction (with elitism 10%) and termination conditions (loop or stop criteria).

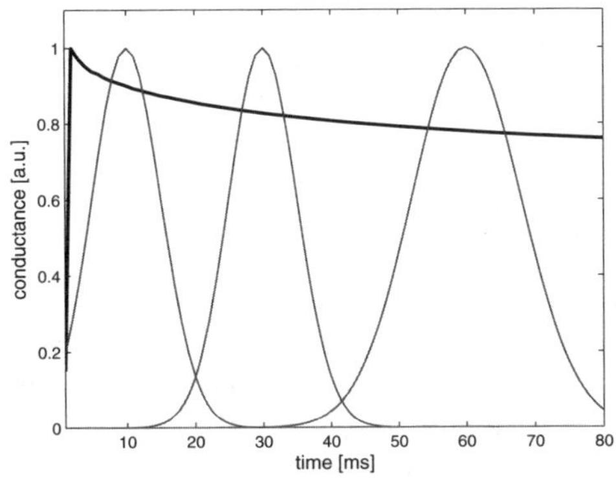

Fig. 3. Gaussian windowing for window time slicing from a conductance curve of gas sensor data

3.2 Particle Swarm Optimization

Particle swarm optimization (PSO) is a non-linear method which also is affiliated to evolutionary computation techniques. Particle swarms explore the search space through a population of particles, which adapt by returning to previously successful regions [4]. The particles then fly over the state space, remembering the best solution encountered. The fitness function is determined by an application-specific objective function. Here, we use overlap assessment measure as the fitness function (see section 3.5). During each iteration, the velocity of each particle is adjusted based on its momentum and influence of the best solutions encountered by itself and its neighbors. The particles then move to a new position, and the process is repeated for a prescribed number of iterations. In the original PSO implementation, the trajectory of each particle is governed by the equations:

$$\mathbf{v}_i(t+1) = \omega \mathbf{v}_i(t) + c_1 \cdot rand() \cdot (\mathbf{p}_i - \mathbf{x}_i(t)) + c_2 \cdot rand() \cdot (\mathbf{p}_g - \mathbf{x}_i(t)) \quad (5)$$

and

$$\mathbf{x}_i(t+1) = \mathbf{x}_i(t) + \mathbf{v}_i(t+1) \quad (6)$$

where $\mathbf{x}_i = (x_{i1}, x_{i2}, ..., x_{id})$ and $\mathbf{v}_i$ are the current vector position and velocity of the i-th particle, $\mathbf{p}_i$ is the position of the best state visited by the i-th particle, $\mathbf{p}_g$ is the particle with the best fitness in the neighborhood of i, and t is the iteration number. The parameter c_1 and c_2 are called the cognitive and social learning rates. The parameter ω is an inertia weight, which used to dampen the velocity during the course of the simulation, and allow the swarm to converge with greater precision.

3.3 Optimizing Feature Computation

In optimizing feature computation, the particle for one member of the population represents the array level values of thresholds for multi-level thresholdings techniques and for Gaussian window function, the particles represent array pairs of mean and standard deviation values. In searching of optimal combination of thresholds, the MLT methods have a constraint, where the rank position of thresholds will not exceed the lower or higher rank threshold.

3.4 Optimizing Feature Selection

The original PSO technique is designed for the real-value problems, whereas the feature selection only uses binary values to represent whether one feature is selected or not. Therefore, the algorithm now has been extended to tackle binary/discrete problems. Kennedy and Eberhart [5] have proposed binary PSO (BPSO), where uses velocity as a probability to determine whether the components of $\mathbf{x}_i$ will be in one or zero state (binary). They squashed $\mathbf{v}_i$ using a logistic function $s(\mathbf{v}) = 1/(1+\exp(-\mathbf{v}))$ while the velocity is calculated using the same equation in Eq. (5). If a randomly generated number within $[0,1]$ is less than $s(v_{id})$, then x_{id} is set to be 1, otherwise it is set to be 0. The minimization of the number of features is not explicitly included in Eq.(7). Instead, we have added an implicit selecting condition, so that in the case of two or more different feature subsets with equal assessment values, the best particle will be set to the feature subset of the smallest cardinality.

3.5 Fitness Function

The process of choosing a subset of features according to certain criteria can be divided into two groups, i.e., wrapper and filter approach [10]. The wrapper approach takes feature subset selection and classification as a whole and selects features based on classification results directly, while the filter approach utilizes statistics underlying training data and operates independently of classification algorithms [11]. The process of feature computation is also based on these approaches as shown in Fig. 4(a) and 4(b). For assessment of feature computation and feature selection in our work, the nonparametric overlap measurement [7] is applied. This measurement is an automatic feature selection of filter approach.

The nonparametric overlap measure q_o, which was inspired by the nearest neighbor concepts, provides a very fine-grained value range. This normalized measure gives values close to one for non-overlapping class regions and decreases towards zero proportional to increasingly overlapping of class regions. The overlap measure is computed by:

$$q_o = \frac{1}{L}\sum_{c=1}^{L}\frac{1}{N_c}\sum_{j=1}^{N_c}\frac{\sum_{i=1}^{k} q_{NN_{ji}} + \sum_{i=1}^{k} n_i}{2\sum_{i=1}^{k} n_i} \tag{7}$$

with

$$n_i = 1 - \frac{d_{NN_{ji}}}{d_{NN_{jk}}} \tag{8}$$

and

$$q_{NN_{ji}} = \begin{cases} n_i & \omega_j = \omega_i \\ -n_i & \omega_j \neq \omega_i. \end{cases} \tag{9}$$

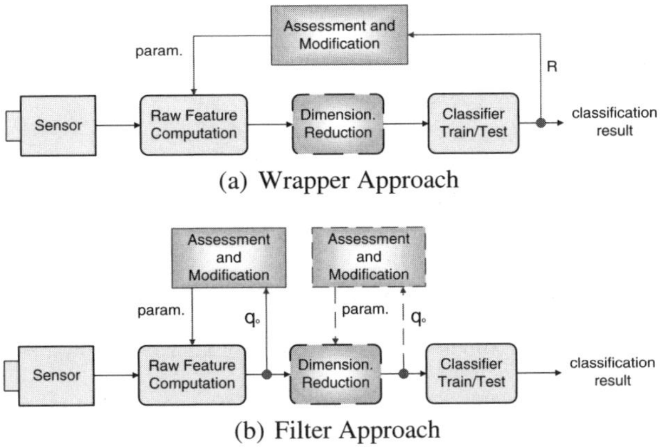

(a) Wrapper Approach

(b) Filter Approach

Fig. 4. The optimization model of recognition system

Here, n_i denotes the weighting factor for the position of the ith nearest neighbor NN_{ji}, $d_{NN_{ji}}$ denotes the Euclidean distance between $\mathbf{z}_j$ and NN_{ji}, $d_{NN_{jk}}$ denotes the distance between $\mathbf{z}_j$ and most distant nearest neighbor NN_{jk}, $q_{NN_{jk}}$ denotes the measure contribution of $\mathbf{z}_j$ with regard to NN_{ji}, L is the number of classes, and ω denotes the class affiliation of $\mathbf{z}$. Typically, the number of nearest neighbors well suited for computation of this measurement are 5 to 10.

4 Experiments and Results

In our experiment, we used a benchmark data of a gas sensor system [6] and [1] obtained from Laboratory for Measurement Technology, University of Saarland. The raw data of sensor response is obtained by using a commercial micro sensor MiCS 5131. These sensors possess a heating that is controlled during measurement in so called temperature cycles. The conductance of the sensors was measured every 10 ms with 16 bits resolution during one high temperature step (500^oC) with a duration of 800 ms, two temperature cycles with four levels (23^oC, 500^oC, 90^oC, and 290^oC) with a duration of 400 ms each level, and temperature off with a duration of 4.1 seconds. There are four test gases (H_2, CH_4, ethanol and CO) used and for each test gas, there are three different concentrations applied. All test gases were performed at two relative humidity values (30% and 70%). The data set consists of 810 measure values and 264 patterns, where each type of gases with different concentration and relative humidity has 11 patterns. Data samples of the four gases were merged to four classes, neglecting underlying differences in concentration and relative humidity.

The available data set was separated into training and testing data set, with 144 and 120 patterns, respectively. This holdout method has been used for classification. We performed feature computation and feature selection only on the training set. Testing set has not been used for feature computation and feature selection because we want to

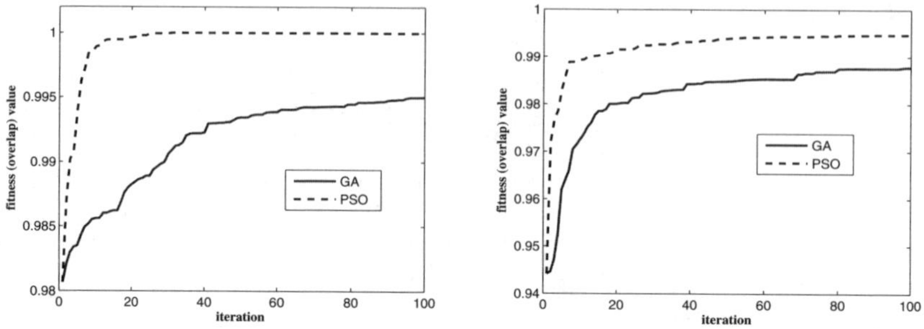

(a) MLT differential mode

(b) MLT cummulative mode

Fig. 5. Comparison of the overlap assessment curves between GA and PSO

make sure that classification results are not biased. In extending our experiments, the leave-one-out cross-validation approach applied on testing set was used for classification estimation.

The control parameters of GA were set as in our prior work [1], where population size is 20 individuals, the number of iteration is 100, recombination rate is 0.8, mutation rate is 0.01 and elitism (reproduction) is 10% of the population size. The parameter settings of PSO in our experiments were determined as following:

- Each experiment is repeated using 10 runs and each run has 100 iterations.
- Population size is 20.
- ω, c_1, and c_2 are 1, 2, and 2 respectively.
- The number of nearest neighbors used is 5 for the *k*NN voting classifier [9] and the overlap measurement.
- In the experiment, the number of window-kernels is used in range of 3 to 10 per temperature step (10 steps).

In our prior work [1], genetic algorithms were used to optimize the multilevel thresholding (MLT) feature computation. In the first step of our experiment, we compared the results between GA and PSO techniques for optimizing the MLT feature computation for both differential and cumulative approaches. The number of thresholdings used for differential mode is nine and for cumulative mode five according to the best results of GA in our prior work [1]. Figure 5(a) and 5(b) show that the PSO performed better than the GA according to the mean and standard deviation of overlap assessment values. Moreover, for the classification accuracies from both training and test data sets, the PSO achieved superior results than GA as shown in Table 1 and Table 2. These conclude the comparison of results between PSO and GA that have been reported alike in relevant references (e.g., [8] and [13]).

In the next experiment, we only used PSO for optimizing Gaussian windowing feature computation due to the PSO performances better than GA. There are no significantly different results for the alteration of number of kernels as shown in Table 3 with regarding to the mean and the standard of the overlap measurement values and classification accuracies. The results achieved by the multilevel thresholding using PSO showed slightly better performances than Gaussian windowing with regarding to

Table 1. Comparison of MLT differential mode result between GA and PSO

MLT CM	overlap q_o mean/std	Recognition accuracy (kNN) train (%) mean/std	test (%) mean/std	test-LOO (%) mean/std
GA	0.9950/0.0035	99.44/0.55	99.67/0.58	99.17/0.79
PSO	1.00/0	100/0	100/0	99.83/0.35

Table 2. Comparison of MLT cummulative mode result between GA and PSO

MLT CM	overlap q_o mean/std	Recognition accuracy (kNN) train (%) mean/std	test (%) mean/std	test-LOO (%) mean/std
GA	0.9878/0.0044	98.89/0.36	99.50/6.36	98.67/1.48
PSO	0.9953/0.0024	99.10/0.34	99.92/0.89	99.83/0.35

Table 3. Results of Gaussian window function using PSO

no. of kernel ($\times$ 10)	overlap q_o mean/std	Recognition accuracy (kNN) train (%) mean/std	test (%) mean/std	test-LOO (%) mean/std
3	0.9806/0.0044	97.91/0.65	99.00/0.66	95.50/2.29
4	0.9791/0.0081	97.78/0.79	99.00/0.77	95.83/1.76
5	0.9794/0.0021	98.13/0.03	99.67/0.43	96.08/1.11
6	0.9797/0.0034	97.71/0.74	98.75/0.90	94.92/2.17
7	0.9795/0.0015	98.13/0.57	99.25/0.73	96.92/1.11
8	0.9786/0.0027	97.92/0.46	99.00/0.53	95.67/0.95
9	0.9786/0.0031	97.92/0.46	99.08/0.61	95.83/1.36
10	0.9787/0.0016	98.13/0.47	99.75/0.40	96.08/0.88

classification accuracies and more efficient in computation effort. The MLT techniques use the summation operator and the number of features only depend on the number of thresholdings used, whereas the Gaussian windowing relies on the multiplication operator and the number of features depends on the number of kernels used and also the temperature steps. In the next step after applying Gaussian windowing feature computation, automated feature selection method was carried out to improve the results of overlap measurement and accuracy in classification as shown in Table 4. Due to the fact that optimization of feature computation and feature selection applies the filter approach, that satisfies the overlap measurement as a criterion function instead of directly using the classification, it can happen that the recognition rates of testing set can be higher than the training set as shown in Table 3 and 4.

Table 4. Results of Gaussian window function after applying feature selection

no. of kernel (x 10)	overlap q_o	selected feature	Recognition accuracy (*k*NN) train (%)	test (%)	test-LOO (%)
3	0.9822	10	99.31	100	99.17
4	0.9854	10	98.61	100	98.33
5	0.9835	36	99.31	100	98.33
6	0.9844	26	99.31	100	97.50
7	0.9805	29	99.31	100	98.33
8	0.9889	34	99.31	100	99.17
9	0.9841	41	99.31	99.17	98.33
10	0.9859	50	99.31	100	98.33

5 Conclusion

In this paper, we contribute to the optimization of the feature computation step in the overall automated design of intelligent sensor systems, investigating multi-level thresholding and Gaussian windowing, which are optimized by evolutionary techniques, i.e., genetic algorithms and particle swarm optimization. The results in the experiments show that PSO performed better and even superior than GA. According to the overlap measurement from filter approach and the classification accuracy, the multi-level thresholding techniques achieved superior results and more efficient computation effort than Gaussian windowing. The next processing step of dimensionality reduction (feature selection) proved the improvement of the recognition system.

In future work, we will consider applications of sensor networks and develop library/toolbox of feature computation techniques for multi-sensor system design intended to increase the choice of the feature processing and apply feature level fusion in order to advance the recognition accuracy of sensor systems. Also, additional feature assessment functions according to wrapper and filter approaches [10] and [11] and the combination of them in multi-objective optimization problems [12] and [13] for assessing feature computation and feature selection with regarding to automated design of intelligent sensor systems will be considered.

Acknowledgment

The providing of the gas sensor benchmark data [1] and [6] by Thomas Fricke, Marc Baumbach and Andreas Schuetze from Laboratory for Measurement Technology, University of Saarland is gratefully acknowledged.

References

1. Iswandy, K., Koenig, A., Fricke, T., Baumbach, M., Schuetze, A.: Towards Automated Configuration of Multi-Sensor Systems Using Evolutionary Computation - A Method and a Case Study. J. Computational and Theoretical Nanoscience, Vol. 2. No. 4. American Scientific Publishers (2005), 574–582.

2. Courte, D. E., Rizki, M. M., Tamburino, L. A., Gutierrez-Osuna, R.: Evolutionary Optimization of Gaussian Windowing Functions for Data Preprocessing. Int. J. Artificial Intelligence Tools, Vol. 12. No. 1. World Scientific (2003) 17–35.
3. Goldberg, D. E.: Genetic Algorithms in Search, Optimization, and Machine Learning. Addison-Wesley, Reading, MA.
4. Kennedy, J., Eberhart, R.C.: Particle Swarm Optimization. Proc. of IEEE Int. Conf. on Neural Networks (ICNN). Vol. 4. (1995) 1942–1948.
5. Kennedy, J., Eberhart, R. C.: A Discrete Binary Version of The Particle Swarm Algorithm. Proc. of Conf. on System, Man, and Cybernetics. (1997) 4104–4109.
6. Baumbach, M., Sossong, A., Delprat, H., Soulantica, K., Schuetze, A., Borrel, H., Maisonnat, A., Chaudret, B.: New Micro Machined Gas Sensors Combined with Intelligent Signal Processing Allowing Fast Gas Identification after Power-Up. Proceedings Sensor 2005. Vol. 2. 91–96.
7. Koenig, A., Gratz, A.: Advanced Methods for the Analysis of Semiconductor Manufacturing Process Data. In: Pal, N. R., Jain, L. C. (eds.): Advanced Techniques in Knowledge Discovery and Data Mining. Springer Verlag (2005) 27–74.
8. Eberhart, R. C., Shi, Y.: Comparison between Genetic Algorithms and Particle Swarm Optimization. In V. W. Porto, N. Saravanan, D. Waagen, and A. E. Eiben, Eds. Evolutionary Programming VII: Proc. 7th Ann. Conf., San Diego, CA. Berlin, Springer Verlag.
9. Raymer, M.L., Punch, W.F., Goodman, E.D., Kuhn, L.A., Jain, A.K.: Dimensionality Reduction Using Genetic Algorithms. IEEE Trans. Evolutionary Computation. Vol. 4. No. 2. (2000) 164–171.
10. Liu, H., Motoda, H.: Feature Selection for Knowledge Discovery and Data Mining. Kluwer Academic Publisher. (1998).
11. Mao, K.Z.: Fast Orthogonal Forward Selection Algorithm for Feature Subset Selection. IEEE Trans. Neural Networks. (2002) 1218–1224.
12. Emmanouilidis, C., Hunter, A., MacIntyre, J.: A Multiobjective Evolutionary for Feature Selection and a Commonality-Based Crossover Operator. In 2000 Congress on Evolutionary Computation (CEC'2000). IEEE Service Center. (2000).
13. Iswandy, K., Koenig, A.: Feature Selection with Acquisition Cost for Optimizing Sensor System Design. In Kleinheubacher Tagung, KH2005, C.1, Integrierte digitale und analoge Schaltungen. Miltenberg, Germany. (2005).

Evaluation of Objective Features for Classification of Clinical Depression in Speech by Genetic Programming

Juan Torres[1], Ashraf Saad[2], and Elliot Moore[1]

[1] School of Electrical and Computer Engineering, Georgia Institute of Technology, Savannah, GA 31407, USA
juan.torres@gatech.edu,emoore@gtsav.gatech.edu

[2] Computer Science Department, School of Computing, Armstrong Atlantic State University, Savannah, GA 31419, USA
ashraf@cs.armstrong.edu

Summary. This paper presents the results of applying a Genetic Programming (GP) based feature selection algorithm to find a small set of highly discriminating features for the detection of clinical depression from a patient's speech. While the performance of the GP-based classifiers was not as good as hoped for, several Bayesian classifiers were trained using the features found via GP and it was determined that these features do hold good discriminating power. The similarity of the feature sets found using GP for different observational groupings suggests that these features are likely to generalize well and thus provide good results with other clinical depression speech databases.

1 Introduction

In studies related to the classification of clinical depression from a patient's speech [1, 2], a database containing speech samples from healthy and depressed subjects was collected and a vast set of features was extracted from the database. In order to facilitate the design of an efficient and robust classifier, it is then desirable to solve the feature selection (FS) problem in order to find a (preferably small) subset of features that maximizes classification performance. Because the size of the feature search space is 2^N, where N is the total number of features, an exhaustive search becomes computationally prohibitive even for moderate values of N. Therefore, for most practical problems, a non-exhaustive FS algorithm must be employed.

Many feature selection algorithms have been proposed and survey of FS under the framework of classification can be found in [3]. FS algorithms can be divided into *filter* and *wrapper* approaches, which differ in the nature of the objective function to be maximized (or minimized). Filter approaches utilize a computationally inexpensive function designed to approximately measure the discriminating ability of the feature subset under consideration, whereas wrapper approaches use the actual performance of the classification algorithm. Thus,

A. Saad et al. (Eds.): Soft Computing in Industrial Applications, ASC 39, pp. 132–143, 2007.
springerlink.com

while wrapper approaches are more computationally expensive, since a classifier must be trained for every feature subset that is evaluated, they can also be more accurate. Certain FS algorithms, such as Branch and Bound [3], are said to be both non-exhaustive and *complete*, meaning that they are guaranteed to arrive at the optimal solution without exploring the entire search space. However, a necessary condition to ensure completeness in these algorithms is that the objective function be monotonic with respect to the size of the feature subset. This condition, at least for finite data sets, precludes the use of the wrapper approach and therefore invalidates the completeness property with respect to classification performance.

A Genetic Programming (GP) approach to classifier design was proposed in [4] and later refined in [5] to perform *simultaneous* feature selection and classifier design by means of a single evolutionary search (GPFS), thus removing the computational expense of classifier training associated with the wrapper approach while allowing classifier performance to be used as a fitness measure for FS. Although this algorithm (like its aforementioned predecessors) is not guaranteed to find a globally optimal solution, the reduced computational burden with respect to wrapper approaches allows the search procedure to evaluate a much larger number of classifiers, which can increase the amount of exploration of the feature space. GPFS has been shown to produce excellent results relative to previous FS methods both in terms of classification performance and feature subset size. The results obtained in [5] using the GENE dataset are particularly motivating. On this dataset, the algorithm achieved an average classification accuracy of 92.55% using an average of 10.45 features out of 7129. In this paper, we present the results of implementing a GPFS-like algorithm using `lilgp` [6] to find a small set of highly discriminating features from a clinical depression speech database. Once this feature subset is found, we then address the question of whether a Bayesian classifier, which is optimal with respect to certain assumptions on the features' probability distributions, can provide better performance over the classifiers designed by the GP's evolutionary search.

The paper is organized as follows. A description of the clinical depression speech database and of the features contained therein is given in Sect. 2. Section 3 provides an overview of Genetic Programming and how it may be used for feature selection. Section 4 consists of a brief introduction to Bayesian classification and to several quantization and Probability Distribution Function (PDF) estimation methods for handling continuous data. Feature selection and classification performance results are given in Sect. 5. Concluding thoughts and directions for future work are offered in Sect. 6.

2 Speech Features

The features used in this paper were obtained from a database populated with the speech data of 18 patients (9 male, 9 female) with no history of mental disorders, and 15 patients (6 male, 9 female) who were undergoing treatment for a depressive disorder at the time of the study [1]. The speech corpus for each speaker consisted of a single recording session of the speaker reading a short story. The 65 sentences contained in the corpus were stored individually. In addition, male and female speech was analyzed separately.

A set of raw features related to vocal tract resonances, the glottal waveform, and the Teager FM component were extracted from each voiced speech frame (approximately 25–30 ms in duration). The extraction of vocal tract and glottal features is based on the source-filter model of speech production [7], which approximates a speech utterance as the convolution of a glottal (vocal fold) excitation signal with an all-pole linear filter representing the resonant frequency response of the vocal tract. Linear Predictive Analysis (LPA) can be applied to a frame of speech in order to approximate the vocal tract filter, from which formant (i.e. resonant) frequencies and bandwidths can be estimated. If done carefully, deconvolution by the LPA filter results in a reasonable approximation of the glottal excitation signal, which can be used to estimate regions of glottal opening and closure. From these regions, several glottal ratio and timing features can be obtained. Further details on the vocal tract and glottal features contained in the database and their extraction are given in [1]. Teager FM Features were extracted using the algorithm given in [8], with the exception that we did not limit ourselves to measuring only the variation in the frequency modulation (FM) component, but instead subjected the Teager FM signal from each speech frame to the statistical measures in Table 2, to obtain a set of 8 raw features. Finally, features related to prosodics (pitch, energy contour, and speaking rate) were extracted from each voiced section of speech within an utterance. The extraction algorithms for prosodic features are described in [2]. Raw features were grouped into the 10 categories listed in Table 1.

Table 1. Raw Feature Categories

Pitch (PCH)	Glottal Ratios (GLR)
Energy Median Statistics (EMS)	Glottal Spectrum (GLS)
Energy Deviation Statistics (EDS)	Formant Locations (FMT)
Speaking Rate (SPR)	Formant Bandwidths (FBW)
Glottal Timing (GLT)	Teager FM (TFM)

For each gender, two separate observation groupings were considered. The first grouping (G1) divided the corpus into 13 observations of 5 sentences each while the second grouping (G2) divided the corpus into 5 observations of 13 sentences each. A set of statistical measures (Table 2) was computed for each feature across each sentence. An additional set of statistics was computed only for pitch and energy features and is given in [2]. The resulting features are denoted as direct feature statistics (DFS). Each direct feature statistic was then subjected to the same set of statistical measures listed in Table 2, this time across all sentences in an observation. The procedure resulted in a large vector (approximately 2000 in size) of observation feature statistics (OFS). To produce an initial reduction in the dimensionality of the feature space, statistical significance tests (using Analysis of Variance - ANOVA) were conducted. Features that did not meet a significance level of $p < 0.001$ were discarded. Table 3 shows the resulting number of OFS as well as the number of observations per gender and observational grouping (G1, G2).

Table 2. Statistical Measures

Statistical Measure	Equation
Average (AVG)	$\frac{1}{N}\sum_{i=1}^{N} x_i$
Median (MED)	$50^{th} percentile$
Standard Deviation (STD)	$\sqrt{\frac{1}{N-1}\sum_{i=1}^{N}(x_i - \bar{x})^2}$
Minimum (MIN)	$5^{th} percentile$
Maximum (MAX)	$95^{th} percentile$
Range (RNG)	$MAX - MIN$
Dynamic Range (DRNG)	$\log_{10}(MAX) - \log_{10}(MIN)$
Interquartile Range (IQR)	$75^{th} percentile - 25^{th} percentile$

Table 3. Observations and Observation Feature Statistics (OFS) per Experiment

Gender	Experiment	Observations	OFS
Male	G1	195	724
	G2	75	298
Female	G1	234	1246
	G2	90	857

3 Genetic Programming and Feature Selection

Genetic Programming is an evolutionary computing method in which an initial population of computer programs, each individual program representing a potential solution to an optimization problem, is evolved by means of biologically-inspired operations that will generally improve the fitness of individuals over generations [9]. Each individual in the population is represented by a set of trees whose nodes are built from a set of functions F and a set of terminals T that are suitable to tackle the problem at hand. Within the context of classification from real-valued data, a suitable choice for F consists of the arithmetic operators $\{+, -, \times, /\}$, where a protected division operator ($\frac{x}{0} = 0$) is used to ensure closeness. The terminal set T was chosen as the set of real number constants in the range of $[0, 10]$ and the OFS set. An advantage of using these node sets is that the resulting classifiers consist of simple expressions that can be easily understood at first glance. A method for evolving N-class classifiers using multiple

GP trees has been presented in [4]. However, for the two-class problem under consideration only one tree per individual is necessary since classification can be performed according to the arithmetic sign of the output value of a tree.

Evolution in GP is performed by a set of operators that include reproduction, crossover, and mutation. These operators use individuals in a given generation to produce a new (and hopefully improved) generation of individuals. During crossover, two parent individuals are selected, and two offspring that are similar to the parents are produced. These offspring are usually created by swapping randomly-selected subtrees across parents. In mutation, a node of a tree is selected and replaced by a new randomly-generated subtree. The modified tree is then copied over to the new generation. Reproduction simply copies a parent to the new generation without modification.

Individuals are selected as inputs for reproduction based on several possible selection schemes, most of which utilize the fitness of an individual as a criterion for selection. The selection schemes considered here are fitness selection, in which individuals are chosen stochastically with a probability that is proportional to their fitness, and tournament selection, in which a pool of individuals is selected at random and the fittest member within this pool is then chosen.

The work reported in [5] has resulted in a new set of fitness functions, reproduction operators and selection criteria that extend the basic classification GP algorithm so that feature selection is simultaneously performed as the classifier population evolves. The algorithm implemented herein is a 2-class version of the one reported in [5], with the main difference being that we use single tree classifiers. A brief overview of the algorithm will be presented below.

Feature selection in the GPFS algorithm begins during the generation of the initial population, where a feature set is randomly generated for each individual. A feature set of size r is chosen with probability p_r linearly decreasing with respect to r, so that the initial population will contain many individuals with small feature sets and few individuals with large feature sets. Once the initial population is created, it is evaluated and the fitness of each individual is computed. In our context, one obvious choice for a fitness function is classification accuracy. GPFS introduces a variation of this fitness function that is biased toward individuals that use smaller feature sets, given as:

$$f_s = f(1 + ae^{-r/n}) , \tag{1}$$

where f is the original normalized fitness function, a is the bias strength, r is the number of features used by the individual, and n is the total number of features. The value of a decreases linearly with generations, so that initial generations highly favor smaller feature sets, but later generations focus on maximizing classification accuracy. As a result of using this fitness function, feature selection is performed mostly in the first few generations. In the current implementation, a can be as much as 0.2, giving a maximum bias of 20%. Finally, we use the two crossover operations introduced in [5]. Homogeneous crossover restricts the selection of a pair of parents to those who share the same feature set. Heterogeneous crossover permits parents to have different feature sets, but has been biased toward selecting parents that use similar features. In heterogeneous crossover, the first parent is selected using tournament selection. During selection of the second parent, the fitness function of each candidate is augmented by a small amount that is proportional to the similarity between its feature set and the feature set of the first parent. This bias is also limited to a maximum of 20%, but its strength remains constant over all generations. During each reproduction phase, heterogeneous and homogeneous crossovers are chosen at random, with the probability of homogeneous crossover P_{hg} given as

$$P_{hg} = \frac{gen}{M} , \tag{2}$$

where gen is the index of the current generation and M is the total number of generations. As such, the probability of using homogeneous crossover increases linearly from 0 to 1 with generations. Heterogeneous crossover is performed with probability $1 - P_{hg}$. As a result, new combinations of features are explored during the first GP generations, while the last few generations focus almost entirely on improving those classifiers that already use a good feature set.

4 Naïve Bayesian Classification

In a probabilistic classifier, the class C with the highest probability of occurrence given the current set of observations is selected. This decision requires knowledge of the posterior probabilities $P(C_j|X_i)$ for $j = 1..N$, where N is the number of classes, and X_i is the feature vector of the i_{th} observation. Using Bayes' rule we can estimate the posterior probabilities as follows:

$$p(C_j|X) = \frac{p(X|C_j)P(C_j)}{p(X)} , \tag{3}$$

where $p(X|C_j)$ is the class-conditional probability density function for class j (also called the likelihood function) and $P(C_j)$ is its *a priori* probability. The denominator term is the same for each class and can be safely ignored. The *a priori* probabilities for each class are usually determined by empirical information, such as the relative frequency of occurrence of each class in nature. In the present context, it is assumed that an incoming patient is equally likely to be depressed or not depressed. Therefore, the *a priori* term can be ignored as well. The task of constructing a classifier is then reduced to estimating the likelihood functions for each class from the training data. To estimate the likelihood functions from continuous data, it is necessary to either quantize the data or assume a known form for the underlying PDF. The naïve Bayes' rule [10] assumes independence between all features in the class-conditional distributions. It has been shown to work fairly well in practice, even in some cases where the data violates the independence assumption [11, 12]. Under this assumption, the likelihood function for class j can be expressed as:

$$p(X|C_j) = \prod_i p(x_i|C_j) , \tag{4}$$

where x_i is the i_{th} component of the feature vector X. Thus, the PDF or discrete probabilities for each feature can be estimated separately. In the following subsections we discuss five quantization and PDF estimation methods, two of which violate the independence assumption of (4).

4.1 Uniform Bins

In this quantization method, the features are scaled individually so that they lie in the range of $[0,1]$. A histogram with N uniformly spaced intervals (bins) is then computed for each feature and each class using the training data. The class-conditional probability can then be estimated as:

$$P(x_i = a|C_j) = f_{ij}(bin(a)) \,, \tag{5}$$

where f_{ij} denotes the normalized histogram for the i_{th} feature and j_{th} class, and the function bin maps the feature value a into the appropriate histogram bin. The parameter N negotiates a tradeoff between quantization bias and variance [12]. The optimum value for N with respect to the product of classification sensitivity and specificity was found for each experiment by exhaustive search and is given in Table 7.

4.2 Optimal Threshold

This method is similar to uniform bins with $N = 2$, with the difference that the cutoff threshold between the two bins is chosen separately for each feature [13]. The optimum threshold for a feature is chosen as the one that maximizes the product of sensitivity and specificity based on classification solely according to that feature. During classification, class-conditional probabilities are computed as in (5), but using a feature-specific bin_j function.

4.3 Gaussian Assumption

Here we estimate the PDF of each feature and each class as a 1-D Gaussian density function whose mean and variance are taken as the sample mean and (unbiased) variance of the training data. The likelihood function for each class can then be evaluated by direct application of (4).

4.4 Gaussian Mixtures

The Gaussian Mixture Model (GMM) is a popular density estimation method in pattern recognition [10, 14]. Each likelihood function $p(X|C_j)$ is modeled as a weighted sum of multivariate Gaussian densities. This approach has the advantage that given a large enough number of mixtures, an arbitrary (and possibly correlated) PDF can be accurately modeled. Training a GMM model consists of estimating the mean, covariance matrix, and weight for each density. The expectation-maximization (EM) algorithm is used to estimate these parameters iteratively. The EM algorithm needs to be initialized with a fixed number of densities and an initial guess for the parameters of each density. This initialization is performed by k-means clustering, where the number of clusters equals the number of densities. The initial mean and covariance matrices are computed from the training points in each cluster. Due to the limited number of observations in our datasets, we use diagonal covariance matrices and limit the number of mixtures to 3 for the G1 experiments and 2 for the G2 experiments in order to reduce the number of parameters to be estimated.

4.5 Multivariate Gaussian

In this method, each (class-conditional) likelihood function is modeled as a single multivariate Gaussian PDF with a full covariance matrix, which is computed from the (unbiased) sample covariance between features in the training data. Like the GMM, this method does not follow the naïve Bayes assumption.

5 Results

The GPFS algorithm was run for each of the experiments shown in Table 3 using the parameters listed in Table 4. Ten iterations of leave-one-out cross-validation (LOOCV) [14] were performed for each combination of gender and observation grouping. Each iteration of cross-validation consisted of a number of GPFS runs equal to the number of observations in the dataset, with a different single observation left-out of the training set during each run. Classification accuracy (Table 5) was computed as the ratio of correctly classified left-out samples across all GPFS runs. In addition, we provide sensitivity and specificity rates, which correspond to the percentage of correctly classified depressed (positive) and non-depressed (negative) samples, respectively. The average feature set size values for each experiment are also given in Table 5.

Table 4. GPFS Parameters

Parameter	Value
Crossover probability	0.8
Reproduction probability	0.05
Mutation probability	0.15
Prob. of selecting int./ext. node during crossover	0.8 / 0.2
Prob. of selecting int./ext. node during mutation	0.7 / 0.3
Tournament size	10
Number of generations	30 for G1 / 20 for G2
Initial height of trees	2 - 6
Maximum allowed nodes of a tree	350
Maximum height of a tree	12
Population size	3000 for G1 / 2000 for G2

Table 5. Average Classification Accuracy and Feature Set Size

Experiment	Male G1	Male G2	Female G1	Female G2	Mean
Classification Accuracy	71.2	71.3	84.9	82.2	77.4
Sensitivity	80.9	74.7	85.4	82.7	80.9
Specificity	64.8	69.1	84.4	81.8	75.0
Feature Set Size	18.5	15.3	16.1	14.2	16.0

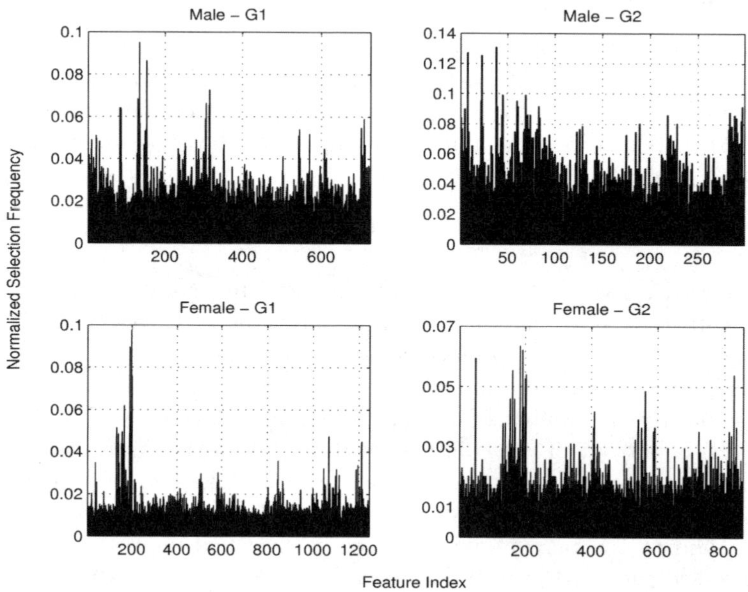

Fig. 1. Feature Selection Histograms

An approximate ranking of features was obtained by computing the frequency with which a feature is selected in the final solution tree of a GPFS run (Fig. 1). The motivation here is that since the GPFS algorithm is stochastic in nature, given a sufficiently large number of runs, the frequency of feature selection in the final solutions should provide a good indication of the discriminating performance of that feature. The 10 most frequently selected features for each experiment are listed in Table 6, with features that appear in more than one experiment in italics. The large similarity in the sets of the top 10 features for the G1 and G2 experiments within the same gender provide a good indication that the GPFS algorithm is consistently selecting certain specific features. It can also be seen in Table 6 that while glottal waveform features appear prominently in the male experiments, we have a large number of energy contour statistics taking the top spots for female subjects.

Table 6. 10 Best Features (Sorted by Rank)

Male - G1	Male - G2	Female - G1	Female - G2
GLT:max(min(cp))	*GLT:max(min(cp))*	*EMS:med(mr)*	EMS:iqr(avg_1)
GLT:drng(iqr(cp))	PCH:med(a_1)	*EMS:med(std_1)*	*EMS:med(std_1)*
GLS:med(max(g_{st}))	*EDS:avg(med)*	*EMS:max(mr)*	PCH:iqr(iqr)
GLT:std(iqr(op))	GLT:iqr(iqr(cp))	*EMS:med(rng)*	EMS:med(std)
GLR:rng(iqr(r_{cpo}))	GLR:min(iqr(r_{opo}))	EMS:max(std_1)	*EMS:max(mr)*
GLS:avg(max(g_{st}))	*EDS:avg(avg)*	EMS:med(avg)	TFM:avg(max(iqr))
EDS:avg(avg)	GLR:med(min(r_{cpop}))	EMS:max(max)	*EMS:med(mr)*
EDS:avg(med)	GLT:std(min(cp))	EMS:avg(std_1)	FBW:med(iqr(bw_{f3}))
EDS:med(med)	GLR:max(min(r_{cpop}))	EMS:avg(med)	EMS:med(max)
GLT:med(min(cp))	*GLS:avg(max(g_{st}))*	EMS:avg(avg)	*EMS:med(rng)*

Table 7. Bayesian Classification Performance

Exp	Method	Acc	Sen	Spec	Exp	Method	Acc	Sen	Spec
	Unif Bin (N=8)	86.7	83.3	88.9		Unif Bin (N=9)	88.0	85.5	90.6
Male	Opt Thresh	82.6	82.1	82.9	Female	Opt Thresh	78.6	65.8	91.5
G1	Gaussian	87.2	88.5	86.3	G1	Gaussian	87.2	91.5	82.9
	GMM	88.7	87.2	89.7		GMM	87.6	88.0	87.7
	MVG	84.1	83.3	84.6		MVG	85.5	83.8	87.2
	Unif Bin (N=2)	90.7	93.3	88.9		Unif Bin (N=5)	93.3	93.3	93.3
Male	Opt Thresh	73.3	50.0	88.9	Female	Opt Thresh	86.7	75.6	97.8
G2	Gaussian	89.3	93.3	86.7	G2	Gaussian	91.1	95.6	86.7
	GMM	90.7	90.0	91.1		GMM	88.0	83.3	91.1
	MVG	86.7	80.0	91.1		MVG	92.2	86.7	97.8

The average classification accuracy of the final GPFS classifiers was not very high. It should be noted that in the male experiments, the classifiers show a bias toward the class with the smaller number of training samples (depressed), which seems counter-intuitive. Nevertheless, the relatively poor performance of these classifiers should not lead to the dismissal of GPFS as a useful feature selection method. Even if GP is unable to consistently find excellent classifiers, the fact that certain features are selected disproportionately more frequently from a uniform initial feature subset population is still an indication of their discriminatory power. To validate this assertion, we used the 16 most frequently selected features for each experiment to train Bayesian classifiers. Because the true form of the probability distribution of the features is unknown, we trained a separate classifier using each of the likelihood estimation methods discussed in Sect. 4.

The results for all likelihood estimation methods are shown in Table 7. Leave-one-out cross-validation was used for all methods. In addition, because the final result from training GMMs with the EM algorithm depends on the initial k-means clustering, which is in turn randomly initialized, the best result out of 10 training episodes is reported. A few interesting results are worth mentioning. For all its simplicity, the uniform quantization method outperforms the optimal threshold method even in the Male-G2 experiment where the number of bins is 2 for both methods. This suggests that setting optimal quantization levels on a per-feature basis may not be beneficial when the features are combined. The uniform quantization method also obtained the best overall accuracy for the female experiments. For all experiments, the naïve Gaussian and GMM results were fairly close. On average, we obtained an improvement in classification accuracy of 18.5% for males (GMM) and 7.1% for females (uniform quantization) relative to the GPFS classifiers.

6 Conclusion and Future Work

By applying the GPFS algorithm, we were able to find a small set of individual speech features that are useful discriminators of clinical depression, as validated by the performance of the Bayesian classifiers that were trained on these features. However, the true goal of FS is the selection of an optimal combination of features. In the context of the present work, this would require that we measure not simply how often single features are selected, but instead how often groups of features are selected together. An algorithm for ranking groups of features based on their joint selection frequency is currently under investigation. Another area to investigate is the convergence rate of the feature selection performed by GPFS. The current implementation is designed to converge in the first few generations, but it might be desirable to instead allow a larger amount of feature set exploration throughout the evolution process. A recent technique [15] involving the use of self-organizing maps to control the amount of exploration of the search space may be useful in this respect.

References

1. Moore, E., Clements, M., Peifer, J., Weisser, L.: Comparing objective feature statistics of speech for classifying clinical depression. In: Proc., 26th IEEE Conf. Eng. in Medicine and Biology. Volume 1., San Francisco, CA (2004) 17–20
2. Moore, E., Clements, M., Peifer, J., Weisser, L.: Analysis of prosodic variation in speech for clinical depression. In: Proc., 25th IEEE Conf. Eng. in Medicine and Biology. Volume 3., Cancún, México (2003) 2925–2928
3. Dash, M., Liu, H.: Feature selection for classification. Intelligent Data Analysis **1**(3) (1997) 131–156
4. Muni, D., Pal, N., Das, J.: A novel approach to design classifiers using genetic programming. IEEE Trans Evol Comput **8**(2) (2004) 183–96
5. Muni, D., Pal, N., Das, J.: Genetic programming for simultaneous feature selection and classifier design. IEEE Trans Syst Man Cybern B Cybern **36**(1) (2006) 106–17

6. Zongker, D., Punch, W.: Lilgp 1.01 User's Manual. Genetic Algorithms and Research Application Group, Michigan State University, East Lansing, MI. (1998) http://garage.cse.msu.edu/software/lil-gp/index.html.
7. Quatieri, T.F.: Discrete-Time Speech Signal Processing: Principles and Practice. Prentice Hall PTR, Upper Saddle River, NJ (2002)
8. Zhou, G., Hansen, J., Kaiser, J.: Nonlinear feature based classification of speech under stress. IEEE Trans Speech Audio Process **9**(3) (2001) 201–16
9. Koza, J.R.: Genetic Programming: On the Programming of Computers by Means of Natural Selection. MIT Press, Cambridge, MA (1992)
10. Duda, R., Hart, P., Stork, D.: Pattern Classification. Wiley, New York, NY (2001)
11. Elkan, C.: Naive bayesian learning. Technical Report Adapted from: CS97-557, Dept. of Computer Science and Engineering, University of California, San Diego, CA (1997)
12. Yang, Y., Webb, G.I.: On why discretization works for naive-bayes classifiers. In: Lect Notes in Artifial Intelligence. Volume 2903., Perth, Australia, 16th Australian Conference on AI (December 2003) 440–452
13. Wiggins, M., Saad, A., Litt, B., Vachtsevanos, G.: Genetic algorithm-evolved bayesian network classifier for medical applications. In: WSC10: 10th Online World Conference on Soft Computing in Industrial Applications. (September 2005)
14. Theodoridis, S., Koutroumbas, K.: Pattern Recognition. Elsevier, San Diego, CA (1999)
15. Amor, H.B., Rettinger, A.: Intelligent exploration for genetic algorithms: Using self-organizing maps in evolutionary computation. In: GECCO Genetic and Evolutionary Computation Conference, Washington, DC, USA (June 2005) 1531–1538

A Computationally Efficient SUPANOVA: Spline Kernel Based Machine Learning Tool

Boleslaw K. Szymanski[1], Lijuan Zhu[1], Long Han[1], Mark Embrechts[1], Alexander Ross[2], and Karsten Sternickel[2]

[1] Rensselaer Polytechnic Institute, Troy, NY, USA
{szymab,zhul4,hanl4,embrem}@rpi.edu
[2] Cardiomag Imaging, Inc., Schenectady, NY, USA
{aross,karsten}@cardiomag.com

Summary. Many machine learning methods just consider the quality of prediction results as their final purpose. To make the prediction process transparent (reversible), spline kernel based methods were proposed by Gunn. However, the original solution method, termed SUpport vector Parsimonious ANOVA (SUPANOVA) was computationally very complex and demanding. In this paper, we propose a new heuristic to compute the optimal sparse vector in SUPANOVA that replaces the original solver for the convex quadratic problem of very high dimensionality. The resulting system is much faster without the loss of precision, as demonstrated in this paper on two benchmarks: the iris data set and the Boston housing market data benchmark.

1 Introduction

Kernel transformations are frequently used in machine learning methods to transform the input domain into a feature domain so that linear methods can be used to find an optimal solution to the learning problem. The most prominent examples of such methods are Support Vector Machines (SVM [1]) and Partial Least Square (PLS) approaches [2]. Despite their predictive power and efficient implementations, they share a fundamental weakness with other machine learning techniques, namely that although they provide answers, they do not give hints on how these answers were produced or on what basis they were reached. To address this weakness, Gunn and Brown [3] and later Gunn and Kandola [4] proposed to use spline kernels and combine them with a full combinatorial decomposition of the feature set. These models explicitly identify feature subsets that are used in producing the answers. Those subsets can then be used to discern the reasons for predictions. The key element of this approach is a sparse solution to the fully decomposed spline kernel prediction function. Therefore, such a sparse solution can be used for hypothesis forming [4]. However, SUPANOVA is not without its challenges, which arise when an efficient and scalable implementation is desired. These challenges were addressed in our initial work on application of SUPANOVA to machine learning for magnetocardiography [5].

A. Saad et al. (Eds.): Soft Computing in Industrial Applications, ASC 39, pp. 144–155, 2007.
springerlink.com

The rest of the paper is organized as follows. Section 2 describes the original SUPANOVA approach published in the literature. Its implementation is discussed in Section 3. Section 4 describes two benchmarks that we used to measure the performance of the implementation and Section 5 provides the results of these measurements. Section 6 offers conclusion and outlines future work in this area.

2 SUPANOVA

Before we discuss the models, we start with some basic definitions and notation. In this paper, we assume that there are N training data points in the form of vectors $x_i = [x_1^i, x_2^i, \ldots, x_n^i] \in R^n$ for $i = 1, 2, \ldots, N$. Each vector represents values of n features and has the corresponding output value, $y_i \in R$. We will denote the matrix containing these vectors (or training data points) as x and the vector of the corresponding output values as y. We assume that the data Mahalanobis scaled [6], that is that each feature has its average and standard deviation computed and then each value is replaced by the difference of the original and average values divided by the standard deviation.

We want to find a function f that is represented by these data points and their values, such that for all training points, as well for any new data point $x_0 = [x_1^0, x_2^0, \ldots, x_n^0]$ with the associated value y_0, we have $y_i \approx f(x_i), i = 0, 1, 2, \ldots, N$. To reconstruct the function f from the training data points we use the following basic kernel model:

$$f(x_0) = \sum_{i=1}^{N} a_i K(x_i, x_0) = K(x, x_0) \cdot a \tag{1}$$

where a kernel function, $K(x_i, x_0)$, yields the value that is a measure of similarity between vectors x_i and x_0 and kernel vector:

$$K(x, x_0) = [K(x_1, x_0), \ldots, K(x_N, x_0)].$$

$a \in R^N$ is the usual weight of data point vector and $\cdot$ denotes the dot product.

The basic idea of the SUPANOVA method based on spline kernels is to represent the solution to a machine learning problem as a sum of kernels that decompose functions of the order n into a sum of terms that are 1-ary, 2-ary,...,n-ary order functions of the original arguments. Each function higher than first order uses a product of spline functions to represent its arguments.

This basic model (1) can be extended by replacing kernel function with a sum of kernels $K_j(x_i, x_0)$ with each one measuring similarity of vectors x_i, x_0 on a subset of features, with $M = 2^n - 1$ we get: $f(x_0) = \sum_{j=0}^{M} c_j K_j(x, x_0) \cdot a$, where $c_j \geq 0$. In this representation, a linear sum of kernels weighted by nonnegative coefficients c_j is used in which each kernel K_j yields the value $K_j(x_i, x_0)$ that defines the jth component of ANOVA decomposition [4] in which an order n function $g(u)$ over an n element vector u is represented as: $g_0 + \sum_{i=1}^{n} g_i(u_i) + \sum_{i<j} g_{i,j}(u_i, u_j) + \ldots + g_{1,2,\ldots,n}(u_1, u_2, \ldots, u_n)$. The appropriate multivariate

ANOVA kernel of two vectors u, v is given by a tensor product of a univariate kernel plus a bias term as follows:

$$K_A(u,v) = \prod_{i=1}^{n}[1 + k(u_i, v_i)] = 1 + \sum_{i=1}^{n} k(u_i, v_i) \ldots + \prod_{i=1}^{n} k(u_i, v_i).$$

As an operator, following Kandola [7], we will use a spline kernel in the form of a piece-wise cubic polynomial: $k_{spline}(u_i, v_i) = u_i v_i + \frac{(u_i+v_i)\min(u_i,v_i)}{2} - \frac{\min(u_i,v_i)^3}{6}$. The resulting kernel is used in a SUPANOVA technique first proposed by Gunn and developed by Kandola [7].

Clearly, the number of potential terms in this representation is very large, $M = 2^n - 1$. However, vector c should be very sparse, so only a few terms in the decomposition will be significant.

The loss function for this kind of representation consists of three (and not two as is the case in traditional kernels) terms:

1. The error of modeling that measures the distance of the prediction from the real results (equal to traditional kernels) and for a quadratic loss can be expressed as $\left|y - \sum_{j=0}^{M} c_j K_j \cdot a\right|_2^2$.
2. The smoothness of the representation, defined by vector a as $\lambda_a \sum_{j=0}^{M} c_j a^T \times K_j \cdot a$. is again the same term as used in traditional kernels to maintain generalizibility of the representation.
3. The sparseness of the representation is specific to the spline kernels, and is defined ideally by the number of non-zeros in vector c, that is by zero-norm of the vector c as $\lambda_c \sum_{j=0}^{M} |c_j|_0$. Since the optimization with zero-norm (that is not differentiable) is difficult, Gunn and Kandola [4] approximated it with the first-norm and used the absolute value of the sum of all vector c elements as a measure of sparseness. Even this first-norm metric leads to the optimization problem that is difficult to solve because of very large numbers of feature subsets that arise even for modest numbers of features. Indeed, for a problem with n features, the corresponding optimization problem would be of 2^n dimensionality. Therefore even for relatively modest numbers of features in the range of 30 to 50, the corresponding optimization problem would have 10^9 to 10^{15} dimensions and would be expensive to solve.

Hence, for the quadratic loss for the entire training data set, the error function is just the sum of the three defined above terms: $\Phi(a,c) = \left|y - \sum_{j=0}^{M} c_j K_j \cdot a\right|_2^2 + \lambda_a \sum_{j=0}^{M} c_j a^T \times K_j \cdot a + \lambda_c \sum_{j=0}^{M} |c_j|_0, \quad c_j \geq 0$. The sparseness and smoothness terms are weighted by regularization parameters λ_a and λ_c, that have to be identified heuristically as they strike the balance between the quality of the predictions on the training data (that is the distance between the predicted and real values for each training data point) and the model's ability to generalize to the testing or new data (that is the distance between the predicted and real values for each testing or new data point).

3 SUPANOVA Implementation

λ_a and λ_c are not initially known but need to be found. In an iterative search for those values, both λ_a and λ_c should initially be set large and reduced gradually (by 2-3 percentage points of the initial value per iteration). Based on Kandola's experience that subsequent iteration steps did not change the solution significantly [7], we follow a solution procedure that consists of an initialization step followed by the single iteration.

Initialize: $c' = 1 \;\; a' = argmin_a(\Phi(a, c'))$. **Single step:** $c^* = argmin_c(\Phi(a', c))$.

The single iteration step used in [4] included also recomputation of the optimal smoothness vector a for the newly found sparseness vector c, but our experience indicates that this additional step does not improve the solution, so we are not executing it in our solution.

In the initialization step, regularization parameter λ_c is not used, so it does not need to be determined in the respective optimization. λ_a is determined by cross-validation (like [4], we use an automatic search procedure to locate a local minimum of the validation error in 8-fold cross-validation runs). The difficult part of the optimization of the loss function is determining the regularization parameters in the single step that computes vector c. Following [4], we use a method based on the best empirical performance. We set $\lambda_a = 0$ and select such λ_c that the loss is equal to the loss of the validation error in the initialization step. As the result, λ_c can be readily computed from the following equation. $\lambda_c = \frac{\lambda_a \sum_{j=0}^{M} c'_j a'^T \times K_j \cdot a'}{\sum_{j=0}^{M} c'_j} = \frac{\lambda_a}{M} \sum_{j=0}^{M} a'^T \times K_j \cdot a'$. This leads to the following revised procedure.

S0: $c' = 1$

S1: $a' = argmin_a(\Phi'(a))$, where $\Phi'(a) = \left| y - \sum_{j=0}^{M} K_j \cdot a \right|_2^2 + \lambda_a a^T \times \sum_{j=0}^{M} K_j \cdot a$, and λ_a is computed by minimizing (possibly locally) validation error in 8-fold cross validation procedure (for problems in which $N < 14$, we use $N/2$-fold cross validation procedure instead).

S2: $c^* = argmin_c(\Phi''(a', c))$, $\Phi''(a, c) = \left| y - \sum_{j=0}^{M} c_j K_j \cdot a \right|_2^2 + \lambda_c \sum_{j=0}^{M} |c_j|$, and λ_c set so the loss is the same as it is in the initialization step.

In step **S1**, by taking the partial derivatives for all elements of vector a, we obtain a system of N linear equations with N unknowns that we represent as $d \times a = f$, where d is a matrix of coefficients for the unknown vector a elements and f is a vector of free terms for those linear equations. Hence, $f_l = 2 \sum_{j=0}^{M} c'_j \sum_{i=1}^{N} y_i K_j(x_i, x_l)$ and each element of matrix d is defined as $d_{l,p} = \sum_{j=0}^{M} c'_j [2 \sum_{i=1}^{N} K_j(x_i, x_l) \sum_{q=0}^{M} c'_q K_q(x_i, x_p) + \lambda_a (K_j(x_l, x_p) + K_j(x_p, x_l))]$.

In step **S2**, Kandola in his Ph.D. thesis [7] replaced the sparseness term with the sum of the elements of vector c (so effectively replaced the zero norm with the first norm measure). This change allowed him to use a convex quadratic problem solver, called BPMPD [8] based on a robust interior point solver technique. However, it is not scalable to a very large number of variables (in our case $M = 2^n - 1$ grows very rapidly with the number of features n). The size of the quadratic problem that we face for realistic data is prohibitively large (when $n > 20$, $M = 2^n - 1 > 10^6$) even for the most modern quadratic problem solvers and computers.

Fortunately, the minimization problem of step **S2** is a very special case of a quadratic problem in which the number of quadratic terms (N) is much smaller than the number of linear terms (M). This fact guided our development of a heuristic that uses a greedy selection of the non-zero entries in vector c, one-by-one, and measures the corresponding error of the solution until this error cannot be further minimized. Indeed, the difficult part of optimization needed in step **S2** is determining which elements of vector c should be non-zero. Once we know that, finding their values is just a matter of solving a set of linear equations of the order equal to the number of non-zero elements chosen. We also know that this number is limited by the number of data points, that is N. To make selection efficient, the heuristic assumes that the subvector of non-zero entries of size $k + 1$ that minimizes the loss function in step **S2** is simply the subvector of non-zero entries of size k that minimizes the loss function plus one more non-zero elements. This approach is often termed a "greedy" selection. The results that we obtained from benchmarks confirmed that a vector selected in such a way is close to the vector that minimizes the loss function in step **S2**.

The additional advantage of the heuristic is that it allows us to use exact value of the sparseness term in the loss function, that is zero norm of the vector c elements. Indeed, as in each step of heuristic the number of non-zero elements is constant, only their values are recomputed, so non-differentiability of the loss function sparseness term does not impact our optimization. Hence, comparing ours with Gunn's approach, we notice that we optimize approximately (by using greedy selection of the vector c non-zero elements) the exact loss function, whereas Gunn optimizes exactly (by using a quadratic problem solver) the approximate expression of the loss function.

The final advantage of the heuristics is its memory efficiency. The size of the sparse kernel array is $O(MN^2)$, however, as shown later, our heuristic requires storing only one element for each column of the spline kernel. Hence, the size of the storage needed for our implementation is just $O(MN)$. This reduction in the memory size by a factor of N enables us to solve much larger problems than solvable on the same machine using the quadratic problem solver. The gain in computational and memory efficiency combined with the quality of the solutions that we obtained justify our approach.

The computation in step **S2** proceeds as follows.

First, we compute λ_c such that the loss is equal to the loss of the validation error in the initialization step, assuming that the maximum number of non-zero elements in vector c is N (as this is the number of independent equations in the corresponding optimization problem which we use to compute the values of non-zero elements of vector c), so $\lambda_c = \frac{\lambda_a}{N} \sum_{j=0}^{M} a'^T \times K_j \cdot a'$.

The loss function contains N approximation error terms of the form

$$\left[y_i - \sum_{j=0}^{M} c_j \sum_{k=1}^{N} K_j(x_i, x_k) a_k \right]^2 = \left[y_i - \sum_{j=0}^{M} c_j P_{i,j} \right]^2 ,$$

where $P_{i,j} = \sum_{k=1}^{N} K_j(x_i, x_k)a_k$. Assuming that only $k \leq N$ positive elements of vector c should be selected and all other elements should be set to 0 (so this is the k-th step of the "greedy" selection in our heuristic), the choice is easy. Taking the derivatives for all non-zero elements selected, each derivative for some non-zero element c_p yields the equation $\sum_{i=1}^{N}\sum_{j=0}^{k} c_j P_{i,j} P_{i,p} = \sum_{i=1}^{N} y_i P_{i,p}$. If solving this set of k equations with k unknowns, we obtain all the non-zero values positive, then we can easily compute the corresponding error as $\sum_{i=1}^{N}\left[y_i - \sum_{j=0}^{M} c_j P_{i,j}\right]^2 + k\lambda_c$. To select the next non-zero value in vector c, we can now substitute the found optimal values of the k elements of vector c so far selected, computing the new vector y^k of approximation error terms in the loss function as $y_i^k = y_i - \sum_{j=1}^{k} c_j \sum_{m=1}^{N} K_j(x_i, x_m)a_m$. Then, we can compute the derivative for each of the $M-k$ unselected elements c_r's of vector c as $\sum_{i=1}^{N} c_r P_{i,r}^2 = \sum_{i=1}^{N} y_i^k P_{i,r}$ to find the optimal value of c_r and the corresponding value of the loss function. Selecting that element c_r that yields the smallest approximation error term of the loss function, we can extend the non-zero value subvector of c with the newly found non-zero element and continue the heuristic.

The more detail description of the implementation of step **S2** follows.

1. **Initialization.** We create initially empty set S of all selected elements of vector c that are positive, and a set E contains all the remaining elements of vector c.
2. **Selection.** We select, one by one, elements e_r in set E and compute, according to Equation 3 the minimum value of the loss function with this selection. Then, we choose that element e_r that yields the smallest minimum among all elements of set E and move this element from set E to set S.
3. **Adjustment.** With the newly created set S, we compute the solution to the set of linear equations obtained by taking derivatives of the error expression for the elements of set S. If all elements of the solution are non-negative, the solution is accepted. Otherwise, the previously found solution is retained and the heuristic stops.
4. **Control loop.** If set E becomes empty, heuristic stops, otherwise step 2 is executed with the extended set S.

In pseudo-code, the heuristic can be written as follows.

$S = \emptyset$, $E = \{c\}$, $k = 1$, toterr=$\sum_{i=1}^{N}\left[y_i - \sum_{j=0}^{M}\sum_{p=1}^{N} K_j(x_i, x_p)a_p\right]^2 + \lambda_c$

do {

  $e_1^{opt} = \sum_{i=1}^{N} y_i^k P_{i,1} / \sum_{i=1}^{N} P_{i,1}^2, kerror = \sum_{i=1}^{N}\left[y_i^k - \sum_{p=1}^{N} e_1^{opt} P_{p,1}\right]^2$

  for r=2 **to** $|E|$ **do**

    $e_r^{opt} = \sum_{i=1}^{N} y_i^k P_{i,r} / \sum_{i=1}^{N} P_{i,r}^2, kerror = \sum_{i=1}^{N}\left[y_i^k - \sum_{p=1}^{N} e_r^{opt} P_{p,r}\right]^2$

    if error<kerror **then** kerror=error, ksel=i **endif**

endfor

$S = S \cup \{e_{ksel}\}$

compute k derivatives and solve a system of k linear equations

if all elements of S are positive **then**

error=$\sum_{i=1}^{N}\left[y_i - \sum_{j=0}^{k} s_j \sum_{m=1}^{N} K_j(x_i, x_m)a_m\right]^2$
continue=(error<toterr)
endif
if continue **then** k=k+1, toterr=kerror+$k\lambda_c$ **endif**
until continue

Let $s(M)$, which we will call a sparse factor, denote the number of non-zero entries in the final sparse vector c of size M. Since we have at most N approximation terms in the loss function, then clearly $s(M) \leq N$. The computational complexity of the entire computation is $O(MN^2)$. Indeed, the spline kernel has $O(MN^2)$ elements, which of each has to be computed. In step **S1**, we solve the set of N linear equations repeatedly to find the optimal value of regularization parameter λ_a, operation of complexity $O(N^3) = O(MN^2)$ for $M > N$. Finally, in step **S2**, we compute coefficients $P_{i,j}$, again an operation of complexity $O(MN^2)$, then we solve $s(M)$ sets of linear equations varying in size from 1 to $s(M)$, so operation of complexity $O(s(M)^3)$. Since $s(M) \leq N < M$ then also the last operation is no worse than $O(MN^2)$. The memory complexity is also reasonable. It is easy to notice that in step **S1**, all elements of vector c are equal to 1, so instead of the full kernel, we need to store only the sum of all values along the vector c dimension, in total $O(N^2)$ memory requirement. In step **S2**, the only arrays $P_{i,j}$ and similarly sized arrays storing the sum of squares of elements $P_{i,j}$ and products of elements $P_{i,j}$ by elements y_i^k need to be stored. So in total only $O(MN)$ data items need to be stored. This is a huge improvement over storing entire spline kernel that would require memory by factor N larger.

4 Performance Metrics and Data Benchmarks

We use two metrics evaluate the prediction performance. The first one is the Root Mean Square Error index or RMSE, which is defined as the average value of the squared error: $RMSE = \sqrt{\frac{1}{n}\sum_i(\hat{y}_i - y_i)^2}$, where $\hat{y}_i = f(x_i)$ is the predicted value for the data vector x_i. The second one is r^2, defined as the correlation coefficient squared between target values and predictions for the response, $r^2 = \frac{(\sum_{i=1}^{ntrain}(\hat{y}_i-\bar{y})(y_i-\bar{y}))^2}{\sum_{i=1}^{ntrain}(\hat{y}_i-\hat{y})^2\sum_{i=1}^{ntrain}(y_i-\bar{y})^2}$. Two related metrics, $q^2 = 1 - r^2$ and $Q^2 = 1 - R^2$, are used to assess the performance of validation or test data. The smaller the values for q^2 the better. Ideally, both values should be similar. Detailed information about these metrics is given in [6].

The model quality was measured on two machine learning benchmarks: the petal iris data and the Boston housing market data. Both data sets are members of the UCI Repository for Machine Learning Data-Bases [9] and are described below.

Petal iris data defines a classification problem with linear relationship. We use this simple data set to verify the implementation. This is dataset is widely used in the pattern recognition literature and was adopted from a classic paper in botany [10] perhaps because of its simplicity for classification. The dataset contains 3 classes of 50 instances each, for a total of 150 data points. Each class refers to a type of iris plant. One class is linearly separable from the other two that are not linearly separable from each other.

The Boston housing dataset has been extracted from the work of Harrison and Rubinfield [11] who were studying the effects of air pollution on housing prices. The data provides the median price for 1970 of owner-occupied houses in 506 census tracts within the Boston metropolitan area. Thirteen features characterizing each census tract are available for use in predicting the median price, including crime rate, mean no. of rooms, distance to job center, etc.

5 Experimental Results

The described above two data sets were subject to many machine learning experiments, including [4, 7, 12]. For this reason, we have chosen them to illustrate the performance of the SUPANOVA method in general and our heuristic in particular.

The results from processing the petal iris dataset using our initial implementation are shown in Figure 1. The available data were split as follows: 50 samples were chosen for training and 30 samples are used for validation. This data set has only four features so only 2^4 elements in vector c were created. The optimal solution uses nine out of these 16 elements. Values of q^2 and Q^2 indicate a high quality of the results, as shown in Figure 1. The positive and negative cases are separated nearly perfectly. Measuring from the ROC curve, the goodness of fit is close to 1.

The results from the Boston housing market data set are shown in Figure 2. They display very similar, high quality results as shown in the previous figures. This data set has a large number of dimensions, 13, and therefore a relatively high number of feature combinations ($2^{13} = 8192$). The non-linearity of the data is limited, so the model that we ran included only empty feature set, single features and pairs of features, yielding 92 feature combinations in total. From this set of 92 potential features, the heuristic selected only 40 features for the solution, as indicted by the number of non-zero elements in vector c. Hence, sparseness of the solution, $40/8192 \approx 0.5\%$, was excellent. Finally, despite of the large size of the model, the results are good, as indicated by the high values of q^2 metrics and Q^2 metrics.

It is interesting to note that in these plots, the largest divergence between the measured and predicted values is at the extremes of the pattern sequence numbers. This is not surprising, because prices of the most expensive houses were censored at the value of $50k$ causing the model to predict wrongly for them.

To simplify the analysis of input variables, we introduce a parameter in the algorithm to control the maximum number of features in feature combinations considered in the solution. In general, it could be expected that the significant elements of vector c may correspond only to single features and their binary and triple combinations. As already mentioned, the results for Boston housing market data set are obtained using at most binary combination of features. To

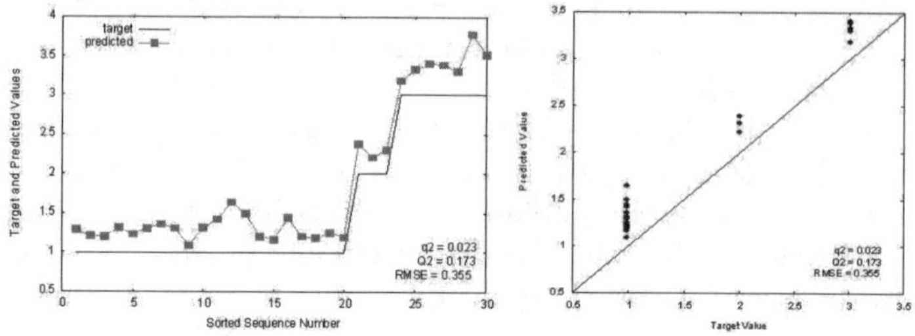

Fig. 1. Predicted vs. target values and scatterplot for the petal iris test data

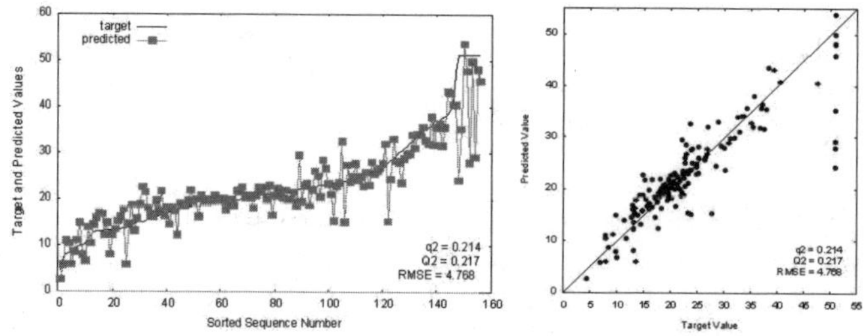

Fig. 2. Processing results for the test Boston housing market data; closeness of the data points to the diagonal in the scatterplot indicates quality of predictions

investigate if adding triple feature combination improves the quality of the prediction, we extend feature space to include 378 feature combinations, including all possible triples, but there are no obvious improvements in the predictions. We may conclude that most of the information for Boston housing market data lies in feature compositions that are generated up to binary combination of features. This result is significant as it yields an extreme reduction in the search space for vector c while retaining good quality of predictions. This is useful for pruning features based on relatively simple binary combination of inputs. In fact, this could be a basis of an efficient and systematic approach to pruning, in which we start with binary feature combinations and then continue with the increasing numbers of features in combinations until no further improvement to the quality of the solution can be obtained. As we expect this iteration to finish after three to five steps, (as it did in the test cases), the size of vectors c is greatly reduced from $M = 2^n - 1$ to M in the range $\frac{n^3}{6}$ to $\frac{n^5}{120}$, making analysis of problems with up to 100 features feasible.

Our results agree with both Husmeier (who used an ensemble of Bayesian neural networks trained using an Expectation-Maximization (EM) algorithm and incorporating automatic relevance determination (ARD) [13]) and Gunn and Kandola (who used SUPANOVA [4]). concluded that Feature 1 in the Boston housing market data, "Crime rate", seems to have a low-impact on house prices, and this conclusion is confirmed in our approach. Likewise, Feature 2, "Percentage of residential land", was also found not to be significant. More interesting is a comparison with the original SUPANOVA analysis provided by Gunn [4]. The paper listed the following single features as significant: mean number of rooms, percentage built before 1940, distance to job center, and percentage of minorities. and binary combination of features with the mean number of rooms and with percentage built pre 1940 as the most important in the 14 non-zero vector c entries selected, but the paper did not provide the prediction quality measurements. On the other hand, the quality of prediction of our implementation of SUPANOVA methods matched other advanced machine learning such as KPLS kernels [6]. Our implementation selected only one single feature: mean number of rooms. It also selected the same binary combinations with mean number of rooms and with percentage built pre 1940 as the Gunn and Kandola's implementation [4] did. In total, both approaches selected 14 non-zero entries in vector c and agreed on 9 of those selections. These results demonstrate that our new heuristic achieves a high accuracy of predictions while preserving the transparency of the solution supported by the original SUPANOVA implementation. We also noticed that the adjustment improves significantly the quality of the results, as measured by the reduction in the total error by over 30% compared with the unadjusted solution. At the same time, the sparseness of the solution can be easily adjusted because of the sequential growth of the size of vector c. For example, the last 20 elements of vector c, that is elements $21 - 40$, improve the final result by less than 3% and therefore can be discarded, improving sparseness of the solution.

Another interesting conclusion is that the values of the c vector element mainly decrease in the order of their selection. For example, the values of the first 20 elements selected by the heuristic vary from 4.86 to 0.36, whereas values of the next 20 elements range from 0.11 to 0.29. This is not surprising, as our heuristic was designed to do exactly that, meaning that it selects the non-zero elements in the order of their importance for the solution. The fact that the order of the selection is very strongly correlated with the magnitude of the impact confirms that the heuristic works correctly, and in fact the difference between the order of selection and the order of magnitude of the value of non-zero elements is a measure of optimality of the heuristics. In fact, at the moment when the heuristic completes, we can compute the largest shift of an element in the vector c in this order to estimate how many additional steps the heuristics should be run to make sure that no better non-zero vector exists. Such an extended vector, if found, can than be truncated using the following observation. Since the magnitude of the entry in the vector c is correlated with its impact on the solution, the corollary is that the elements with small magnitude can be dropped without decreasing the quality of the approximation error but improving the sparseness of the solution. This is important, because the more sparse the solution is the more transparent the model becomes, thereby increasing the value of spline kernel solutions to the users.

Finally, it should be noted that the presented heuristic is capable of eliminating linearly or nearly linearly dependent feature combinations (including single features). Indeed, if two feature sets are linearly dependent, only one of them would be selected as having a non-zero coefficient in vector c, whereas the other (or others if there are more than one) would have their coefficients kept at zero, as the inclusion of any of them would not improve the solution.

6 Conclusion and Future Work

So far we have used the benchmark data to examine a new implementation of the SUPANOVA technique. our current work focuses on applying the developed sparse kernel implementation with the new heuristic presented above to biological data, such as magnetocardiograms. In this special case, the number of data points, N, (the number of patients tested) is now relatively modest (on the order of a few hundred, but it is expected to grow quickly to a few thousand). This range of values of N is not a problem from a computational point of view for the current implementation to handle. The challenge is that the number of features that dictates the number of the feature subsets defining the size of the sparseness vector, c. There is a need to limit the number of features considered to $100 - 200$, as larger numbers of features would result in our implementation exceeding the current limits of memory in modern workstations (for example, with triples of 200 features and $1,000$ data points in the training set, our implementation requires about 8 GB of memory). Our heuristic is instrumental here, as without it, the same 8 GB memory workstation would be able to process at most 23 feature problem with $1,000$ data points in the training set and triples of features using the quadratic problem solver [14].

Acknowledgement

Content of this paper does not necessarily reflect the position or policy of the U.S. Government - no official endorsement should be inferred or implied.

References

1. Vapnik, V.: The Nature of Statistical Learning Theory. Springer, New York, NY (1995)
2. Wold, S., Sjstrm, M., Eriksson, L.: PLS-Regression: a Basic Tool of Chemometrics. Chemometrics and Intelligent Laboratory Systems **58** (2001) 109–130
3. Gunn, S., Brown, M.: Supanova - a sparse, transparent modelling approach. Proc. IEEE Workshop on Neural Networks for Signal Processing, Madison, WI (1999) 21–30
4. Gunn, S., Kandola, J.: Structural Modelling with Sparse Kernels. Machine Learning **48**(1) (2002) 137–163
5. Embrechts, M., Szymanski, B., Sternickel, K., Naenna, T., Bragaspathi, R.: Use of Machine Learning for Classification of Magnetocardiograms. Proc. IEEE Conference on System, Man and Cybernetics Washington DC (2003) 1400–1405

6. Embrechts, M., Szymanski, B., Sternickel, K.: Introduction to Scientific Data Mining. Computationally Intelligent Hybrid Systems: The Fusion of Soft Computing and Hard Computing. Wiley, New York, NY (2005)
7. Kandola, J.: Structural Modelling with Sparse Kernels. Department of Electronics and Computer Science, University of Southampton, U.K (2001)
8. Meszaros, C.: The BPMPD interior point solver for convex quadratic problems. TR WP 98-8, Computer and Automation Institute, Academy of Sciences, Budapest, Hungary (1998)
9. Merz, C.J., Murphy, P.M.: UCI repository for machine learning data-bases. http://www.ics.uci.edu/ mlearn/MLRepository.html (1996)
10. Fisher, A.: The use of multiple measurements in taxonomic problems. Annual Eugenics **7** (1936) 179–188
11. Harrison, D., Rubinfield, D.: Hedonic housing prices and the demand for clean air. Journal of Enviromental Economics and Management **5** (1978) 81–102
12. Neal, R.: Bayesian Learning for Neural Networks. Springer, New York, NY (1995)
13. Husmeier, D.: Neural Networks for Conditional Probability Estimation. Springer, New York, NY (1999)
14. Szymanski, B., Han, L., Embrechts, M., Ross, A., Sternickel, K., Zhu, L.: Using efficient supanova kernel for heart disease diagnosis. Proc. 16th Conf. Artificial Neural Networks Industrial Engineering, ANNIE (2006)

Part IV

Classification

Multiobjective Genetic Programming Feature Extraction with Optimized Dimensionality

Yang Zhang and Peter I Rockett

Department of Electronic and Electrical Engineering, University of Sheffield, Sheffield, S1 3JD, UK
hegallis@gmail.com,p.rockett@shef.ac.uk

Summary. We present a multi-dimensional mapping strategy using multiobjective genetic programming (MOGP) to search for the (near-)optimal feature extraction pre-processing stages for pattern classification as well as optimizing the dimensionality of the decision space. We search for the set of mappings with optimal dimensionality to project the input space into a decision space with maximized class separability. The steady-state Pareto converging genetic programming (PCGP) has been used to implement this multi-dimensional MOGP. We examine the proposed method using eight benchmark datasets from the UCI database and the Statlog project to make quantitative comparison with conventional classifiers. We conclude that MMOGP outperforms the comparator classifiers due to its optimized feature extraction process.

1 Introduction

Normally, the discriminability between classes in a pattern space can be enhanced by feature extraction to obtain more desirable classification performance. Feature extraction as a pre-processing step in a pattern recognition system needs to target the extraction of discriminatory information from the input patterns in order to optimize the classification performance. This is implemented by projecting the pattern space into a new *decision space* in which the discriminability between classes is maximized.

Although a large amount of research has focused on designing ever-more sophisticated and powerful classifiers, such as support vector machines and neural networks, feature extraction has remained an important step in the design of pattern recognition systems. Due to the lack of a generic feature extraction methodology, conventional feature extraction is usually problem-specific and designed largely on the basis of domain-dependent knowledge or assumptions about class distributions. Some general purpose methods such as principal component analysis (or the Karhunen-Loève transform) are mostly constrained to linear transformations. Non-linear mappings like supervised learning networks (*e.g.* multilayer perceptrons and radial basis functions) provide model-free or semi-parametric methods to design non-linear mappings/discriminant functions but in such paradigms, the feature selection/extraction process is subsumed into a single entity within the classifier. Hence the optimality is hard to guarantee.

A. Saad et al. (Eds.): Soft Computing in Industrial Applications, ASC 39, pp. 159–168, 2007.
springerlink.com

Depending on the criteria used, feature extractions can be grouped into two categories: *representational* and *classificational*. In the first approach, the objective is to maintain the fidelity between the original input and projected spaces. The major concern of these methods is to maintain the discriminatory information of the original input space during what is typically, dimensionality reduction. For the classificational approach, the objective is to enhance the discriminability between classes in the new projected decision space. The present work is targeted at optimizing the discriminability in the transformed space. We show that the determination of an optimal multi-dimensional mapping does not necessarily imply dimensionality reduction.

Typically, a prospective solution in GP can be interpreted as a sequence of operations [1]. Langdon [2] developed a multi-tree program with redefined genetic operations to reduce the extent of building block disruption; see also Koza's work involving automatically defined functions (ADFs) [1]. The CoacH system [3] specifies multiple trees within a single individual in the GP population. Sherrah [4] has attempted to address the issue of generating n-to-m mappings ($m > 1$) by applying a multi-tree chromosome structure but had to impose some severe restrictions to compensate for the limitation of his single objective function. Bot [6] described a hybrid algorithm using GP to evolve a (*near-*)optimal feature extraction stage and added extra transformed features, one-at-a-time. Each new feature was retained only if it yielded an improvement in classification error above a pre-defined threshold. Bot's approach is a greedy algorithm and hence suboptimal. Muni *et al.*. [5] applied a multi-tree representation to produce an q-class classifier ($q > 2$) by simultaneously evolving q independent dichotomizers. Recent work on GP-evolved feature extraction/classification has been reviewed in [8] where two distinct research strands have been identified: Either GP is used to evolve the whole classifier or the tree produces a mapping to a real-valued decision space which forms the input to a conventional classifier.

Zhang & Rockett [7][8] proposed a generic framework to produce optimal feature extractors independent of domain-specific knowledge and class distributions using multiobjective genetic programming (MOGP). Through a multiobjective optimization process, their 1D mappings comprised a series of mathematical transformations projecting input patterns into a one-dimensional decision space. Classification performance was markedly enhanced. In this paper, we seek to extend that work on evolving feature extraction stages to produce n-to-m mappings of optimal dimensionality. For a given classification problem, generally an optimal value of m exists due to the well-known *peaking phenomenon* [17] – here m becomes a free parameter within the optimization. In Section 2 we describe our method to find the (*near-*)optimal set of m extracted features. In order to substantiate the effectiveness of the proposed method, in Section 3 we report results of quantitative comparison with eight conventional classifiers on two groups of datasets from the UCI and Statlog databases. After analysis of the multi-dimensional feature extraction process, we offer conclusions in the final section.

2 Multi-dimensional Feature Extraction

To distinguish between the present work and our earlier investigations using a 1-dimensional decision space [7], we term the present method *multi-dimensional multiobjective genetic programming* (MMOGP). As in [7], we optimize a vector of multiple objectives within a Pareto framework.

2.1 The Objective Vector

2.2 Single Tree Representation

We have used a single, vectorizing tree representation in this feature extraction application. Note that the single tree representation scheme used here differs from the multiple-output structure of *MRtree* [15] and Sherrah's *EPrep* [4] approach in that the output of a candidate tree from MMOGP is a multi-dimensional feature vector in the projected decision space, not a set of tentative class labels as in *MRtree*. Furthermore, *MRtree* results are very sensitive to the defined *Modi* parameter [15]. In *EPrep*, the output of the evolved feature pre-processing unit is formed by a special function node – an *Output Point* - operating on the tree [4]. One of the disadvantages of *EPrep* is that the features extracted by Output Points from the lower levels of the tree may be severely correlated with those collected at higher levels. Notably, *EPrep* is able to improve performance only on some, not all problems.

In this work we have added two special node types - *Root* and *Dummy* nodes - to our single tree representation: A tree contains a single instance of a *Root* node, unsurprisingly at the root of the tree where its role is to assemble the scalar outputs from child sub-trees into a (decision space) vector. A *Dummy* node is a special kind of terminal node which allows trees to change dimensionality. If a child of the Root node evolves to a Dummy node, this effectively reduces the dimensionality of the tree mapping by one. If the immediate child of the Root node is a Dummy node, the resulting element in the decision space vector is taken to be zero – which, of course, has no discriminatory power. Similarly, the dimensionality of a tree can increase if the child of a Root node which was previously a Dummy node is replaced by a sub-tree. Since they are just a form of terminal node, Dummy nodes are free to occur anywhere in a tree, can be swapped and counted as normal terminal nodes but during evaluation, they return a constant zero value.

In actuality, there are two simultaneous search processes occurring during the optimization – one locates the optimal dimensionality of the decision space vector (the number and/or the positions of the non-dummy *subtrees*). The second, together with the optimization of the dimensionality, finds the optimal transformations to effect the mappings for each element of the decision space vector. The vectorizing tree structure employed here is illustrated in Figure 1.

Tree Complexity Measure: In mapping to a 1D decision space, Zhang & Rockett [7] employed the total number of tree nodes as a straightforward measure of tree complexity to provide a selection pressure which favored simpler

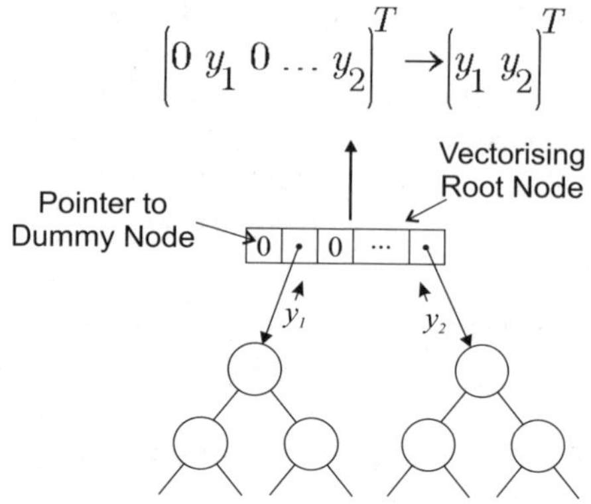

Fig. 1. Illustration of the vectorizing tree chromosome structure

solutions during evolution; this approach has been demonstrated to control tree-bloat in GP applications. With the multi-tree representation here, using total node count as a complexity measure will introduce an undesirable 'hidden' selection pressure which implicitly favors trees with lower dimensionality. Higher-dimensional solutions with more subtrees will tend to contain more nodes. Thus we have used the *mean* tree size as a complexity measure in this work:

$$Complexity = \frac{1}{m}\sum_{i=1}^{m} Size_i$$

where the $Size_i$ denotes the number of nodes in *subtree* i, and m is the number of non-Dummy sub-trees, that is, the dimensionality of the evolved decision space vector.

Misclassification Error: A Fisher linear discriminant (FLD) [16] has been trained during the fitness evaluation phase of each tree to perform classification in the m-dimensional decision space. The misclassification error (0/1 loss) from the Fisher discriminant over the training set is used as the second objective. The training of an FLD can be done quickly in closed-form, thus making a negligible contribution to the time of an iteration; we determine the decision threshold within the FLD classifier [16] by golden section search which is terminated when there is no further improvement in error rate.

Lower Bound on Misclassification Error: Following on from the ideas presented in the one-dimensional MOGP work [7][8], we have found that a further objective is needed to enhance selective pressure in the early stages of evolution when that from misclassification error alone is not sufficient. Partly because the misclassification error turns-out *initially* to be a rather insensitive gage of class separation, we have added the overlap area of the class-conditioned densities along the 1D Fisher projection direction [16] as our third and final objective. This overlap is a measure of the lower bound on the misclassification error, independent of the decision-making mechanism. We estimate this using a straightforward histogramming procedure. The overlap objective is analogous to, although not equivalent to Zhang & Rockett's use of an estimate of Bayes error in the 1D setting [7].

Overall, we have found experimentally that the combination of the three objectives is necessary for the algorithm to rapidly generate a Pareto set of parsimonious solutions which generalize well. Without the Complexity measure the trees bloat and the optimization tends to stagnate. Without the Lower Error Bound objective, convergence is very slow or non-existent. Each of the three objectives thus has a key role to play during the evolutionary process although since we are ultimately considering the classification domain, after we have generated a set of non-dominated solutions whose properties are 'shaped' by the multiple objectives, we select the solution which has the lowest (mean) validation error. This is a critical distinction between the current area and most other uses of multiobjective optimization: The multiple objectives are vital *constraints* during the evolutionary process but do not form part of any natural trade-off at the end of the optimization.

Table 1. MMOGP (PCGP) Settings

Terminal set	Input pattern vector elements
	Dummy nodes
	10 floating point numbers $\in \{0 \ldots 1\}$
Function set	sqrt, log, pow2, -, sin, not
	-, +, *, /, max, min, xor, or, and
	if-then-else
Max. dimensionality	50
Sub-tree preservation probability	0.2
Max. no. of tree evaluations	20,000
Stopping criterion	Max generations exceeded

2.3 Extended Breeding Operators

Crossover: In our single tree representation, the Root node is a special type of output node which only appears at the root of the GP tree. Hence the depth-fair crossover used for one-dimensional MOGP [7] is modified to avoid selecting the root node. Furthermore, we have used a second modification to depth-fair crossover designed to preserve useful genetic building blocks within the feature transformations associated with each dimension. With some probability, the *sub-tree preservation probability*, we perform crossover only within corresponding pairs of sub-trees. For example, given two parents, A and B, we perform cross over within sub-tree 1 of parent A and sub-tree 1 of parent B, then within sub-tree 2 of parent A and sub-tree 2 of parent B, and so on up to sub-tree p such that $p = \min[m_A, m_B]$ where $m_{A,B}$ is the dimensionality of the parents.

Mutation: Like [7], we use depth-fair, size-dependent mutation except that the mutation operator is modified to allow mutation on the whole tree *including* the

Root node. If the Root node is selected based on depth-fair conditions, a new random GP tree of random dimensionality will be created. After the population has been initialized, the mutation operation is responsible for introducing new dummy nodes in the GP tree.

Steady-State MOGP Implementation

In a typical multiobjective optimization problem, a family of equivalent solutions exists, the members of which are superior to all the other feasible solutions. None of this family of *non-dominated* solutions can be considered 'better' than any other one from the point of view of the simultaneous optimization of multiple objectives; this family is termed the *Pareto-optimal set.* A traditional way to obtain multiobjective optima is to aggregate the multiple objectives into a single, weighted objective using linear or non-linear combination based on domain knowledge or experience, although this is far from satisfactory. Here we apply steady-state Pareto converging genetic programming (PCGP) to find the Pareto-optimal set for our multiobjective feature extraction problem since this method has been shown to yield smaller trees than competitor methods [8]. Further details of PCGP can be found in [8] and [9]. The population size for all problems in this paper was 200 and the maximum tree depth used to create the random initial population was 5. To create some diversity in the initial population, half of each initial population was created at full tree depth (5, here) and half with random tree depth (≤ 5) although thereafter, no limit was imposed on tree depth during the evolutionary optimization; bloat was controlled solely by the tree complexity objective discussed above. The full list of MMOGP settings used in this work are listed in Table 1.

3 Experiments

We re-emphasize: The method we present here is not a classifier, rather a feature extraction framework to design a classifier adapted to an individual problem without prior knowledge or any distributional assumptions. To justify our method, we investigated the design process on a representative range of benchmark datasets from the UCI Machine Learning database [10] and the Statlog project [11]. The key issue is that the generation of the feature extraction stage as well as the dimensionality of the projected decision space are driven by the notion of optimality.

Datasets

We make statistical comparisons of classification performance between the results of our MMOGP algorithm and a range of established classifiers on 8 two-class learning problems. From the Statlog database [11] we use: (a) Australian credit approval, (b) German credit and (c) the heart disease datasets. From the UCI Machine Learning database [10] we use: (d) Glass (classifying between float

Table 2. The Eight Datasets Used in this Work

Name	Features	Size and Distributions	Projected Dimensionality
Glass	9	163 = 87 (float) + 76 (non-float)	19
BUPA	6	345 = 200 (Benign) + 145 (Malignant)	19
PID	7	532 = 355 + 177 (Diabetic)	7
WBC	10	699 = 458 (Benign) + 241 (Malignant)	6
WDBC	30	569 = 357 (Benign) + 212 (Malignant)	23
AUS	14	690 = 383 (Positive) + 307 (Negative)	20
GER	24	1,000 = 700 (Positive) + 300 (Negative)	34
HEA	13	270 = 120 (Diseased) + 150 (Benign)	13

and non-float glasses), (e) BUPA - liver disorder prediction, (f) Wisconsin Diagnostic Breast Cancer, (g) Pima Indians Diabetes and (h) Wisconsin Breast Cancer datasets. Table 2 lists details of the eight datasets studied in this paper. The last column contains the decision space dimensionalities optimized by MMOGP for each dataset - these will be discussed below.

Classification Algorithms

As the basis for comparison with MMOGP, we have used eight existing classification algorithms with implementations from the Weka Machine Learning system [12]; we used the default parameter settings except where noted below. The classifiers used were: (i) Radial Basis Functions (RBF)– a normalized Gaussian radial basis function network using the k-means clustering algorithm. We estimated the number of clusters, k for a given dataset by considering a random split of the dataset, training the classifier on the first half and calculating a validation error on the second half. We adopted the value of k which gave the lowest validation error for each dataset by this method; (ii) Logistic - a modified multinomial logistic regression model with a ridge estimator; (iii) NNge – Nearest-neighbor-like algorithm using non-nested generalized exemplars; (iv) BayesNet – Bayes Network classifier using the K2 learning algorithm; (v) IB1 – Instance-based learning algorithm using a simple distance measure to find the training instance closest to the given test instance and predict the same class as this training instance; (vi) ADTree – the alternating decision tree learning algorithm; (vii) SMO – Sequential minimal optimization algorithm for training a support vector classifier; (viii) C4.5 – The well-known decision tree algorithm. (This is called J48 in Weka.)

Dietterich [13] has pointed-out that the commonly-used practice of classifier comparison by N-fold cross-validation followed by a t-test is unsound due to the implicit assumptions about independence being violated and has proposed an empirical 5×2 cv test. Alpaydin [14] has modified Dietterich's test to remove the unsatisfactory aspect of the result depending on the ordering of the folds: it is Alpaydin's F-test which we use here to statistically compare classifier performance. We perform five repetitions of splitting the dataset into two folds, treating one fold as the training set and the other as the test set to compute an F-statistic with which to decide whether to reject the hypothesis that the performances of the two classifiers are identical. (See [14] for full details.) Throughout this work we have used a 95% confidence level to infer a statistical difference.

Table 3 shows the mean test errors from the 5×2 *cv* test for all nine classifiers (8 conventional + MMOGP) for all eight datasets. As explained above, we have used the MMOGP solution which displays the smallest mean error since in a classification problem we are (generally) only interested in the lowest classification rate – classifier structure is generally an unimportant detail. The outcomes of the F-test comparisons are also shown in Table 3 where a tick denotes that MMOGP is statistically superior to the conventional comparator classifier for that particular dataset and a dash denotes no statistical difference. It is noteworthy that MMOGP has the smallest error rates over all dataset/classifier combinations and the F-test results show that over the $8 \times 8 = 64$ paired comparisons, MMOGP is superior in 57 of these comparisons at the 95% confidence level. On the remaining 7 pairwise comparisons, MMOGP is statistically identical to the respective conventional classifiers. Most significantly, MMOGP is not bettered by any of the comparator classifiers on any dataset.

The *optimal* dimensionalities extracted by multi-dimensional mapping for each dataset are list in Table 2. There is a reduction in dimensionality for WBC and WDBC while the dimensionalities of the projected spaces are identical to that of the original input space for HEA and PID. (The two spaces are not the same, of course.) Optimal performance on the other datasets is obtained with an *increase* in dimensionality over the original pattern spaces. The MMOGP algorithm is constructing new, typically non-linear features which maximize the class discriminability.

Table 3. Mean Error Comparisons of the 9 Classifiers on 8 Datasets (*5* × *2 cv* Test); F-test comparisons between algorithms for each dataset at 95 % confidence level. A tick represents superiority of MMOGP over the comparator classifier/dataset combination; a dash denotes no statistical difference.

Datasets	Classifiers							
	RBF	LOG	NNge	Bayes Net	IB1	ADTree	SMO	C4.5	MMOGP
GLASS	0.354 √	0.364 √	0.322 √	0.311 √	0.300 √	0.317 √	0.392 √	0.338 √	0.135
BUPA	0.442 √	0.383 √	0.449 √	0.420 √	0.388 √	0.343 √	0.423 √	0.391 √	0.215
PID	0.255 √	0.233 √	0.249 √	0.249 √	0.301 √	0.248 √	0.222 √	0.263 √	0.203
WBC	0.048 √	0.045 √	0.038 –	0.026 –	0.042 √	0.043 √	0.030 –	0.057 √	0.024
WDBC	0.061 √	0.068 √	0.077 √	0.054 √	0.046 √	0.052 √	0.030 –	0.067 √	0.024
AUS	0.182 √	0.130 –	0.176 √	0.139 –	0.202 √	0.153 √	0.162 √	0.162 √	0.121
GER	0.288 √	0.270 √	0.250 √	0.274 √	0.324 √	0.272 √	0.266 √	0.304 √	0.230
HEA	0.178 –	0.185 √	0.233 √	0.184 √	0.244 √	0.235 √	0.179 √	0.242 √	0.140

Feature Selection and Extraction

Due to space limitations, we show only one representative MMOGP tree in Figure 2 to illustrate the outcome of the multi-dimensional mapping process. This tree is the solution with the lowest error over the HEA dataset. The leaf nodes are labeled as Xn, $n \in \{1 \dots N\}$ where N is the number of raw attributes in the input pattern vector. The "if-then-else" node returns the value of the second child if the first child value is greater than 0, otherwise the third child value is returned. The "max" node returns the larger value from its leaves while "min" return the smaller. If both leaf node values are larger than 0, "xor" returns 0, otherwise it returns 1.

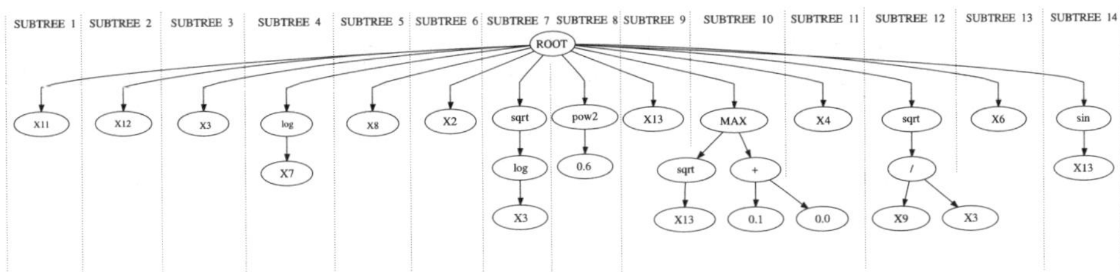

Fig. 2. Example GP tree to perform multi-dimensional mappings on the HEA datasets

The nominal dimensionality of this tree is 14, the number of the sub-trees. Actually, sub-tree 8 in Figure 2 returns a constant (0.6^2) to the final feature vector and hence will not provide any discrimination power. Clearly the objective which minimizes tree size does not exert quite enough selective pressure to remove minor redundancies from the trees – this is an area for future work. Thus, the dimensionality of the output feature vector is 13 and the mapping (listed in Table 2) is 13-to-13. Interestingly, the raw attributes: $X_1, X_5$and X_{10} have not been used at all while some of the new features are constructed from non-linear combinations of other raw attributes which are used repeatedly. During the multiobjective optimization process, feature selection has been *implicitly* conducted together with explicit feature extraction to design/search for the optimal multi-dimensional mappings.

4 Conclusions

In this paper we have proposed a generic, multi-dimensional, multiobjective genetic programming method to design an *optimal* feature extraction pre-processing stage for pattern recognition with the optimal dimensionality of the decision space being determined. The method is domain-independent and makes no assumptions about the class distributions or prior information. The single-tree structure is shown to be an effective way to represent the multi-dimensional mapping problem. We have carried-out an extensive range of comparisons with 8 conventional classifiers over a 8 benchmark datasets and conclude that MMOGP yields error rates which, in the vast majority of cases are statistically superior to the best of the conventional classifiers or, at worst, statistically identical.

References

1. J.R.Koza. (1994) Genetic Programming II, Automatic Discovery of Reusable Programs. *The MIT Press,* Cambridge, Massachusetts.
2. W.B. Langdon, (1998) Genetic Programming and Data Structures: Genetic programming + Data Structures = Automatic Programming. *Kluwer Academic Publishers*, London.
3. S.Raik & B.Durnota, (1994) *The Evolution Of Sporting Strategies*, in R Stonier and X Yu (eds), *Complex Systems '94: Mechanisms of Adaption, IOS Press.*
4. J.R.Sherrah, R.E.Bogner & A.Bouzerdoum. (1997) The Evolutionary Pre-Processor: Automatic Feature Extraction for Supervised Classification using Genetic Programming. *Genetic Programming 1997: Proc.of the 2nd Ann. Conf.* pp 304-312.
5. D.P.Muni, N.R.Pal, & J.Das, (2004) A Novel Approach to Design Classifiers using Genetic Programming, *IEEE Trans. on Evolutionary Computation*, vol. 8, no. 2, pp.183–196.
6. M.C.J.Bot, (2001) Feature Extraction for the k-Nearest Neighbor Classifier with Genetic Programming, In *Genetic Programming, Proc.of EuroGP 2001*, pp. 256–267.
7. Y.Zhang & P.I.Rockett, (2005) Evolving Optimal Feature Extraction using Multi-objective Genetic Programming: A Methodology and Preliminary Study on Edge Detection," In *GECCO 2005*, pp. 795-802.
8. Y.Zhang & P.I.Rockett, (2006) Feature Extraction using Multi-objective Genetic Programming, in *Y.Jin (ed), Multi-Objective Machine Learning*, pp.79-106. Springer.
9. R.Kumar & P.I.Rockett, (2002) Improved Sampling of the Pareto-Front in Multi-objective Genetic Optimization by Steady-State Evolution: A Pareto Converging Genetic Algorithm, In *Evolutionary Computation*, vol.10, no. 3, pp. 283-314.
10. C.L.Blake & C.J.Merz, (1998) UCI Repository of Machine Learning Databases [`http://www.ics.uci.edu/~mlearn/MLRepository.html`]. Irvine, CA: University of California, Department of Information and Computer Science.
11. D. Michie, D.J. Spiegelhalter & C.C.Taylor, (1994) Machine Learning, Neural and Statistical Classification, *Ellis Horwood.*
12. I.H.Witten & E.Frank, (2005) Data Mining: Practical Machine Learning Tools and Techniques (2nd Edition). *Morgan Kaufmann.*
13. T.Dietterich, (1998) Approximate Statistical Tests for Comparing Supervised Classification Learning Algorithms, *Neural Computation*, vol. 10, no. 7, pp. 1895-1923.
14. E.Alpaydin, (1999) Combined 5×2 cv F-test for Comparing Supervised Classification Learning Algorithms, *Neural Computation*, vol. 11, no. 8, pp. 1885-1892.
15. Y.Zhang & M.Zhang, (2004) A Multiple-Output Program Tree Structure in Genetic Programming, *Tech. Report CS-TR-04/14*, Victoria University, New Zealand.
16. R.O.Duda, P.E.Hart & D.G.Stork. (2000) Pattern Classification (2nd ed.), *Wiley Interscience.*
17. F.Camastra. (2003) Data Dimensionality Estimation Methods: A Survey. *Pattern Recognition* vol. 36, no. 12, pp.2945-2954.

A Cooperative Learning Model for the Fuzzy ARTMAP-Dynamic Decay Adjustment Network with the Genetic Algorithm

Shing Chiang Tan[1], M.V.C. Rao[2], and Chee Peng Lim[3]

[1] Faculty of Information Science & Technology, Multimedia University, Melaka Campus, Jalan Ayer Keroh Lama, Bukit Beruang, 75450 Melaka, Malaysia
[2] Faculty of Engineering & Technology, Multimedia University, Melaka Campus, Jalan Ayer Keroh Lama, Bukit Beruang, 75450 Melaka, Malaysia
{sctan, machavaram.venkata}@mmu.edu.my
[3] School of Electrical & Electronic Engineering,University of Science Malaysia, Engineering Campus, 14300 Nibong Tebal, Penang, Malaysia
cplim@eng.usm.my

Abstract. In this paper, combination between a Fuzzy ARTMAP-based artificial neural network (ANN) model and the genetic algorithm (GA) for performing cooperative learning is described. In our previous work, we have proposed a hybrid network integrating the Fuzzy ARTMAP (FAM) network with the Dynamic Decay Adjustment (DDA) algorithm (known as FAMDDA) for tackling pattern classification tasks. In this work, the FAMDDA network is employed as the platform for the GA to perform weight reinforcement. The performance of the proposed system (FAMDDA-GA) is assessed by means of generalization on unseen data from three benchmark problems. The results obtained are analyzed, discussed, and compared with those from FAM-GA. The results reveal that FAMDDA-GA performs better than FAM-GA in terms of test accuracy in the three benchmark problems.

Keywords: Fuzzy ARTMAP, Dynamic Decay Adjustment, Genetic Algorithms, Cooperative Learning, Classification.

1 Introduction

Over the last few years, combination between artificial neural networks (ANNs) and evolutionary algorithms (EAs) has attracted a lot of attention. In fact, both ANNs and EAs cross-fertilize each other; ANNs provide a framework of accurate and exact computation whereas EAs provide a robust and efficient approach for undertaking complex optimization problem (Fogel 1995; Yao 1999). The focus of the combination between ANNs and EAs is highlighted on the accuracy rate produced by the resulting hybrid system (Cantú-Paz and Kamath 2005).

EAs, which encompass genetic algorithms (GAs), genetic programming, evolutionary programming and evolution strategies, are stochastic in nature, and they are based on populations made up of individuals with a specific behaviour similar to biological phenomenon. They are search algorithms that are less likely to be trapped

A. Saad et al. (Eds.): Soft Computing in Industrial Applications, ASC 39, pp. 169–178, 2007.
springerlink.com

in a local minimum. Among existing EAs, the most well-known branch is GAs, which is essentially a mechanism of natural selection, genetic and evolution (Goldberg 1989). Compounded by a capability of evaluating many points in the search space simultaneously, GAs are likely to find near-optimum solution of a given problem. They are general-purpose optimization methods that require no derivative information for solving search problems. GAs have been employed primarily in two main research areas; optimization and machine learning. In machine learning, GAs can be used to aid the learning of ANNs (Whitley 1995). On the other hand, learning of ANNs is actually a process of compressing codes and refining the connection weights from a large pool of incoming data within a specific structure. In this regards, the role of GAs (or general EAs) is to aid in the learning process of ANNs by searching for a set of "optimized" weights of ANNs. Another way of hybridization between GAs and ANNs is that GAs are used to design the structure of ANNs. However, the second approach is not the focus of this paper. Instead, the main objective of our work is to improve the learning process of ANNs by searching for the weights that would eventually lead to a better network generalization performance.

Among many ANN models, the multilayer perceptron (MLP) network, which is trained with the backpropagation algorithms, has often been used as the numerical model in a hybrid system with the GAs (Lam and Leung 2004; Tsai et al. 2006). The GAs have been used to search for the number of hidden nodes as well as for the connection weights of the network. As pointed out in Taha and Ghost (1999), one major limitation of the standard MLP network is that it is difficult to adopt an incremental learning scheme in its structure. On the other hand, one ANN model that is able to overcome the issue of incremental learning and has demonstrated good performance especially for pattern classification problems is Fuzzy ARTMAP (FAM) (Carpenter et al. 1992). Indeed, FAM is a supervised model of Adaptive Resonance Theory (ART) (Carpenter and Grossberg 1987). The FAM network is an incremental learning model that has overcome the stability-plasticity dilemma (Carpenter and Grossberg 1987), i.e., it is stable enough to preserve information learned from previous data and, at the same time, flexible enough to learn new information from new data. Combination between GA and FAM is not new. However, most of the work leans towards the use of GA in selecting relevant features for assisting the learning of the FAM network (Palaniappan and Raveendran 2002; Palaniappan et al. 2002), and there is relatively little work in the literature that reports fusion between GA and FAM in machine learning and optimization.

In our work, we aim to develop a neuro-genetic system that can perform machine learning for undertaking classification problems. Unlike MLP, FAM performs incrementally learning for which it does not require handcrafting of network architecture prior to training process. Nevertheless, like MLP, FAM might likely to produce sub-optimal solutions upon the completion of the training process. Meanwhile, GAs require the initialization of population, which is predefined with arbitrary size and is of random basis, before performing the search procedure. The combination of FAM and GA amplifies the strengths and, at the same time, ameliorates the disadvantages of each constituent.

In this work, FAM learns incrementally the information by means of prototypes and the resulting templates are employed by the GA as a set of sub-optimal weights that would guide the future search and adaptation. Such a cooperative learning scheme can improve the search efficiency of the hybrid system for obtaining good solutions, as will be demonstrated in the later section of this paper.

Learning in the original FAM network involves the formation of new prototypes and/or invariable boundary expansion of the existing prototypes. In this regards, FAM does not impart a learning scheme that resolves conflicts resulting from overlapping among the prototypes of different classes in the attribute space. In our previous work (Tan et al. 2004a), FAM is integrated with the Dynamic Decay Adjustment (DDA) algorithm for resolving conflicts during training period. In Tan et al. (2004b), an investigation of using an *ordering algorithm* (Dagher et al. 1999) to identify a fixed order of training data presentation for network learning is reported. The sequence of the training patterns is determined prior to network learning, and it is independent of any permutations of the training patterns. The resulting network (i.e., Ordered FAMDDA) learns the *ordered* training patterns for improving generalization performance.

In this paper, the focus is on the weight reinforcement of the FAMDDA network, for better generalization. A GA (Baskar et al. 2001) is proposed to co-assist in searching and adapting the weights that are learned by FAMDDA. To show the effectiveness of the hybrid system FAMDDA-GA, a performance comparison between FAM and FAMDDA, which are trained respectively with the GA (i.e., FAM-GA and FAMDDA-GA), is made. The main focus of the investigation is to ascertain that the cooperative machine learning scheme of FAMDDA-GA is more effective than that of FAM-GA towards good solutions.

This paper is organized as follows. Section 2 briefly describes FAM, FAMDDA and the GA that are applied in this work. The fusion of FAMDDA-GA is explained in section 3. To demonstrate the effectiveness of the proposed FAMDDA-GA, its performance is evaluated with three benchmark datasets and the results are compared with those from FAM-GA. All of the experimental results are analyzed and discussed in section 4. A summary of the work is given in section 5.

2 Fuzzy ARTMAP-Based Networks

The section presents a brief description on the operations of FAM and FAMDDA. For full description, readers can refer to Carpenter and Grossberg (1987), Carpenter et al. (1991), and Carpenter et al. (1992) for FAM; and Tan et al. (2004a) for FAMDDA.

2.1 Fuzzy ARTMAP (FAM)

FAM is a neural network that is constructed by a pair of Fuzzy ART modules (i.e., Fuzzy ART_a and Fuzzy ART_b) interconnected through a mapping field. On presentation of input patterns, FAM self-organizes and self-stabilizes information and network configuration. Such ability of FAM is reliant on the recruitment of a feedforward pass (that is usually employed in most ANN models) and a feedback pass for pattern matching. On the other hand, the network growth is also controlled by the vigilance parameter (Carpenter et al. 1992), which is a user-defined threshold. The vigilance test evaluates the similarity between the prototype patterns of the network and the input pattern. If none of the existing neurons could satisfy the vigilance test, a new neuron is introduced, and the input is coded as its prototype pattern. Hence, subject to the vigilance criterion, the number of prototype patterns grows with time and this would engender the formation of an autonomous network topology for a problem in hand.

2.2 Fuzzy ARTMAP with the Dynamic Decay Adjustment Algorithm (FAMDDA)

The architecture of FAMDDA is actually similar to that of FAM. The difference is that FAMDDA incorporates a conflict-resolving facility for dealing with overlapping among prototypes of different classes. Such conflict-resolving facility is provided by the DDA algorithm (Huber and Berthold 1995), which comprises a width shrinking procedure for rectangular basis prototypes. Width adjustment of the prototype is class dependent, which distinguishes the prototype from different neighbours. Several modifications have been made in FAMDDA to perform width shrinking. In order to avoid the formation of a very thin rectangle of a prototype in FAMDDA, a user-defined minimum width threshold, $\varepsilon_{\min}$ (Huber and Berthold, 1995), is enforced on each dimension. The learning procedure of FAMDDA is attached in the Appendix.

3 Reinforcement Learning of FAMDDA with GA

GAs are adaptive heuristic search algorithm premised on the evolutionary ideas of natural selection and genetic (through crossover and mutation operations). The basic concept of GAs is designed to simulate processes in natural system necessary for evolution. The combination of FAMDDA and GA (i.e., FAMDDA-GA) amplifies the strengths of both constituents and at the same time, it complements the weaknesses of each individual entity. In our work, the GA participates in enhancing the learning capability of the FAMDDA network, by searching and adapting its weights. Fig. 1 shows the overall procedure of the cooperative learning model of FAMDDA-GA. Training starts with the weight learning process of FAMDDA within a predefined number of maximum epochs. FAMDDA performs supervised learning based on the given training patterns. The architecture as well as the weights of the network are identified autonomously. It is noted that all weights of Fuzzy ART$_a$ that represent the essence of input data space are grouped as a Pittsburg chromosome z_0 for search mechanism and adaptation in the GA environment. In fact, a chromosome in the GA environment represents the weights of different classes in the data space in the network environment. On the other hand, both weights of Fuzzy ART$_b$ and mapping field are secured from evolution. Such arrangement is intended to reinforce weights of Fuzzy ART$_a$ in the GA environment without affecting its behavior link with Fuzzy ART$_b$, upon returning to the network environment. A finite number of chromosomes are generated in the GA environment according to a modified function (Baskar et al. 2001).

$$z_i = z_0 + \text{RMF} . \text{rand}(1, n) \tag{1}$$

where z_i is the i-th replicated chromosome of z_0; RMF$\in$ [0,1] stands for the range multiplication factor;. represents scalar multiplication operation; and, rand(1,n) $\in$ [0,1] is a uniformly distributed random vector with n represents the length of z_0.

1. Weight initialization and nodes self-organization using FAMDDA. 2. Generate chromosomes (Pittsburgh Approach). Repeat 3. Compute fitness value of each chromosome. 4. Apply Roulette-Wheel selection. 5. Generate new generation through crossover and mutation. Until terminating condition has been satisfied.

Fig. 1. The overall training procedure of FAMDDA-GA

Each chromosome is converted into the Fuzzy ART$_a$ module and the fitness of the network is evaluated in terms of accuracy, by presentation of training patterns. Once the fitness of all chromosomes is calculated, the GA applies Roulette-Wheel to select chromosomes. The genetic operators (i.e., crossover and mutation) are applied to process the selected chromosomes. In this study, the arithmetic crossover (Baskar et al. 2001) that can deal with a pair of selected real-coded chromosomes (i.e., z_a, z_b) is applied, as follows.

$$z' = \lambda z_a + (1 - \lambda) z_b \tag{2}$$

$$z'' = (1 - \lambda) \lambda z_a + \lambda z_b \tag{3}$$

where $\lambda \in [0,1]$ is a uniformly distributed random variable; z' and z'' are the offsprings.

A dynamic mutation as described in Baskar et al. (2001) is adopted in our work. For a given chromosome z, if the element z_k is selected for mutation, the resulting chromosome is $z' = \lfloor z_1 \cdots z_k' \cdots z_n \rfloor$, where z_k' is randomly selected from two possible choices

$$z_k' = z_k + \left(z_k^U - z_k\right) . \ r \ . \ (1\text{-}t/T)^d \ \text{ or } \ z_k' = z_k - \left(z_k - z_k^L\right) . \ r \ . \ (1\text{-}t/T)^d \tag{4}$$

where z_k^U / z_k^L are the upper/lower bounds of the element z_k; $r \in [0,1]$ is a random number; T is the maximum generation number; and, d is a parameter that determines the degree of non-uniformity (usually assumed as 2).

After crossover and mutation have been applied, new set of chromosomes (offspring) are obtained. They are new set of individuals that forms a new population. The process of fitness evaluation, selection, and genetic operations on a population of chromosomes is repeated until one of the stopping criteria has been satisfied, i.e., either an achievement of 100% recognition rate of training patterns by one of the chromosome or a maximum setting of 10 generations (i.e., T=10), has been reached. If second stopping criterion is fulfilled, the chromosome of the current population that has the highest fitness value is identified as the weights of the Fuzzy ART$_a$ module.

4 Experimental Results and Discussion

To demonstrate the effectiveness of the proposed FAMDDA-GA model, a series of experiments were conducted using three benchmark datasets; pima Indian diabetes (PID), Australian credit approval (AUS), and heart (HEA). All these datasets are available from the UCI machine-learning repository (Blake and Merz 1998). The PID, AUS and HEA datasets are binary-class problems with 768, 690, and 270 samples, and have 8, 14, and 13 input attributes, respectively.

In each experiment, the dataset is divided into training and test sets. The FAMDDA-GA model was trained using 50% randomly selected data while the remaining data were used for evaluating the network's performance. The FAMDDA-GA model underwent two different modes of training: FAMDDA and GA. In the FAMDDA environment, the system was trained at 1 epoch with the "default" parameter settings: fast learning rate, $\beta = 1$; minimum width, $\varepsilon_{\min} = 0.1$; and ART$_a$ baseline vigilance, $\bar{\rho}_a = 0.0$. On the other hand, in the GA environment, the number of replicate chromosomes is 10, and the search and adaptation of the system are driven by parameters which include RMF=0.30, T=10, crossover rate 0.50, and mutation rate 0.01. For each classification task, eight independent experiments were carried out. The network performances, in terms of the test accuracy rate and the number of nodes, were averaged. To compare the performance of FAMDDA-GA with that of its counterpart (FAM-GA) as well as with the ordinary classifiers FAMDDA and FAM, the same experimental procedure was adopted. Tables 1 and 2 summarize the overall results obtained from the three datasets.

Table 1. Performance comparison between FAM-GA and FAMDDA-GA, in terms of test accuracy and number of nodes. The results are averages of eight runs, and their respective standard deviation is given in parenthesis. The p-values are the results from the bootstrap hypothesis tests between FAM-GA and FAMDDA-GA.

Dataset	Accuracy (%)			Size (Nodes)		
	FAMDDA-GA	FAM-GA	p-value	FAMDDA-GA	FAM-GA	p-value
PID	72.55 (2.24)	68.53 (2.45)	0.006	15.6 (1.4)	13.9 (1.9)	0.03
AUS	78.89 (2.57)	72.61 (2.27)	0	15.2 (3.1)	13.0 (2.6)	0.081
HEA	77.02 (2.33)	71.50 (4.63)	0.010	9.3 (1.5)	8.5 (1.4)	0.151

Table 2. Performance comparison between the FAM and FAMDDA classifiers, in terms of test accuracy and number of nodes. The results are averages of eight runs, and their respective standard deviation is given in parenthesis. The *p*-values are the results from the bootstrap hypothesis tests between the classifier and its GA-based network (*a* denotes FAM vs. FAM-GA; and *b* denotes FAMDDA vs. FAMDDA-GA).

Dataset	Accuracy (%)			Size (Nodes)		
	FAMDDA	FAM	*p*-value	FAMDDA	FAM	*p*-value
PID	70.11	64.36	0.008^{a}	16.2	13.9	0.172^{a}
	(2.26)	(3.35)	0.007^{b}	(2.6)	(1.9)	0.403^{b}
AUS	72.25	67.61	0.006^{a}	15.2	13.0	0.513^{a}
	(2.42)	(3.58)	0^{b}	(3.1)	(2.6)	0.496^{b}
HEA	69.28	62.97	0.003^{a}	8.7	9.0	0.685^{a}
	(5.33)	(4.46)	0.018^{b}	(1.9)	(2.0)	0.271^{b}

From Table 1, we can see that FAMDDA-GA produces higher accuracy rate than FAM-GA. The comparison between the generalization performance of FAMDDA-GA and FAM-GA can be made by a bootstrap hypothesis test (Efron, 1979) with a significance level of 0.05. One advantage of the bootstrap method is that it does not rely on the assumption that the samples must be drawn from a normal distribution. In this regards, the null hypothesis states no difference between the test accuracy rate of FAM-GA and FAMDDA-GA whereas the alternative hypothesis claims that the test accuracy rate of FAM-GA is lower than that of FAMDDA-GA. The *p*-values of the corresponding tests are tabulated Table 1. We can observe that all *p*-values of the test for the accuracy between FAMDDA-GA and FAM-GA are less than 0.05. This indicates that the classification performance of FAM-GA is statistically lower than that of FAMDDA-GA.

Table 1 also shows the comparison between the network size of FAMDDA-GA and FAM-GA. The alternative hypothesis of the bootstrap hypothesis test for network size claims that FAM-GA has a smaller network size than that of FAMDDA-GA in the respective benchmark studies. Note that the *p*-values for the network size in the AUS and HEA case studies are greater than 0.05, whereas the *p*-value for the PID case study is less than 0.05. This indicates that, statistically, there is no significant difference between the network size of FAM-GA and FAMDDA-GA in both AUS and HEA studies. But, it is on the contrary for the PID case study. However, this observation from the PID case study does not affect the finding that the cooperative learning of FAMDDA-GA has yielded better generalization performance (than FAM-GA) in these three studies.

The performances between the FAMDDA/FAM classifiers and their respective counterparts (FAMDDA-GA/FAM-GA) are also compared with the same claims of alternative hypothesis aforementioned. From Table 2, we can observe that all *p*-values of the test for the accuracy between FAMDDA (FAM) and FAMDDA-GA (FAM-GA) are smaller than 0.05 whereas all *p*-values of the test for comparing network size FAMDDA (FAM) and FAMDDA-GA (FAM-GA) are greater than 0.05. These results indicate that the generalization performances of individual classifiers are, respectively, lower than their GA-version counterparts, and the difference in the network sizes is statistically insignificant.

5 Summary

In this paper, a neuro-genetic system which integrates FAMDDA and the GA for performing classification is proposed. The performance of the propose FAMDDA-GA was accessed with three benchmark datasets. In order to show the effectiveness of FAMDDA-GA, its performance is compared with that of FAM-GA, as well as the ordinary classifier of FAMDDA and FAM. The results reveal that the generalization performance of FAMDDA-GA is better than that of FAM-GA. As for future work, a study on the possibility of integrating FAMDDA with another variant of EA-based global search algorithm, e.g. differential evolution, will be conducted. Another direction of future work is to investigate the possibility of including a local search into the existing framework for refining the solutions that have been obtained via global search.

References

Baskar, S., Subraraj, P., Rao, M.V.C. (2001), "Performance of hybrid real coded genetic algorithms," *International Journal of Computational Engineering Science*, vol. 2, pp. 583-601.

Blake, C. and Merz, C. (1998), UCI Repository of Machine Learning Databases, URL http://www.ics.uci.edu/~mlearn/MLRepository.html

Cantú-Paz, E., Kamath, C. (2005), "An empirical comparison of combinations of evolutionary algorithms and neural networks for classification problems," *IEEE Trans. Systems, Man, and Cybernetics – Part B: Cybernetics*, vol. 35, pp. 915-927.

Carpenter, G. A. and Grossberg, S. (1987), "A massively parallel architecture for a self-organizing neural pattern recognition machine," *Computer Vision, Graphics and Image Processing*, vol. 37, pp. 54-115.

Carpenter, G. A., Grossberg, S., and Rosen, D. (1991), "Fuzzy ART: Fast stable learning and categorization of analog patterns by an adaptive resonance system," *Neural Networks*, vol. 4, pp. 759-771.

Carpenter, G. A., Grossberg, S., Markuzon, N., Reynolds, J., and Rosen, D. (1992), "Fuzzy ARTMAP: A neural network architecture for incremental learning of analog multidimensional maps," *IEEE Trans. Neural Networks*, vol. 3, pp. 698-713.

Dagher. I, Georgiopoulos. M, Heileman. G.L., Bebis, G. (1999), "An ordering algorithm for pattern presentation in fuzzy ARTMAP that tends to improve generalization performance," *IEEE Trans Neural Networks*, vol. 10, pp. 768-778.

Efron, B. (1979), "Bootstrap methods: another look at the jackknife," *The Annals of Statistics*, vol. 7, pp. 1-26.

Fogel, D. B. (1995), *Evolutionary Computation: Toward a New Philosophy of Machine Intelligence*. Piscataway, NJ: IEEE Press.

Goldberg, D.E. (1989), *Genetic Algorithms in Search Optimization and Machine Learning*. Reading, MA: Addison-Wesley.

Huber, K.-P. and Berthold, M.R. (1995), "Building precise classifiers with automatic rule extraction," *in Proc. of the IEEE Int. Conf. Neural Networks,* vol. 3, pp. 1263-1268.

Lam, H.K., and Leung, F.H.F. (2004), "Digit and command interpretation for electronic book using neural network and genetic algorithm," *IEEE Trans. Systems, Man, and Cybernetics – Part B: Cybernetics*, vol. 34, pp. 2273 – 2283.

Lim, C.P. and Harrison, R.F. (1997), "An incremental adaptive network for on-line supervised learning and probability estimation," *Neural Networks*, vol. 10, 1997, pp. 925-939.
Palaniappan, R., and Raveendran, P. (2002), "Individual identification technique using visual evoked potential signals," *Electronics Letters*, vol. 28, pp. 1634-1635.
Palaniappan, R., Raveendran, P., and Omatu, S. (2002), "VEP optimal channel selection using genetic algorithm for neural network classification of alcoholics," *IEEE Trans. Neural Networks*, vol. 13, pp. 486-491.
Taha, I.A., and Ghosh, J. (1999), "Symbolic interpretation of artificial neural networks," *IEEE Trans. Knowl. Data Eng.*, vol. 11, pp. 448–463.
Tan, S.C., Rao, M.V.C., and Lim, C.P. (2004a), "An adaptive conflict-resolving classifier," *WSEAS Trans. on Information Science and Applications*, vol. 1, pp. 113-117.
Tan, S.C., Rao, M.V.C., and Lim, C.P. (2004b), "An adaptive fuzzy min-max conflict-resolving classifier," in Proceedings of the 9th Online World Conference on Soft Computing in Industrial Applications, *WSC9*, 20 September – 8 October 2004.
Tsai, J.-T, Chou, J.-H., and Liu, T.-K. (2006), "Tuning the structure and parameters of a neural network by using hybrid taguchi-genetic algorithm," *IEEE Trans. Neural Networks*, vol. 17, pp. 69-80.
Whitley, D. (1995), "Genetic algorithms and neural networks," in *Genetic Algorithms Engineering and Computer Science*, G. Winter, J. Periaux, M. Galan, and P. Cuesta, Eds. New York: Wiley, pp. 191–201.
Yao, X. (1999), *Evolutionary Computation: Theory and Applications*, Singapore: World Scientific.

Appendix

In FAMDDA, the operation of Fuzzy ART$_a$ module is modified for resolving conflicts resultant from the overlapping among prototypes of different classes. A set of *reference vector* $\boldsymbol{w}^r$ (Lim and Harrison1997) is introduced to each prototype j in F_2^a (Carpenter et al. 1992). Initially, $\boldsymbol{w}_j^r$ is equal to zero. In general, as learning takes place, $\boldsymbol{w}_j^r$ is updated. In addition, the width condition of all dimension of each F_2^a prototype is initialized as zero (i.e., $S_{j.} = 0$) to indicate "infinite volume". If $\boldsymbol{x}$, which represents the weights of the M – dimension (or $2M$ – dimension with complement coding) winning prototype falls in the region formed by the prototype of conflicting classes, a shrinking procedure then ensues. In this regard, the width of the conflicting prototypical region at arbitrary dimension is shrunk. In this regard, shrinking of an existing finite dimension is preferred for not losing "infinite volume" of other infinite dimensions. Nevertheless, to avoid the formation of a very thin rectangle, a user-defined minimum width threshold, $\varepsilon_{\min}$, is enforced on each finite dimension. The shrinking procedure is applied successively between the winning prototype and other conflicting prototypes. Three cases of width shrinking as in Huber and Berthold (1995) are considered. First, if the existing finite dimensions of the conflicting prototype q can be shrunk without falling below a pre-set $\varepsilon_{\min}$, the one with the smallest loss in volume ($\gamma_{best,k}$) is chosen, as follows.

$$\gamma_{best,k} = \min\{\left|w^{r}_{qk} - x_k\right| : \forall 1 \le i \le n, i \ne k : \left(\frac{\lambda_{qk} - \left|w^{r}_{qk} - x_k\right|}{\lambda_{qk}} \le \frac{\lambda_{qi} - \left|w^{r}_{qi} - x_i\right|}{\lambda_{qi}} \right) \wedge \left(\lambda_{qk} \ge \varepsilon_{k,\min}\right) : S_{qi}, S_{qk} = 1\} \quad (5)$$

where λ_q represents the current width of the prototype q. If the first option is not satisfied, the following options are considered, i.e., either one of the remaining infinite dimensions is shrunk $(\gamma_{\max,l})$,

$$\gamma_{\max,l} = \max\left\{\left|w^{r}_{ql} - x_l\right| : S_{ql} = 0\right\} \quad (6)$$

or, the width of a conflicting prototype is shrunk

$$\gamma_{\min,m} = \min\{\left|w^{r}_{qm} - x_{qm}\right| : \forall 1 \le i \le n, i \ne m : \left(\frac{\lambda_{qm} - \left|w^{r}_{qm} - x_{qm}\right|}{\lambda_{qm}} \le \frac{\lambda_{qi} - \left|w^{r}_{qi} - x_{qi}\right|}{\lambda_{qi}} \right) : S_{qi}, S_{qk} = 1\} \quad (7)$$

Width shrinking using Eq. (6) is selected if $\gamma_{\max,l} > \gamma_{\min,m}$. The learning algorithm of FAMDDA in one epoch is described in Fig. 2.

```
%Training of one epoch
%1.Consider all ordered complement-coded (Carpenter et al. 1992) samples (A,c) c - class
   for all samples A do
   if ART_a vigilance test = TRUE & ART_b vigilance test = TRUE
      if Mapfield link established = TRUE then
   % 2. Sample covered by wining prototype J of the F_2^a layer (Carpenter et al. 1992)
         w_J^a(new) = β(A ∧ w_J^a(old)) + (1 − β) w_J^a(old);
         N_J = N_J + 1;
         (w_J^r)^new = (w_J^r)^old + (1/N_J) (A − (w_J^r)^old)
         % β - learning rate; w_J - the weight of prototype J
   % 3. Commit new neuron
      else
         N_J = 1; % N_J - number of sample covered by prototype J
         w_J^r(new) = A ; (w_J^r)^new = A ;
      end
   % 4. Shrink width of conflicting neurons: (according to Eq. (5), (6), or (7))
   end
   end %for all samples A
```

Fig. 2. The operation of the FAMDDA network

A Modified Fuzzy Min-Max Neural Network and Its Application to Fault Classification

Anas M. Quteishat and Chee Peng Lim*

School of Electrical & Electronic Engineering, University of Science Malaysia
Engineering Campus, 14300 Nibong Tebal, Penang, Malaysia
cplim@eng.usm.my

Abstract. The Fuzzy Min-Max (FMM) network is a supervised neural network classifier that forms hyperbox fuzzy sets for learning and classification. In this paper, we propose modifications to FMM in an attempt to improve its classification performance in situations when large hyperboxes are formed by the network. To achieve the goal, the Euclidean distance is computed after network training. We also propose to employ both the membership value of the hyperbox fuzzy sets and the Euclidean distance for classification. To assess the effectiveness of the modified FMM network, benchmark pattern classification problems are first used, and the results from different methods are compared. In addition, a fault classification problem with real sensor measurements collected from a power generation plant is used to evaluate the applicability of the modified FMM network. The results obtained are analyzed and explained, and implications of the modified FMM network in real environments are discussed.

1 Introduction

Researches on formulating a classifier that can accurately classify input patterns have been conducted decades ago [2,3,4,10]. There are two types of learning classifiers: supervised and unsupervised learning, which differ in the way they are trained. In supervised classifiers, the class labels are provided along with the input patterns. In unsupervised classifiers, the training patterns are unlabeled and the classifiers have to split the data into homogenous clusters based on some similarity measure.

In this paper, we focus on the Fuzzy Min-Max [11] (hereafter refer to as FMM) neural network as the pattern classification system. FMM can be used for tackling clustering (unsupervised) or classification (supervised) problems, as proposed in [5]. The approach finds the decision boundary between classes, and clusters patterns that cannot be said to belong to any of the existing classes. However, in this paper, our focus is on supervised FMM for classification problems.

FMM is constructed using hyperbox fuzzy sets, each of which is an n-dimensional box defined by a set of minimum and maximum points. Each input pattern is classified based on the degree of membership to the corresponding boxes. The size of hyperboxes is a user defined value bounded between zero and one. A smaller hyperbox size means that the hyperbox can contain only a smaller number of patterns, which

* Corresponding author.

A. Saad et al. (Eds.): Soft Computing in Industrial Applications, ASC 39, pp. 179–188, 2007.
springerlink.com

will increase the network complexity. A larger hyperbox size means that the hyperbox can contain a larger number of patterns, and will decrease the network complexity. However, large hyperboxes may lead to a low classification performance as the hyperboxes may not capture salient features of the input patterns.

The objective of this paper is to propose methods to improve the classification performance of FMM in situations when large hyperboxes are formed by the network. This improvement is achieved by introducing the Euclidean distance into the classification procedure of FMM. Both the membership value of the hyperbox fuzzy sets and Euclidean distance are used to classify the input patterns. The effectiveness of the modified FMM network is first evaluated using benchmark problems, i.e. Pima Indian Diabetes (PID) and IRIS data sets [9]. The practical application of the network is then demonstrated using a real-world fault classification problem. The results are compared, analyzed, and discussed.

The paper is organized as follows: Section 2 gives a brief description of FMM. Section 3 describes the proposed method of modified FMM. The experimental studies and the results obtained are explained and discussed in section 4. The conclusion of the paper is presented in section 5.

2 The Fuzzy Min-Max Neural Network (FMM)

The FMM classification network is formed using hyperbox fuzzy sets. A hyperbox defines a region of the n-dimensional pattern space that has patterns with full class membership. The hyperbox is completely defined by its minimum and maximum points. The membership function is defined with respect to these hyperbox min-max points, and describes the degree to which a pattern fits in the hyperbox. For an input pattern of n-dimensions a unit cube I^n is defined. In this case the membership value ranges between 0 and 1. A pattern which is contained in the hyperbox has the membership value of one. The definition of each hyperbox fuzzy set B_j is:

$$B_j = \{X, V_j, W_j, f(X, V_j, W_j)\} \qquad \forall\, X \in I^n \tag{1}$$

where V_j and W_j are the min and max points, respectively. Figure 1 illustrates the min and max points in a three dimensional box.

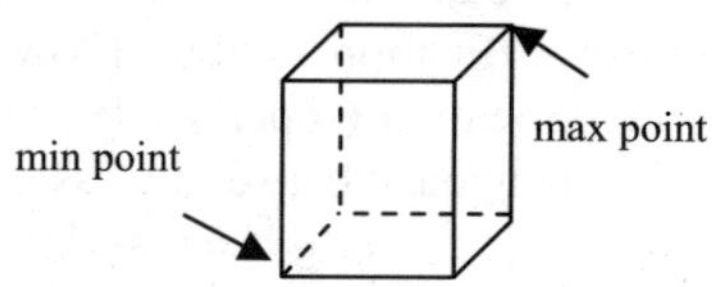

Fig. 1. A min-max hyperbox $B_j=\{V_j,W_j\}$ in I^3

Applying the definition of a hyperbox fuzzy set, the combined fuzzy set that classifies the Kth pattern class, C_k, is defined as:

$$C_k = \bigcup_{j \in K} B_j \tag{2}$$

where K is the index set of those hyperboxes associated with class k. One important property of this approach is that the majority of the processing is concerned with finding and fine-tuning the boundaries of the classes, as shown in Figure 2.

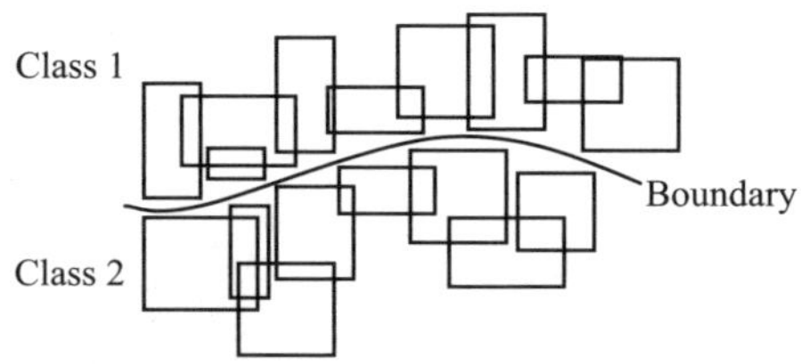

Fig. 2. An example of FMM hyperboxes placed along the boundary of a two-class problem

The learning algorithm of FMM allows overlapping of hyperboxes of the same class while eliminating overlapping among different classes. The membership function for the jth hyperbox $b_j(H_h)$, $0 \le b_j(H_h) \le 1$, measures the degree to which the hth input pattern A_h falls outside the hyperbox B_j. On a dimension-by-dimension basis, this can be considered as a measurement of how far each component is greater (or lesser) than the max (or min) point value along each dimension that falls outside the min-max bounds of the hyperbox. In addition, as $b_j(H_h)$ approaches 1, the point should be more "contained" by the hyperbox. The function that meets all these criteria is the sum of two complements–the average amount of max point violation and the average amount of min point violation. The resulting membership function is:

$$
\begin{aligned}
b_j(A_h) = \frac{1}{2n}\sum_{i=1}^{n} [&\max\left(0,1-\max\left(0,\gamma\min\left(1,a_{hi}-w_{ji}\right)\right)\right) \\
&+\max\left(0,1-\max\left(0,\gamma\min\left(1,v_{ji}-a_{hi}\right)\right)\right)]
\end{aligned}
\tag{3}
$$

where, $A_h = (a_{h1}, a_{h2}, \ldots, a_{hn}) \in I^n$ is the hth input pattern, $V_j = (v_{j1}, v_{j2}, \ldots, v_{jn})$ is the min point for B_j, $W_j = (w_{j1}, w_{j2}, \ldots, w_{jn})$ is the max point for B_j, and γ is the sensitivity parameter that regulates how fast the membership values decrease as the distance between A_h and B_j increases.

FMM is a three layer network, as shown in Figure 3. The first layer F_A is the input layer which contains input nodes equal in number to the number of dimensions of the input pattern. Layer F_C is the output layer. It contains nodes equal in number to the number of classes. The hidden layer is called the hyperbox layer F_B. Each F_B node represents a hyperbox fuzzy set, where F_A to F_B connections are the min-max points. The F_B transfer function is the hyperbox membership function defined by (3). The min and max points are stored in matrices V and W, respectively. The connections between the F_B and F_C nodes are binary valued and are stored in matrix U. The equation for assigning the values from F_B to F_C connections is

$$
u_{jk} = \begin{cases} 1 & \text{if bj is a hyperbox for class } \mathrm{C_k} \\ 0 & \text{otherwise} \end{cases}
\tag{4}
$$

where b_j is the jth node and C_k is the kth node. Each F_C node represents a class. The output of the F_C node represents the degree to which input pattern A_h fits within class k. The transfer function for each of the F_C nodes performs the fuzzy union of the appropriate hyperbox fuzzy set values. This operation is defined as

$$c_k = \max_{j=1}^{m} b_j u_{jk} \quad (5)$$

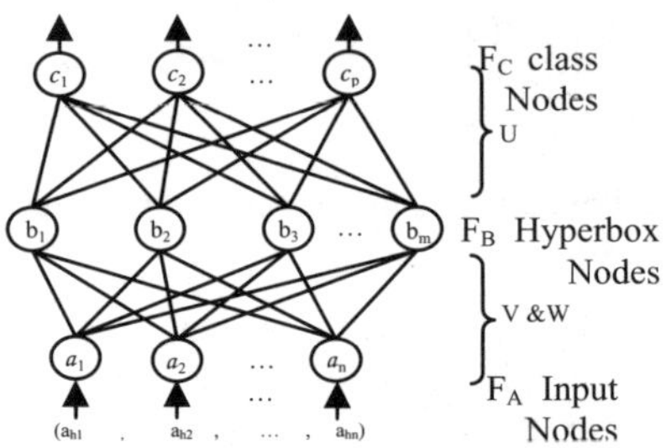

Fig. 3. A three layer FMM network

There are two main ways in which the output of F_C class nodes can be utilized. If a soft decision is required, the outputs are utilized directly. If a hard decision is required, the F_C node with the highest value is located, and its node value is set to 1 to indicate that it is the closest pattern class, while the remaining F_C node values are set to 0, i.e., winner-take-all [8].

In FMM, the fuzzy min-max learning methodology is an expansion/contraction process. The training set D consists of a set of M ordered pairs $\{X_h, d_h\}$, where, $X_h = (x_{h1}, x_{h2}, ..., x_{hn}) \in I^n$ is the input pattern and $d_h \in \{1,2,..., m\}$ is the index of one of the m classes. The learning process begins by selecting an ordered pair from D and finding a hyperbox of the same class that can expand, if necessary, to include the input. The expansion criteria has a constraint that must be met, and is defined as:

$$n\theta \geq \sum_{i=1}^{n} \left(\max\left(w_{ji}, x_{hi}\right) - \min\left(v_{ji}, x_{hi}\right) \right) \quad (6)$$

where $0 \leq \theta \leq 1$ is the hyperbox size. If a hyperbox that meets the expansion criteria cannot be found, a new hyperbox is formed in the network. This growth process allows classes to be formed that are nonlinearly separable. It also allows new classes to be added without retraining. When hyperboxes expand there is a possibility of overlap among these boxes. So, an overlap test is introduced to check if the overlap is among the same or different classes. If overlapping occurs among the same classes nothing is done. But if overlapping occurs among different classes, the contraction process is activated and the overlap is eliminated. Basically, the fuzzy min-max learning process comprises of series of activities that create and expand hyperboxes, and fine-tune these hyperboxes by the overlap test and the contraction process.

3 Modifications on the FMM Network

3.1 Prediction Using the Euclidean Distance

After FMM is trained, the network is tested to evaluate its classification accuracy. In the prediction phase, the patterns are classified based on the membership value calculated using equation (3). The degree of membership for the pattern under consideration is calculated against all hyperboxes created in the learning process. The pattern is so classified that it belongs to the class associated with the hyperbox that has the highest membership value.

We propose another method in the prediction phase of the FMM network. This method is based on the Euclidean distance between the input pattern and the centroid of the hyperbox. In addition to the min and max points, the centroid of patterns falling in each hyperbox is computed, as follows

$$C_{ji} = C_{ji} + \frac{|a_{hi} - C_{ji}|}{N_j} \tag{7}$$

where C_{ji} is the centroid of the jth hyperbox in the ith dimension, and N_j is the number of patterns included in the jth hyperbox. The Euclidean distance between the centroid and the input pattern is calculated using

$$E_{jh} = \sqrt{\sum_{i=1}^{n} (C_{ji} - a_{hi})^2} \tag{8}$$

where E_{jh} is the Euclidean distance between j^{th} hyperbox and the hth input pattern. Now, when a pattern is to be classified the Euclidean distances between the pattern and the centroids of the existing hyperboxes are first calculated. The hyperbox with the smallest distance is selected as the winner, and the pattern is so classified that it belongs to this hyperbox. Figure 4 shows the classification process of a two-dimensional input pattern using the described method. In Figure 4, E1 and E2 are the distances between the input and the centroids of hyperboxes 1 and 2, respectively. Since E1 is smaller than E2, the input pattern is classified as class 1.

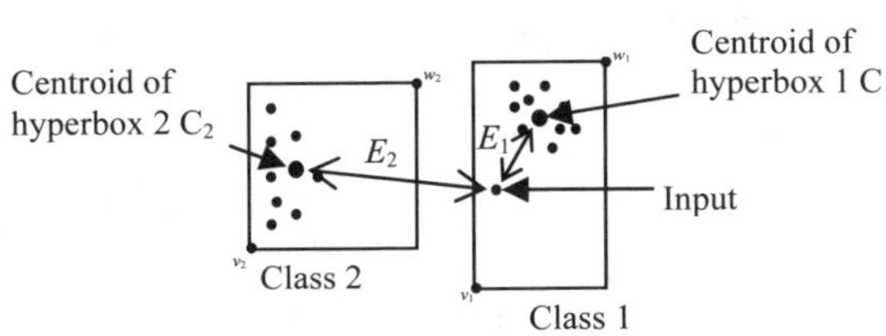

Fig. 4. The classification process of an input pattern using the Euclidean distance

3.2 Prediction Using Both the Membership Function and Euclidean Distance

The size of a hyperbox is controlled by θ, which varies between 0 and 1. When θ is small, more hyperboxes are created. When θ is large, the number of hyperboxes is small, and the membership function values are high. A lot of hyperboxes may end up with the same value of membership function (e.g., unity). In this case, depending on the membership value, inaccurate prediction may result. As such, we propose to use both the membership value and Euclidean distance for classification. The hyperboxes with the highest membership value are selected and then the Euclidean distance between the centroid of these boxes and the input pattern is calculated. The hyperbox with the smallest distance is used to classify the input pattern. This method, as will be shown, gives improvement in the classification results when θ is large.

4 Experiments and Results

4.1 Benchmark Problems

The proposed method was first evaluated using two benchmark problems, i.e., the Pima Indian Diabetes (PID) and IRIS data sets. The data sets were obtained from the UCI machine learning repository [9].

The first experiment was conducted using the PID dataset. This data set contains 768 cases which belong to two classes, in which 268 cases (35%) are from patients diagnosed as diabetic and the remaining as healthy. Following the procedure as described in [12], an experiment was conducted using 75% of samples in each class as a training set (576 samples) and the remaining 25% as a test set (192 samples). The second experiment was conducted using the IRIS data set. This data set consists of 150 samples with four continuous attributes from three classes. The data set was used after normalizing each attribute value to a real number in the unit interval [1,0]. Following the procedure as described in [6], an experiment was conducted using 80% of the data set (120 samples) for training and the remaining 20% (30 samples) for testing.

The experiments started with varying the hyperbox size θ from 0.01 to 0.95 (with an increment of 0.05) to determine the classification accuracy rate and the number of hyperboxes created. Since the proposed modifications do not affect the learning process of FMM, the number of hyperboxes created remained the same for both original and modified FMM networks.

Figure 5 (a) show the classification accuracy rates using three different methods for the PID and IRIS data sets. Curve 'A' represents the results of using the membership value only (i.e. original FMM); 'B' represents the results of using both the membership

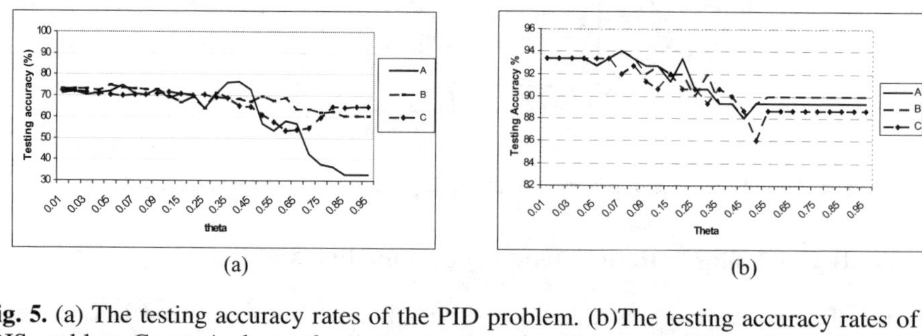

(a) (b)

Fig. 5. (a) The testing accuracy rates of the PID problem. (b)The testing accuracy rates of the IRIS problem Curve A shows the accuracy rate using the membership value only; curve B shows the accuracy rate using both the membership value and Euclidean distance; curve C shows the accuracy rate using the Euclidean distance only.

value and Euclidean distance; and 'C' represents the results of using the Euclidean distance only.

Table 1 shows the best results of Curves A, B and C along with a list of results from different classification algorithms. The results for PID and IRIS data sets are as reported in [12] and [6] respectively.

Table 1. A comparison of accuracy rates of various methods for the PID data set and IRIS data set

PID data set		IRIS data set	
Methods	Accuracy (%)	Methods	Accuracy (%)
LDA	77.5	C4.5	91.60
C4.5	73.0	OC1	93.89
CART	72.8	LMDT	95.45
K-NN	71.9	LVQ	92.55
Curve A (best result)	76.6	Curve A (best result)	94.00
Curve B (best result)	74.9	Curve B (best result)	94.00
Curve C (best result)	72.4	Curve C (best result)	93.33

In Figure 5 (a) one can see that as θ increased, the performance of all three methods decreased. However, the testing accuracy rate of the membership value only approach (curve A) deteriorated faster than those from the other two methods especially when large hyperboxes were formed in the network (high values of θ). In Figure 5 (b), again one can see that as θ increased, the performance of all three methods decreased. Notice that curve 'B' gave better performance than curve 'A' and curve 'C' when θ assumed a large value.

4.2 Fault Classification

An important industrial application of pattern classification system is fault detection and classification. Fault detection and classification is a research area that is getting increasingly important owing to complexity of modern industrial systems and growing demands for quality, cost efficiency, reliability and safety [1,7]. A fault detection and classification system predicts failures, and when a failure occurs, it identifies the reason(s) of failures.

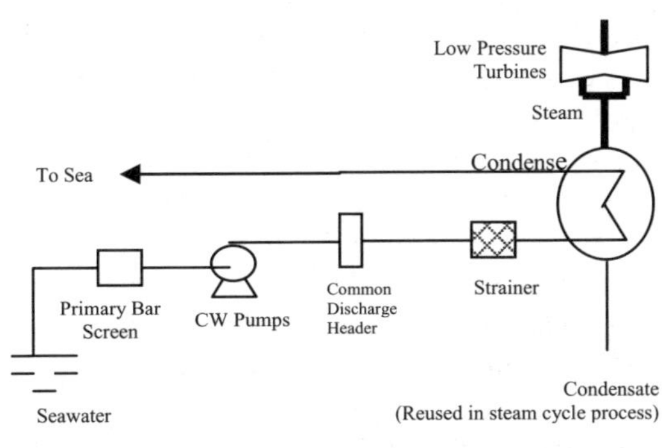

Fig. 6. The circulating water system

In this study, we investigate the applicability of modified FMM using a set of sensor measurements collected from a power generation plant in Malaysia. The system under consideration is a circulating water (CW) system, as shown in Figure 6. The main function of the CW system is to provide a sufficient and continuous supply of cooling water to the main turbine condensers to condense steam from the turbine exhaust and the steam then flows into the condenser [13].

A data set of 2439 samples was collected. As shown in Table 2, each data sample consisted of 12 features comprising of temperature and pressure measurements at various inlet and outlet points of the condenser, as well as other important parameters. Two case studies were conducted: first, the heat transfer conditions and second, the tube blockage conditions.

(a) Heat Transfer Conditions

The heat transfer conditions were classified into two categories: efficient or inefficient. From the data set, 1224 data samples (50.18%) that showed inefficient heat-transfer condition, whereas 1215 data samples (49.82%) showed efficient heattransfer condition in the condenser. The data set (excluding one sample) was divided into two equal sets, each containing 1219 samples, one for training and the other for testing. Both data sets contained 50% of the data samples belonging to each class.

Table 2. List of sensor parameters used in the experiments

No	Parameter	Description
1	LPT A	Low Pressure Cylinder Exhaust Temperature A
2	LPT B	Low Pressure Cylinder Exhaust Temperature B
3	GEN	Generator
4	CWIT A	Condenser Circulating Water Inlet Temperature A
5	CWIT B	Condenser Circulating Water Inlet Temperature B
6	CWOT A	Condenser Circulating Water Outlet Temperature A
7	CWOT B	Condenser Circulating Water Outlet Temperature B
8	CWIP A	Condenser Circulating Water Inlet Pressure A
9	CWOP A	Condenser Circulating Water Outlet Pressure A
10	CWIP B	Condenser Circulating Water Inlet Pressure B
11	CWOP B	Condenser Circulating Water Outlet Pressure B
12	VAC	Condenser vacuum

A series of experiments was conducted by varying the value of θ from 0.01 to 0.95, with an increment of 0.05. The results obtained are shown in Figure 7 (a). One can see that the performance of curves A and B is more stable than that of curve C. However, when θ is high, the proposed modifications (curves B and C) outperform the original system (curve A). Overall, the performance of curve B is the most stable and the best. These results justify the usefulness of using both the membership value and Euclidean distance in the prediction phase of the FMM network. Since the objective of this work is focused on situations when the values of θ are large, Table 3(a) summarizes the testing accuracy rates and the number of hyperboxes created using the three described methods from 0.5 to 0.95, with an increment of 0.05. It can be seen that the proposed modifications produce better performances than original FMM when $\theta \geq 0.75$. The test accuracy rate for both the membership value and Euclidean distance was stable at the level of 92.23% (with only two hyperboxes). This performance is better than the best result produced by original FMM (membership value alone), i.e. 90.9% accuracy (with 226 hyperboxes) at $\theta \geq 0.50$.

(b) Tube Blockage Conditions

In this experiment, the objective was to predict the occurrence of tube blockage in the CW system. The conditions of the condenser tubes were categorized into two classes: significant blockage and insignificant blockage. The data set used in the first experiment was again employed. A total of 1313 samples (53.83%) showed significant blockage and the remaining samples showed insignificant blockage in the condenser tubes. The data samples were divided into two sets for training and testing.

Again, the data set (excluding one sample) was divided into two equal sets, each containing 1219 samples, one for training and the other for testing. Both data sets contained 50% of the data samples belonging to each class.

The same procedure as in the previous experiment was adopted. Figure 7(b) and Table 3(b) summarize the overall results. As shown in Figure 7 (b), all the methods were able to produce perfect performance when θ was low. When θ increased, the performance, as expected, dropped. The performance of using the Euclidean distance only, deteriorated sharply. However, it was able to achieve good results when θ was high, i.e., $\theta \leq 0.9$. On the other hand, the membership value and Euclidean distance approach was able to maintain a good performance of over, in general, 90% accuracy. When $\theta \leq 0.85$, modified FMM with the membership value and Euclidean distance was able to produce an accuracy rate of over 96%.

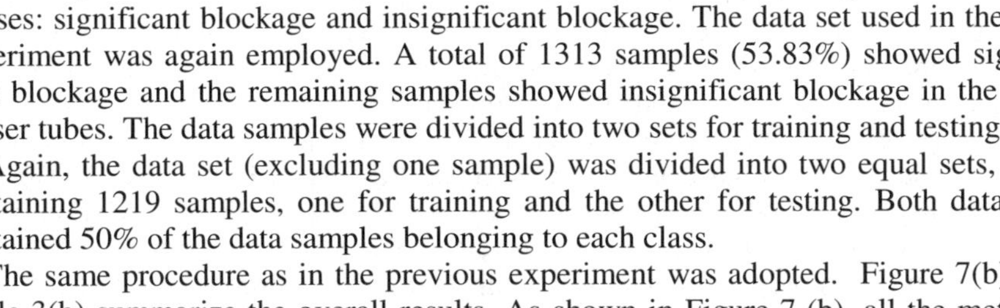

(a) (b)

Fig. 7. (a) The testing accuracy rates of the Heat Transfer Conditions. (b) The testing accuracy rates of the Tube Blockage Conditions. Curves A, B, and C show the testing accuracy rates using the membership value only, both the membership value and Euclidean distance, and the Euclidean distance only, respectively.

Table 3. Testing accuracy for the heat transfer data set and tube blockage conditions

(a) Heat Transfer Conditions					(b) Tube Blockage Conditions				
	Testing accuracy					Testing accuracy			
θ	Membership value (%)	Euclidean distance (%)	Membership value and Euclidean distance (%)		θ	Membership value (%)	Euclidean distance (%)	Membership value and Euclidean distance (%)	
0.5	90.9	92.82	93.36	226	0.5	99.88	96.61	100.00	264
0.55	90.53	85.56	92.03	187	0.55	100	85.56	99.68	155
0.6	88.56	77.32	89.94	101	0.6	100	86.79	95.28	78
0.65	90.53	72.05	90.63	81	0.65	91.88	83.39	93.75	62
0.7	90.41	79.19	92.57	38	0.7	92	85.85	92.13	81
0.75	87.08	90.43	92.23	2	0.75	91.88	85.36	93.01	21
0.8	87.08	90.43	92.23	2	0.8	92	86.35	89.25	42
0.85	87.08	90.43	92.23	2	0.85	92	88.98	96.41	6
0.9	87.08	90.43	92.23	2	0.9	92	93.75	96.75	2

5 Summary

In this paper, modifications to the FMM network have been proposed to tackle the problem when the number of hyperboxes created by the network is small, i.e. size of the hyperboxes is large. A prediction method using the Euclidean distance combined with the membership value was employed. The proposed method was applied to FMM after its training stage; thus the incremental learning property and other salient learning features of original FMM remain undisturbed. The effectiveness of the proposed modifications has been evaluated using PID and IRIS benchmark data sets. Applicability of the modified FMM network to a real-world fault classification task in a power generation plant has been demonstrated. The results obtained reveal the usefulness of the proposed modification in improving the performance of FMM when large hyperboxes are formed by the network.

Although the results obtained from the fault classification problem are encouraging, more investigations with data sets from different problem domains need to be done in order to further ascertain the effectiveness of the proposed modifications in making the FMM network an accurate and useful classifier in a variety of application areas.

References

1. Al-Najjar, B. (1996) Total quality maintenance: An approach for continuous reduction in costs of quality products, Journal of Quality in Maintenance Engineering, vol. 2, pp. 2-20.
2. Devijver, P. and Kittler, P. (1982) Pattern Recognition: A Statistical Approach. Englewood Cliffs, NJ: Prentice-Hall.
3. Duda, R.O., and Hart, P.E. (1973). Pattern Classification and Scene Analysis. New York: Wiley.
4. Fukunaga, K. (1972) Introduction to Statistical Pattern Recognition. New York Academic Press.
5. Gabrys, B. and Bargiela, A. (2000) General Fuzzy Min-Max Neural Network for Clustering and Classification. IEEE Trans on Neural Networks, vol. 11, pp 769-783.
6. Hoang, A., (1997) Supervised Classifier Performance on the UCI Data Set. M.Sc. Thesis, Department of Computer Science, University of Adelaide, Australia.
7. Huang,Y. C., Yang, H. T., Huang, K.Y. (2002) Adductive network model-based diagnosis system for power transformer incipient fault-detection. IEE Proceedings- Generation, Transmission and Distribution, vol. 149, pp. 326-330.
8. Kohonen, T. (1984) Self-Organization and Associative Memory. Berlin: Springer-Verlag.
9. Murphy, P.M. and Aha, D.W. (1995). UCI repository of machine learning databases, (Machine-readable data repository). Department of information and computer Science. University of California, Irvine, CA.
10. Nakashima T., Nakai G., and Ishibuchi H. (2003) Constructing Fuzzy Ensembles for Pattern Classification Problems. IEEE SMC Int. Conf. Vol. 4, pp. 3200 – 3205.
11. Simpson, P. (1992) Fuzzy Min-Max Neural Networks–Part 1: Classification. IEEE Trans on Neural Networks, vol. 3, pp 776-786.
12. Ster, B. and Dobnikar, A. (1996). Neural Network in medical diagnosis: Comparison with other methods, In: A. Bulsari et al., eds, Proc. Int. Conf. EANN'96. pp. 427-430.
13. System description and operating procedures (1999). Prai Power Station Stage 3, vol. 14.

AFC-ECG: An Adaptive Fuzzy ECG Classifier

Wai Kei Lei[1], Bing Nan Li[1], Ming Chui Dong[1,2], and Mang I Vai[2]

[1] Institute of System and Computer Engineering, Taipa 1356, Macau
[2] Dept. Electrical & Electronic Engineering, FST, University of Macau, Taipa, Macau
bingoon@ieee.org, {ma46530, dmc, fstmiv}@umac.mo

Abstract. After long-term exploration, it has been well established for the mechanisms of electrocardiogram (ECG) in health monitoring of cardiovascular system. Within the frame of an intelligent home healthcare system, our research group is devoted to researching/developing various mobile health monitoring systems, including the smart ECG interpreter. Hence, in this paper, we introduce an adaptive fuzzy ECG classifier with orientation to smart ECG interpreters. It can parameterize the incoming ECG signals and then classify them into four major types for health reference: Normal (N), Premature Atria Contraction (PAC), Right Bundle Block Beat (RBBB), and Left Bundle Block Beat (LBBB).

Keywords: ECG classifier; fuzzy sets; medical advisory system; health prognosis.

1 Introduction

Most people do not care about their health condition until they fall in illness. Then, it is an excruciating and cost-expensive procedure for the subsequent therapy. Moreover, early prevention and healthcare has been proven as an effective measure to prevent the sudden death due to heart diseases. So, the mode of contemporary healthcare is experiencing an ultimate revolution from disease recovery to health prevention. Nowadays home healthcare, including home health monitoring, has been widely accepted to improve the quality of our life. The major advantage of home health monitoring is to provide a cost-effective way for health prognosis with various physiological signals collected remotely [1].

In the pilot project - "Intelligent e-Home Healthcare System", we propose a series of embedded medical advisory systems to enhance the intelligence of conventional biomedical transducers, such as intelligent sphygmogram analyzers (*SGA*) and smart electrocardiogram interpreters (*ECGI*) [2], [3], [4]. Then, people can master their health condition better at home. Beyond collecting and submitting physiological signals, these intelligent healthcare apparatus, benefited from the embedded medical advisory system, can report the health condition in a real-time manner. Meanwhile, the embedded-link mode of medical advisory systems enable home subjects to submit the collected signals to health centers for further analysis.

Coming to cardiovascular system monitoring, electrocardiogram (ECG) is most competent because it can reflect the complete cycle of subtle heart beating. An ECG signal is the record of changing bioelectric potentials with respect to time as the

A. Saad et al. (Eds.): Soft Computing in Industrial Applications, ASC 39, pp. 189–199, 2007.
springerlink.com

human heart beats. Confirmed by numerous exploration and clinical trails, ECG monitoring and analysis has been widely accepted as a useful manner to diagnose heart disorder. The problem is, tremendous amount of data are generally involved in the procedure of ECG monitoring and analysis. Consequently, a smart tool for ECG interpretation is indispensable for cardiologists, let alone home subjects.

In former exploration, quite a few methods have been proposed to implement computerized ECG analysis and diagnosis. Among them, neural networks and fuzzy sets are two most popular scenarios. Neural networks are in consideration due to their self-adaptation and robustness. In other words, a neural network is a nonlinear map between the input vectors and the desired outputs. Its knowledge is stored in the weight net between nodes. The problem is, neural networks often require a completed dataset with independent and identical distribution (i.i.d.) for training and self-adaptation [5], [6]. Nevertheless, in practical, it is hardly to find such datasets. Meanwhile, physicians do not favor it because it lacks a structural knowledge base for review and reference.

The method of fuzzy sets is another popular solution for ECG analysis and diagnosis because it uses smooth variables with membership functions for medical inference. In the first side, it can represent imprecise concepts within a linguistic form, such as "maybe", "likely", and "absolutely", etc. Therefore, fuzzy sets are particularly suitable to describe health status because it is impossible to assure which kind of diseases a patient certainly has. Secondly, a classifier based on fuzzy sets is built with a structural knowledge like "if-then" production rules [7]. Such classical fuzzy classifiers are generally based on the priori knowledge and experience of domain experts. Different from neural networks as non-linear mapping, fuzzy sets not only describe the mapping relations but also incorporate the inference justifications. However, it has been recognized that knowledge acquisition is an annoying bottleneck of system implementation. In other words, conventional fuzzy classifiers are restricted to the self-learning capability, an underlying mechanism of intelligent ECG classifiers.

Recently, some advanced solutions, such as adaptive neuro-fuzzy inference networks, have been proposed to combine the advantages of neural networks for self-learning and fuzzy sets for legible inference [8]. In this paper, we propose a different adaptive fuzzy ECG classifier (AFC-ECG) based on statistical learning theory. Its adaptation is oriented to online modification in accordance with the input ECG signals. Then, the performance of the proposed AFC-ECG can be enhanced in the sight of any specific home subject. In the following, the section 2 introduces ECG data and patterns necessary for auto classification; and afterward, a conventional fuzzy classifier, including its performance, is presented in the section 3; then, AFC-ECG with enhanced self-adaptation will be detailed in the section 4; the final part is for discussion and conclusion.

2 ECG Data and Patterns

Our research is based on the datasets from the MIT-BIH arrhythmia database [9], which includes recordings of many common and life-threatening arrhythmias along with examples of normal sinus rhythm. The database contains 48 half-hour two-channel ambulatory ECG recordings and measures from 47 subjects at BIH Arrhythmia Laboratory. The data are sampled at 360Hz and band pass filtered at 0.1-100Hz. All practical issues occur in ECG classification can be found in this database, i.e. baseline drift, power noise, etc.

During an ECG cycle, the fiducial points with physiopathological significance are often marked as P, Q, R, S, and T wave, as shown in Fig. 1. From the perspective of cardiology, these points correspond to the action potentials of different myocardial chambers. In generally, P wave corresponds to the contraction of the atria; QRS complex (composed by Q, R and S wave) corresponds to the contraction of left ventricle; and T wave corresponds to relaxation of the ventricles. Their morphologies will vary in accordance with the physiological condition of cardiovascular system. Hence, it is possible to infer out the health condition of cardiovascular system inversely based on the features of ECG signal.

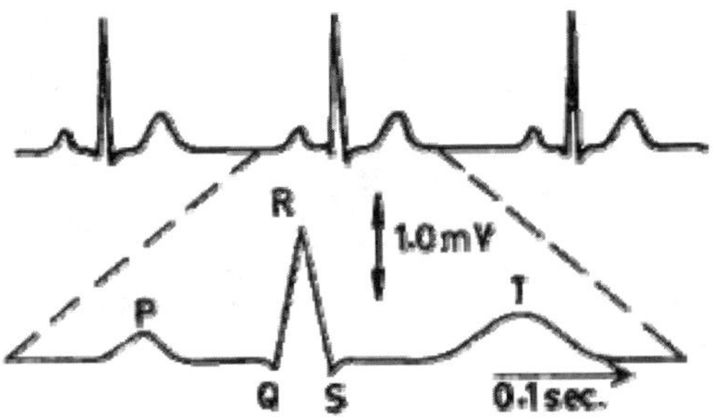

Fig. 1. Typical patterns of ECG signals

Nowadays the timing or morphological ECG features in cardiology, as shown in table 1, have been well defined by clinical professionals. Here the prior-HR is further defined as the time between the prior and current R waves, and the post-HR is the time difference of the current and next R waves. On the other hand, the morphological features of ECG signals are usually described by the words of "upward or downward", "early or late" and "narrow or broad" in medical literatures. The cardiologists often make effective diagnosis based on these characteristic features.

Table 1. ECG patterns and linguistic variables

ECG features	Medical linguistic variables
Prior-Heart Rate (RR_0)	{Short; Normal; Long}
Post-Heart Rate (RR_1)	{Short; Normal; Long}
P wave	{Early; Normal; Disappear}
QRS Complex	{Upward; Downward}
R Wave Amplitude	{High; Normal; Low}
T wave	{Upward; Downward; Disappear}

3 Fuzzy Classifier

3.1 Classifier Structure

Shown as in Fig. 2, a practical fuzzy ECG classifier is generally comprised of two major function blocks: ECG Parameterizer and Fuzzy Classifier. The first function block is used to detect the fiducial points, including P, Q, R, S and T, of ECG signals. Our proposed system is based on the methodology of Wavelet Transform (WT) [10]. The derived parameters including amplitudes and durations are then exported to the latter function block for ECG classification.

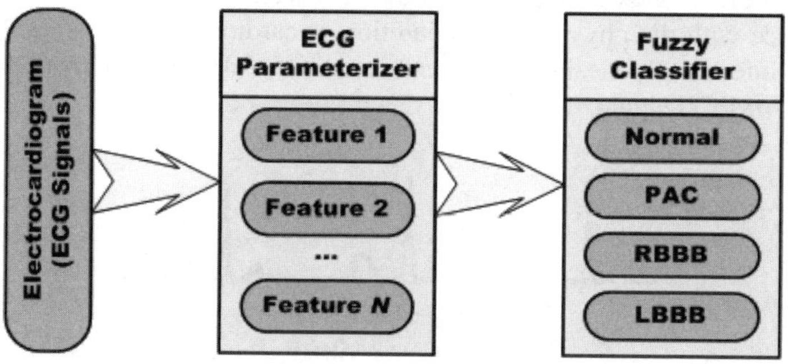

Fig. 2. The structure of a fuzzy ECG classifier

3.2 Fuzzy Inference

The proposed fuzzy classifier achieves ECG classification with those features listed in table 1. A set of characteristic vectors are selected from those features for ECG classification. In general, any fuzzy ECG classifier has to undergo iterative adjustment in terms of fuzzy variables, for example, the choice of membership functions and the definition of fuzzy rules in knowledge base. These adjustments are referring to medical literatures and fuzzy set theory. Then, the infrastructure of a fuzzy ECG classifier can be illustrated as shown in Fig. 3.

During fuzzification stage, 3 kinds of membership functions are adopted in the proposed classifier, that is, *S*-type, *Z*-type and Gaussian functions whose mathematical forms are $f_S(a,b)$, $f_Z(a,b)$ and $f_G(a,b)$ respectively. The parameter "*a*" and "*b*" in both *S* and *Z* function represent the values of lower and upper boundary. In contrast, "*a*" and "*b*" of Gaussian function represent its mean value and standard deviation respectively.

In clinical practice, the cardiologists usually make empirical diagnosis based on the timing and morphological features of ECG signal. Take Normal Beat (*N*) for example. It can be delineated with the fuzzy rules comprised of a series of ECG patterns, such as "upward *P* wave", "upward *QRS* complex", and "upward *T* wave", etc. Then, the most challenging issue in fuzzy ECG classifiers stems from how to convert the

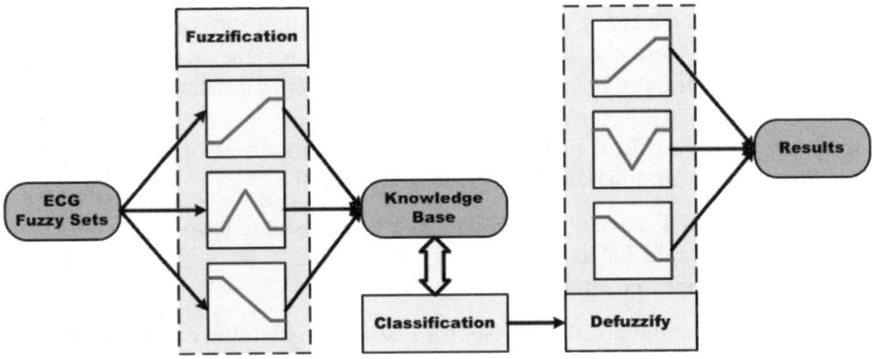

Fig. 3. The infrastructure of a fuzzy ECG classifier

subjective linguistic terms to precise numerical values suitable for computer processing. The following table 2 is to illustrate the possible timing characteristics of membership functions in fuzzy rules for classifying the type of normal heart beat. Here the choice of membership functions is based on medical literatures and so are the parameters "*a*" and "*b*". For example, it is reported that the normal range of heart rate varies from 60 to 100 bpm. So the "*a*" and "*b*" of Gaussian function are set as 80 and 20 respectively.

Table 2. Definition of membership functions for Normal Beat (*N*)

Characteristic	Feature	Function Type	Parameter (*a*)	Parameter (*b*)
P upward	*P* peak value	*S*	0.10 mV	0.15 mV
QRS upward	*R* peak Value	*S*	0.70 mV	0.80 mV
T upward	*T* peak Value	*S*	0.10 mV	0.15 mV
RR_0	Prior-HR	Gaussian	80 bpm	20 bpm
RR_1	Post-HR	Gaussian	80 bpm	20 bpm

It has been recognized that one of main bottlenecks of building a conventional fuzzy classifier lies in finding the boundary values of membership functions. Here an adapting mechanism will be detailed in the following section 4. After the fuzzification step, all incoming features are described by a membership value [0, 1]. These values form the hypotheses of fuzzy inference. The inference mechanism of the proposed fuzzy ECG classifier obeys the following production rules:

IF	("Feature 1" is "Linguistic Variable 1")	**AND**
	("Feature 2" is "Linguistic Variable 2")	**AND**
	…	
	("Feature *N*" is "Linguistic Variable *N*")	
THEN	(Class Name)	

The hypotheses of a fuzzy rule are consisted of *N* linguistic variables with logical "AND" operation. Here the rule will become true once the accumulated result is larger than an empirical threshold value, that is, 0.6. This threshold is from medical literatures or data analysis, and can be modified further.

According to the medical literature, the detecting properties of Left Bundle Block Beat (*LBBB*) can be summarized by following statements:

- The direction of T wave and QRS complex is opposite to each other.
- The prior-heart rate RR_0 is small.
- P wave is disappeared.

Base on above statements, the rule set for *LBBB* classification can be defined as:

<RULE 1 :>		
IF	"*P* wave is disappear"	**AND**
	"Prior-Heart Rate is small"	**AND**
	"*QRS* is downward"	**AND**
	"*T* is upward"	
THEN	"Left Bundle Block Beat (*LBBB*)"	
<RULE 2 :>		
IF	"*P* wave is disappear"	**AND**
	"Prior-Heart Rate is small"	**AND**
	"*QRS* is upward"	**AND**
	"*T* is downward"	
THEN	"Left Bundle Block Beat (*LBBB*)"	

The linguistic variables, including "upward", "small" and so on, are fuzzified by calculating their membership grades. The product of their membership grades forms the firing hypothesis of production rules.

3.3 Experiment Results

The data used for system evaluation of the proposed fuzzy ECG classifier are from the MIT/BIH arrhythmia database [9]. Each dataset comes from the 2-channel trace of ECG signals in 30 minutes. A prototyping fuzzy ECG classifier was programmed with the help of Matlab© for "ECG Parameterizer" and Visual C++© for "Fuzzy Classifier". As mentioned in previous, its membership functions and production rules of that fuzzy classifier have been carefully tuned in accordance with medical literatures and physician instructions. The following table 3 demonstrates its performance in different ECG morphological segments.

4 An Adaptive Fuzzy Classifier of ECG (AFC-ECG)

As shown in the table 3, the accurate rate as a performance index averages around 77.0% even though the fuzzy ECG classifier has been manually optimized. It is

Table 3. The performance of a conventional fuzzy ECG classifier on MIT/BIH dataset

Signal Segment	MIT/BIH Annotation		Fuzzy Classifier	Correct Rate (%)
Sig 106	Normal	1507	1208	80.2%
	PAC	0	7	-
	RBBB	0	30	-
	LBBB	0	0	-
	Unclassified	0	262	-
Sig 118	Normal	0	1	-
	PAC	96	49	51.0%
	RBBB	2166	1602	74.0%
	LBBB	0	0	-
	Unclassified	0	610	-
Sig 207	Normal	0	19	-
	PAC	107	127	81.3%
	RBBB	86	71	82.6%
	LBBB	1457	1357	93.1%
	Unclassified	0	76	-

understandable because any ECG signal is physiologically and pathologically unique in accordance with the specific subject's health condition. In the proposed fuzzy ECG classifier, its membership functions in essence come into being from generalized medical knowledge and datasets. Hence, predictably, its fixed membership functions will impair the performance once it is extended to the specific ECG segments.

Therefore, to succeed in real-life applications, the self-adaptation should be a necessary step for any fuzzy ECG classifier. In fact, it has been well recognized that generalization and self-adaptation serve as two indispensable steps to any intelligent system. A competent intelligent system often comes into being with the tradeoff of generalization and self-adaptation to the specific application. Coming to fuzzy classifiers for ECG analysis, a paradigm based on neural computation has been recommended for self-adaptation. Nevertheless, it requires a batch of identical and independent datasets for learning and updating. Unfortunately, it is often a mission impossible for the fuzzy classifier embedded in biomedical transducers.

Consequently, in this paper, we propose a method of online self-adaptation based on statistical learning theory. Such adaptive fuzzy ECG classifier (AFC-ECG) is as shown in Fig. 4. Before self-adaptation, all parameters are set as default values in accordance with medical knowledge. Such default system runs as it is for a while. Then, AFC-ECG will try to adapt itself from historical data. After adaptation, it will run as it is again for a while. Such mode of online self-adaptation is particularly suitable for our proposed embedded-link medical advisory systems: AFC-ECG implements fuzzy classification in biomedical transducers while accomplishing self-adaptation at the site of central servers.

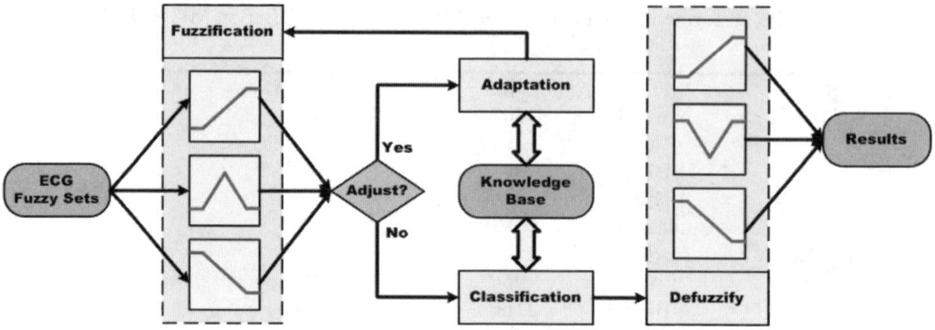

Fig. 4. Flowchart of an adaptive Fuzzy ECG Classifier

4.1 Statistical Learning Method

It has been pointed out that AFC-ECG entails a self-learning stage so as to optimize its system parameters, for example, membership boundaries and threshold values. During the self-learning (pre-classification) stage, the boundary values in membership function are firstly preset in accordance with each type of heart beats. When a specific ECG signal comes, it will be classified to the appropriate group. Then, all significant values of the pre-classified heart beats are taken advantages for the modification of the boundary values in membership functions. Here statistical learning method is adopted for such self-adaptation procedure. In essence, it is based on the population distribution of beat features.

By using equation 1 and 2, the mean value (μ) and the standard deviation (σ) of a feature can be evaluated. The lower, upper boundary value of *S* and *Z* function is defined as $(\mu - 2\sigma)$ and $(\mu + 2\sigma)$ respectively. In statistics, it is pointed out that about 95.45% of the elements are within two standard deviations. Using this manner, the parameters "*a*" and "*b*" in Gaussian function are set as μ and 2σ respectively. An illustrative example in Fig 5 exhibits the variation of characteristic curve (RR_0) before and after self-adaptation. By using the proposed learning method, the adapted membership curve is able to represent both characteristic patterns of medical literature and pathophysiological condition of the specific subject.

$$\mu = \left(\sum_{i=1}^{N} x_i \right) / N \tag{1}$$

$$\sigma \sqrt{\left[\sum_{i=1}^{N} (x_i - \mu)^2 \right] / N} \tag{2}$$

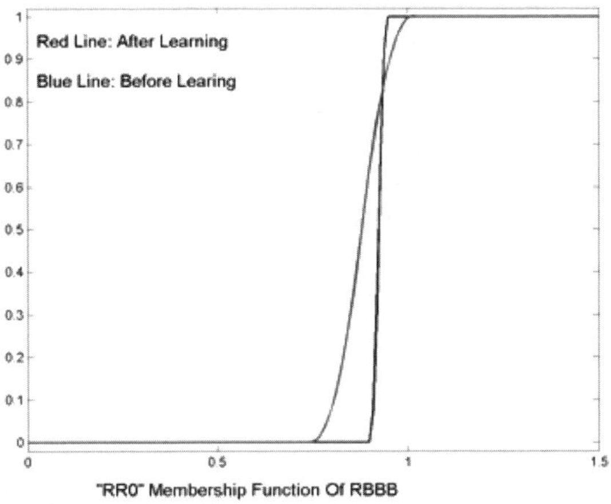

Fig. 5. An illustrative adaptation by statistical learning method

4.2 Experiment Results

For performance comparison, the data used for training and testing are based on the same ECG segments from MIT/BIH arrhythmia database. That is, each dataset comes from the 2-channel trace of ECG signals in 30 minutes. Differently, here the first 10 minutes of each record is reserved for pre-classification and self-adaptation. The following 20-minute segment is for evaluating the enhanced AFC-ECG.

Table 4. Performance evaluation of the proposed AFC-ECG

			AFC-ECG			
Record	MIT/BIH Annotation		Before Learning		After Learning	
			Result	Accuracy	Result	Accuracy
Sig 106	Normal	1507	1208	80.2%	1258	83.5%
	PAC	0	7	-	1	-
	RBBB	0	30	-	27	-
	LBBB	0	0	-	1	-
	Unclassified	0	262	-	220	-
Sig 118	Normal	0	1	-	0	-
	PAC	96	49	51.0%	72	75.0%
	RBBB	2166	1602	74.0%	1831	84.5%
	LBBB	0	0	-	0	-
	Unclassified	0	610	-	359	-
Sig 207	Normal	0	19	-	2	-
	PAC	107	127	81.3%	105	98.1%
	RBBB	86	71	82.6%	80	93.0%
	LBBB	1457	1357	93.1%	1389	95.3%
	Unclassified	0	76	-		-

The average correct rate of the proposed AFC-ECG is around 88.2%, which is a substantial improvement in comparison with the result of conventional fuzzy ECG classifier, that is, 77.0%. The worst result is 75.0% for "PAC" in the record "Sig 118". That is because there are only 20 characterisitc beats of "PAC" during the first 10-minute learning procedure. Such amount is definitely not enough to train the system effectively. Hence, similar as those neuro-fuzzy paradigms, the pattern distribution of historical data for self-adaptation is of vital importance for a successful AFC-ECG.

5 Conclusion

In this paper, an adaptive method based on statistical learning theory is proposed to enhance conventional fuzzy ECG classifiers. Experimental results approve its effectiveness in comparison with those conventional fuzzy ECG classifiers with fixed system parameters. Within the frame of embedded-link medical advisory systems, the AFC-ECG can accomplish self-adaptation dynamically in accordance with the incoming ECG signals. However, it is noteworthy that the proposed AFC-ECG still depends on the pattern distribution of former classification results.

In future research, we will try to improve the proposed AFC-ECG from the following aspects: in the first, it can deal with 4 types of heart beats at present, notwithstanding there are over 10 types of heart beats in MIT/BIH arrhythmia database; secondly, the computational efficiency of AFC-ECG should be further optimized because it is oriented to the embedded-link medical advisory systems. At last, definitely, the performance of "ECG parameterizer" is of vital importance for the following classification. It is necessary to enhance the capability of our ECG parameterizer to extract ECG characteristic parameters.

Acknowledgements. The authors would like to thank the financial support from the Research Committee of University of Macau under grants RG071/04-05S/ DMC/FST and RG074/04-05S/VMI/FST.

References

1. Adler, A. T.: A Cost-Effective Portable Telemedicine Kit for Use in Developing Countries. Master Thesis, Massachusetts Institute of Technology (2000)
2. Li, B.N., Dong, M.C., Vai, M.I.: An embedded medical advisory system for mobile cardiovascular monitoring devices. Proceedings of 2004 IEEE International Workshop on Circuit and Systems. IEEE Press, New York (2004) 1-4
3. Li, B.N., Dong, M.C., Vai, M.I.: A novel intelligent sphygmogram analyzer for health monitoring of cardiovascular system. Expert Systems with Applications 28 (2005) 693-700
4. Li, B.N., Dong, M.C., Vai, M.I.: The application of soft computing in embedded medical advisory systems for pervasive health monitoring. In Abraham, A., Baets, B.D., Köppen, M., Nickolay, B. (Eds.): Applied Soft Computing Technologies: The Challenge of Complexity, Springer Verlag, Germany (2006)
5. Silipo, R., Marchesi, C.: Artificial neural networks for automatic ECG analysis. IEEE Transactions on Signal Processing 46 (1998) 1417-1425

6. Bortolan, G., Degani, R., Willems, J.L.: Neural networks for ECG classification. Proceedings of Computers in Cardiology. IEEE Press, New York (1990) 269-272
7. Donna, L.H.: Fuzzy Logic in medical expert systems. IEEE Engineering in Medicine and Biology 13 (1994) 693-698
8. Guler, J., Ubeyli, E.D.: Application of adaptive neuro-fuzzy inference system for detection of electrocardiographic changes in patients with partial epilepsy using feature extraction. Expert Systems with Applications 27 (2004) 323-330
9. Moody, G.B., Mark, R.G., Goldberger, A.L.: PhysioNet: a web-based resource for the study of physiologic signals. IEEE Engineering in Medicine and Biology Magazine 20 (2001) 70-75
10. Chan, W.C.: Parameter Extractor of ECG Signals for The Intelligent Home Healthcare Embedded System. Master Thesis, University of Macau (2005)

A Self-organizing Fuzzy Neural Networks

Haisheng Lin[1], X.Z. Gao[2], Xianlin Huang[1], and Zhuoyue Song[1]

[1] Department of Control Science and Engineering,
Harbin Institute of Technology, Harbin, P.R. China
haishenglin318@yahoo.com.cn,
xlinhuang@hit.edu.cn,
zhuoyuesong@yahoo.com.cn

[2] Institute of Intelligent Power Electronics,
Helsinki University of Technology, Espoo, Finland
gao@cc.hut.fi

Abstract. This paper proposes a novel clustering algorithm for the structure learning of fuzzy neural networks. Our clustering algorithm uses the reward and penalty mechanism for the adaptation of the fuzzy neural networks prototypes at every training sample. Compared with the classical clustering algorithms, the new algorithm can on-line partition the input data, pointwise update the clusters, and self-organize the fuzzy neural structure. No priori knowledge of the input data distribution is needed for initialization. All rules are self-created, and they grow automatically with more incoming data. There are no conflicting rules in the created fuzzy neural networks. Our approach also shows that supervised clustering algorithms can be used for the structure learning of the self-organizing fuzzy neural networks. The identification of several typical nonlinear dynamic systems is developed to demonstrate the effectiveness of this learning algorithm.

1 Introduction

It is well known that fuzzy logic provides human reasoning capabilities to capture uncertainties, which cannot be described by precise mathematical models [1]. Neural networks offer remarkable advantages, such as adaptive learning, parallelism, fault tolerance, and generalization. They have been proven to be very powerful techniques in the discipline of system control, especially when the controlled system is difficult to be modeled accurately, or when the controlled system has large uncertainties and strong nonlinearities. Therefore, fuzzy logic and neural networks have been widely adopted in model-free adaptive control of nonlinear systems [2].

There are numerous kinds of neural fuzzy systems proposed in the literature, and most of them are suitable for only off-line cases [3]. Some on-line learning methods for the neural fuzzy systems are studied as well. In [4], Fuzzy Adaptive Resonance Theory (ART) and backpropagation method were employed to the structure and parameter learning of the Fuzzy Adaptive Learning Control Network (FALCON), respectively. The FALCON-ART can on-line partition the input /output spaces, tune membership functions, find proper fuzzy logic rules, and dynamically annihilate redundant rules with on-line training data. E. Lughofer and E. P. Klement [5] adopted

A. Saad et al. (Eds.): Soft Computing in Industrial Applications, ASC 39, pp. 200–210, 2007.
springerlink.com

the combination of vector quantization and Adaptive Resonance Theory (ART) as an on-line clustering algorithm for the structure learning. This open-loop clustering method has the feature of pointwise adaptation of latest recorded data points.

In this paper, we proposed a novel on-line clustering algorithm of structure learning for our fuzzy neural networks. This new clustering algorithm employs the mechanism of reward and penalty action used in the Learning Vector Quantization (LVQ) [6]. Our fuzzy neural networks with the on-line structure and parameter learning is a suitable candidate for real-time applications due to its fast convergence.

This paper is organized as follows. Section II introduces the principle and structure of the fuzzy neural networks. Section III describes the new clustering algorithm for the structure identification. In Section IV, parameter learning algorithm is described. In Section V, these two algorithms are used to control the inverted pendulum to demonstrate their effectiveness. Finally, Section VI concludes this paper with some remarks and conclusions.

2 Structure of Fuzzy Neural Networks

In this section, we describe the principle and structure of the fuzzy neural networks. As we can see in Fig. 1, the fuzzy neural networks have five layers with nodes. Each node consists of one input integration function $f(\cdot)$ and one output activation function $a(\cdot)$ (see Fig. 2). For Layer l, the node input is expressed as

$$node_{(in)} = f(u_1^l, u_2^l, ..., u_p^l; w_1^l, w_2^l, ..., w_p^l)\,, \tag{1}$$

where $u_1^l, u_2^l, ..., u_p^l$ are the inputs to the node, $w_1^l, w_2^l, ..., w_p^l$ are the associated weights. The output activation function $a(\cdot)$ is given as

$$node_{(out)} = a^l(node_{(in)}) = a^l(f) = u^{(l+1)}\,. \tag{2}$$

The link weight is treated as unity unless we specify it individually in each layer.

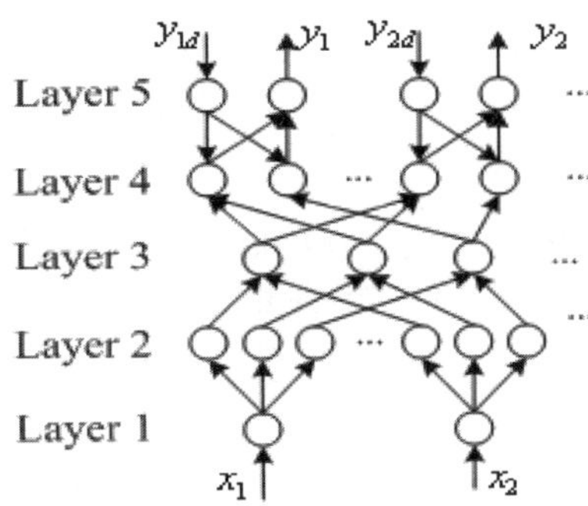

Fig. 1. Structure of fuzzy neural networks

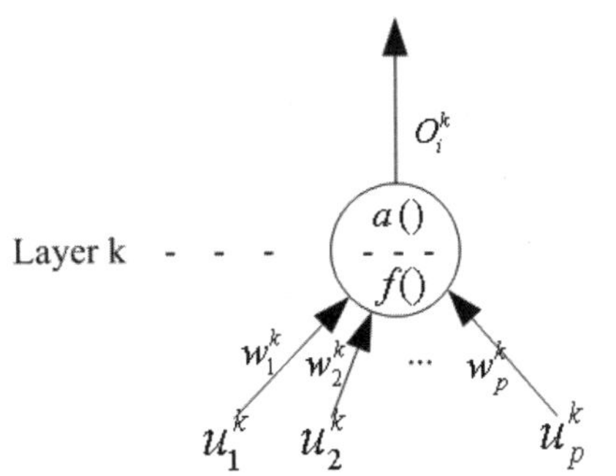

Fig. 2. Basic structure of a node in neural network

Layer 1: Each node in this layer, which corresponds to one input variable, only transmits input values to the next layer directly. Thus, the function of the i th node is defined as

$$f = u_i^1 = x_i \text{, and } a = f \text{ .} \tag{3}$$

Layer 2: Each node in this layer corresponds to one linguistic label of one of the input variables in Layer 1. In other words, the membership value, which specifies the degree to which an input value belongs to a fuzzy set is calculated in Layer 2. In our paper, a Gaussian membership is employed in this layer, and the operation performed in this layer is

$$f = -\frac{1}{2}(\frac{u_i^{(2)} - m_{ij}}{\sigma_{ij}})^2 \text{, and } a = e^f \text{ ,} \tag{4}$$

where m_{ij} and σ_{ij} , respectively, are the center and width of the Gaussian membership of the j th term of the i th input variable x_i . The link weight is interpreted as $[m_{ij}, \sigma_{ij}]$.

Layer 3: Nodes in this layer are rule nodes, and constitute the antecedents of the fuzzy rule base. Each node is equivalent to a cluster in the input space. The links of the rule nodes implement the precondition matching of fuzzy logic rules. The rule nodes perform the product operation to calculate the firing strength. Thus, the input and output functions of the j th rule node are

$$f = \prod_{i=1}^{n} u_i^{(3)} \text{, and } a = f \text{ .} \tag{5}$$

Layer 4: The nodes in this layer are called "output-term nodes"; each has two operating modes: down-up and up-down transmission modes. In the down-up transmission mode, the links in Layer 4 perform the fuzzy OR operation on activated rule nodes that have the same consequent

$$f = \sum_{j=1}^{J} u_j^4 \text{, and } a = \min(1, f) \text{ .} \tag{6}$$

In the up-down transmission mode, the nodes in this layer and links in Layer 5 function exactly the same as those in Layer 2, except that only a single node is used to act as a membership function for output linguistic variables.

Layer 5: There are two kinds of nodes in this layer. The first kind of nodes perform the up-down transmission for feeding the training data into the network. For this kind of nodes,

$$f = y_k \text{, and } a = f \text{ .} \tag{7}$$

The second kind of nodes perform the down-up transmission for the decision signal output. These nodes and Layer 5 links attached act as the defuzzifier. If $m_{ij}^{(5)}$ and $\sigma_{ij}^{(5)}$ are the centers and widths of membership functions, respectively, the following functions can perform the Center Of Area (COA) defuzzification method:

$$f = \sum w_{ij}^5 u_i^5 = \sum (m_{ij}\sigma_{ij}) u_i^5 \text{ , and } a = \frac{f}{\sum \sigma_{ij} u_i^5} . \quad (8)$$

Here, the link weight at Layer 5 is $m_{ij}\sigma_{ij}$.

Based on the above structure, an on-line learning algorithm will be proposed to determine the proper centers (m_{ij} 's) and widths (σ_{ij} 's) of term nodes in Layer 2 and Layer 4. Also, it will learn fuzzy logic rules by deciding the existence and connection types of the links at Layer 3 and Layer 4; that is, the precondition links and consequence links of the rule nodes.

3 Learning Algorithm for Structure Identification

In this section, we present a new clustering algorithm for the structure learning of the fuzzy neural networks more precisely. In our fuzzy neural networks, for every on-line incoming training pattern, we first use the novel cluster algorithm to identify the structure, and next apply the backpropagation algorithm to optimize the parameters. In our learning method, only the training data is need. The input /output-term nodes and rule nodes are created dynamically as learning proceeds upon receiving on-line incoming training data. During the learning process, novel input-term and output-term nodes and rule nodes will be added.

Before we introduce our novel clustering algorithm, we illustrate some conventional methods: Learning Vector Quantization (LVQ) and Fuzzy ARTMAP [7]. Learning Vector Quantization (LVQ) is a supervised learning technique that rewards the winners by adapting the synaptic vectors with a positive learning coefficient, and penalizes the losers with a negative learning coefficient. The fuzzy ARTMAP system incorporates two fuzzy ART modules, fuzzy ART_a and fuzzy ART_b, which are connected by an inter-ART module. This inter-ART module includes a map field that controls the learning of an associative map from fuzzy ART_a recognition categories to fuzzy ART_a recognition categories. A mismatch at the map field between the fuzzy ART_a categories activated by an input a and the fuzzy ART_b category activated by the input b increases fuzzy ART_a vigilance by the minimum amount needed for the system to search for, and if necessary, learn a novel fuzzy ART_a category, whose prediction matches fuzzy ART_b category. This inter-ART vigilance resetting signal is a form of "back propagation" of information.

Here, we adopt the reward and penalty mechanism of the LVQ algorithms and the concept of the "back propagation" of information through the inter-ART reset in fuzzy ARTMAP, to develop our novel learning algorithm for the fuzzy neural networks, The main idea of our clustering algorithm is that for every input data, we first find the winner clusters in the input and output space respectively. Next, just like the mechanism in the fuzzy ARTMAP, we check that if the winner cluster in the input space is connected to the winner cluster in the output space. If so, we assume that the winner cluster in the output space is the correct prediction of the winner cluster in the input space, which is analogous to the fact that the fuzzy ART_b category activated by the input b is the correct prediction of the fuzzy ART_a categories activated by an input a in the fuzzy ARTMAP. If not, we assume that the mismatch occurs between the winner cluster in the input space and the winner cluster in the output space and we will begin to search for another cluster in the input space which will match the winner cluster in the output space. The reward and penalty mechanism is employed in our clustering algorithm. We can describe the novel clustering algorithm particularly as follows.

Step 1: Initialize the fuzzy system with zero cluster: $In = 0$, $On = 0$.

Step 2: For the first input and output training vectors, they are selected as the centers of the first clusters in the input and output space, respectively, that is, $Im_1 = x_1$, $I\sigma_1 = 0$ and $Om_1 = y_1$, $O\sigma_1 = 0$. We connect the first cluster in the input space to the first cluster in the output space, and set the number of data belonging to the cluster as $Ic_1 = 1$, $Oc_1 = 1$.

Step 3: For an input data point $[x_i, y_i]$ in the input space, we compute the distances between the input vector and the existing input space clusters using the Euclidean metric function:

$$d_p = \sum_{i=1}^{q} \left\| x_i^l - I\_m_p^l \right\|^2, \quad 0 \le p \le n \tag{9}$$

where q is the dimension of the input training vector, x_i is the i th input training vector, n is the number of the existing clusters, and l is the l th dimension of the input training vector. The nearest cluster j (winner neuron j) is chosen by selecting the minimum d_j. Next, the following algorithms are used:

If d_j is larger than a certain value $d_{vigilance}$, we assume that this input data does not belong to any existing cluster, we form a novel cluster, set $In = In + 1$, $Im_j = x_j$ and $I\sigma_j = 0$. The newly added cluster is the winner cluster in the input space. If d_j is smaller than $d_{vigilance}$, we assume that cluster j is the winner cluster in the input space. The procedure in the input space in Step 3 is also adopted in the output space. We can also find the winner cluster in the output space.

Step 4: We check the mapping process from input clusters to the output clusters.

(1). If the winner cluster in the input space is a novel cluster, we connect this novel cluster to the winner cluster in the output space, and update the parameters of the winner cluster in the output space. That is, the centers (Om_{winner}), variances ($O\sigma_{winner}$) and the counter (Oc_{winner}) of this cluster are given by the following algorithm:

$$eph\_Om_{winner} = \frac{Om_{winner} \times Oc_{winner} + x}{Oc_{winner} + 1}, \tag{10}$$

$$O\sigma_{winner}^2 = \frac{Oc_{winner} \times (O\sigma_{winner}^2 + Om_{winner}^2) + y^2}{Oc_{winner} + 1} - eph\_Om_{winner}^2, \tag{11}$$

$$Om_{winner} = eph\_Om_{winner}, \tag{12}$$

$$Oc_{winner} = Oc_{winner} + 1, \tag{13}$$

where $eph\_O\mathrm{m}_{winner}$ is the updated center of the winner cluster in the output space.

(2). If the winner cluster of the input space is connected to the winner cluster of the output space originally, we adopt the following algorithm to update the centers

(Im_{winner}), variances ($I\sigma_{winner}$) and the counters (Ic_{winner}) of the winner cluster in the input space. For the input space, we have

$$eph\_Im_{winner} = \frac{Im_{winner} \times Ic_{winner} + x}{Ic_{winner} + 1}, \tag{14}$$

$$Im_{winner} = eph\_Im_{winner}, \tag{15}$$

$$I\sigma^2_{winner} = \frac{Ic_{winner} \times (I\sigma^2_{winner} + Im^2_{winner}) + x^2}{Ic_{winner} + 1} - eph\_Im^2_{winner}, \tag{16}$$

$$Ic_{winner} = Ic_{winner} + 1, \tag{17}$$

where eph_Im_{winner} is the updated center of the winner cluster in the output space. Similarly, for the output space, we use the same algorithm to update the parameters of the winner cluster there.

(3). If the winner cluster of the input space is not connected to the winner cluster of the output space yet, we use the following algorithm to punish the winner cluster of the input space

$$eph\_Im_{winner} = \frac{Im_{winner} \times Ic_{winner} - x}{Ic_{winner} - 1}, \tag{18}$$

$$Im_{winner} = eph\_Im_{winner}, \tag{19}$$

$$I\sigma^2_{winner} = \frac{Ic_{winner} \times (I\sigma^2_{winner} + Im^2_{winner}) - x^2}{Ic_{winner} - 1} - eph\_Im^2_{winner}, \tag{20}$$

$$Ic_{winner} = Ic_{winner}. \tag{21}$$

After that, we return to Step 3 to search for another cluster in the input space which will match the winner cluster in the output space.

The unsupervised clustering algorithms are usually employed for the on-line structure learning of the self-organizing fuzzy neural networks. Our structure learning algorithm, which is a fusion of the LVQ and fuzzy ARTMAP, is actually a supervised clustering algorithm. In this paper, we proposed that supervised clustering algorithms are also suitable candidates. Based on the "back propagation" of information through the inter-ART reset in the algorithm of fuzzy ARTMAP, supervised clustering algorithm LVQ can be used to learn the structure of the fuzzy neural networks. Adopting the supervised clustering algorithms, the fuzzy neural networks get the initial structures faster, and the training algorithms are more effective. In summary, our clustering algorithm provides a new way to use a class of supervised clustering algorithms for the on-line structure learning.

4 Parameter Learning of Fuzzy Neural Networks

After the fuzzy neural network structure has been trained according to the training patterns, we use the backpropagation algorithm to tune the parameters of the fuzzy neural networks. The goal is to minimize the error function

$$E = \frac{1}{2}[y1(t) - y(t)]^2, \tag{22}$$

where $y1(t)$ is the desired output, and $y(t)$ is the current output. For the current training data pair starting at the input nodes, a forward pass is first used to compute the activity levels of all the nodes in the network. The output of the node j in Layer 3 is computed by

$$f_j^3(t) = \prod_{i=1}^{Id} e^{-\frac{(x_i - Im_j^i(t))^2}{(I\sigma_j^i(t))^2}}, \tag{23}$$

where Id is the dimension of the input data.

If node k in Layer 4 is connected to node j in Layer 3, the output of node k in Layer 4 is computed by

$$f_k^4(t) = \sum f_j^3(t). \tag{24}$$

Starting at the output nodes, a backward pass is used to compute $\frac{\partial E}{\partial y}$ for all the hidden nodes. We simply introduce two parameters in advance to be utilized, $f_1(t)$ and $f_2(t)$:

$$f1(t) = \sum_{k=1}^{Od} O\sigma_k(t) \times f_k^4(t), \tag{25}$$

$$f2(t) = \sum_{k=1}^{Od} Om_k(t) \times O\sigma_k(t) \times f_k^4(t), \tag{26}$$

where Od is the dimension of the output data.

The centers and variances of the cluster in Layer 5 are updated by

$$Om_k(t+1) = Om_k(t) + \eta \times [y1(t) - y(t)] \times \frac{O\sigma_k(t) \times f_k^4(t)}{f1(t)}, \tag{27}$$

$$O\sigma_k(t+1) = O\sigma_k(t) + \eta \times [y1(t) - y(t)] \times \frac{Om_k(t) \times f_k^4(t) \times f1(t) - f2(t) \times f_k^4(t)}{f1^2(t)}. \tag{28}$$

The error to be propagated to the Layer 4 is

$$error_k^4(t+1) = [y1(t) - y(t)] \times \frac{Om_k(t) \times f_k^4(t) \times f1(t) - f2(t) \times f_k^4(t)}{f1^2(t)}. \tag{29}$$

If node k in Layer 4 is connected to node j in Layer 3, the error to be propagated to node j in Layer 3 is

$$error_j^3(t+1) = error_k^4(t+1) = [y1(t) - y(t)] \times \frac{Om_k(t) \times f_k^4(t) \times f1(t) - f2(t) \times f_k^4(t)}{f1^2(t)}. \tag{30}$$

The centers and variances of the cluster in Layer 2 are updated by

$$Im_j^i(t+1) = Im_j^i(t) + \eta \times error_j^3(t) \times f_j^3(t) \times 2 \times \frac{x_i - Im_j^i(t)}{(I\sigma_j^i(t))^2}, \tag{31}$$

$$I\sigma_j^i(t+1) = I\sigma_j^i(t) + \eta \times error_j^3(t) \times f_j^3(t) \times 2 \times \frac{(x_i - Im_j^i(t))^2}{(I\sigma_j^i(t))^3}. \tag{32}$$

The structure and parameter learning algorithms of the fuzzy neural networks will be examined using simulations in the next section.

5 Simulations

In this section, we use three examples to demonstrate the effectiveness of the above proposed learning algorithms. The first example is to identify a Single-Input-Single-Output (SISO) dynamic system [8], the second example is to identify a Multiple-Input-Single-Output (MISO) dynamic system, and the last one is to identify a Multiple-Input-Multiple-Output (MIMO) dynamic system [8].

Example 1 — Identification of an SISO dynamic system: The plant to be identified is described by the following difference equation [8]:

$$y(t+1)=\frac{y(t)}{1+y^2(t)}+u^3(t). \tag{33}$$

The training input patterns are generated with $u(t)=\sin(\frac{2\pi t}{100})$, and there are 10000 training iterations. Figure 3 shows the Root-Mean-Square (RMS) errors during learning. Each point on the curve is the average RMS of 100 training time steps. Figure 4 shows the outputs of the plant and the identification model. In this figure, the outputs of the fuzzy neural networks are presented as dotted curve, while the plant output values are represented as solid curve. Figure 5 shows the distribution of the output membership functions in $y(t+1)$ domain (output space). Figure 6 illustrates the distribution of input membership functions in $[u(t), y(t)]$ plane (input space). There are five hyperellipsoids (IR1, IR2, IR3, IR4, IR5) formed in the input space. Four Gaussian membership functions (OR1, OR2, OR3, OR4) are generated in the output space. Proper mapping between the input clusters and output clusters is constructed. There are five rules formed as follows:

Rule 1: IF x is IR1, THEN y is OR1

Rule 2: IF x is IR2, THEN y is OR2

Rule 3: IF x is IR3, THEN y is OR1

Rule 4: IF x is IR4, THEN y is OR3

Rule 5: IF x is IR5, THEN y is OR4

The effectiveness of our adaptive learning algorithm and powerful approximation capability of the fuzzy neural networks are demonstrated in this example.

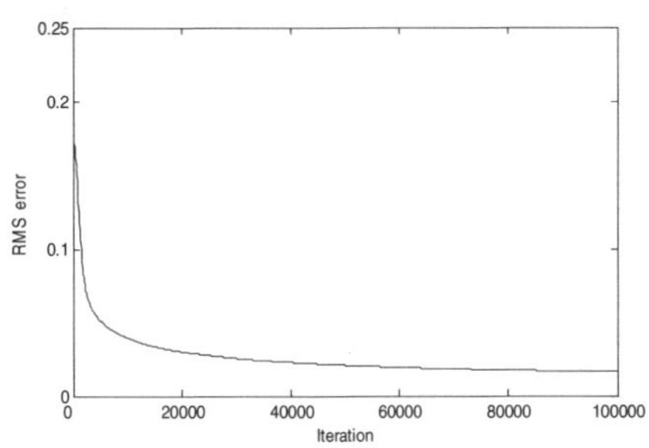

Fig. 3. Root-mean-square errors during learning

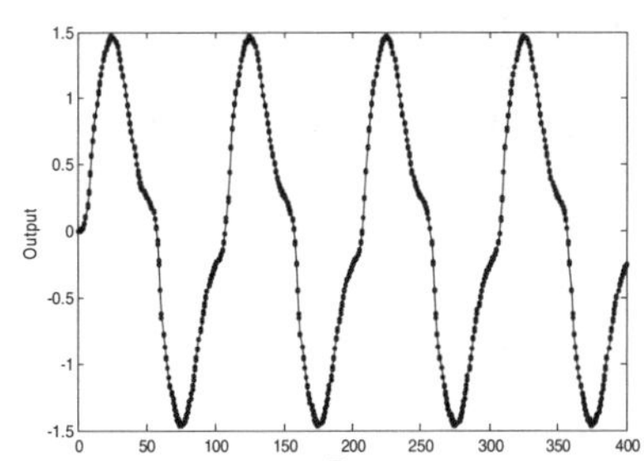

Fig. 4. Outputs of SISO system and identification model

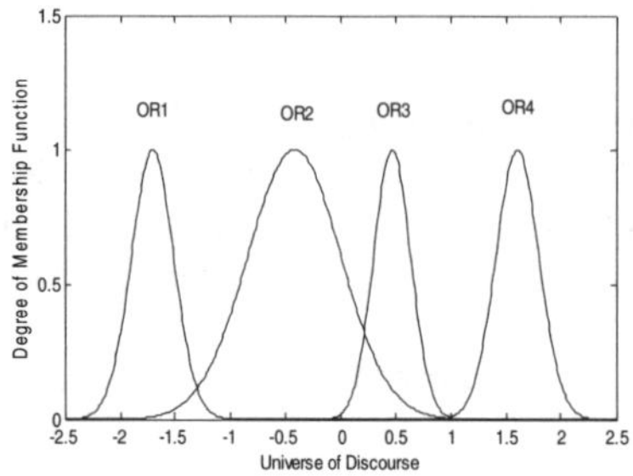

Fig. 5. Distribution of output membership functions

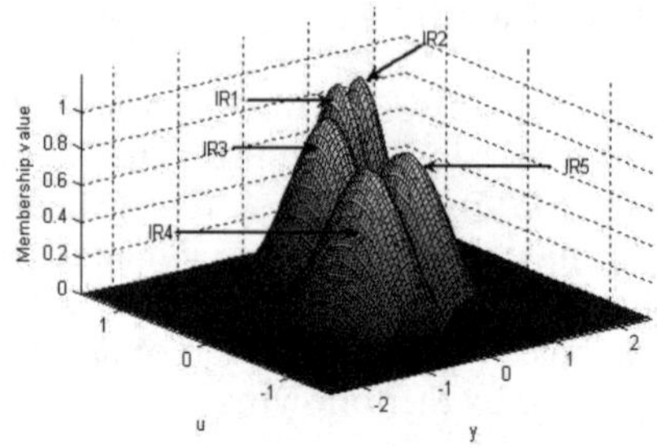

Fig. 6. Distribution of input membership functions

Example 2 — Identification of an MISO dynamic system: the plant to be identified is a two-input and one-output dynamic system described by the following equations:

$$y = (x_1^2 - x_2)^2 + (1 - x_1)^2 . \tag{34}$$

In the training procedure, the training data is randomly distributed in the interval $[-1,+1]$. In the verification procedure, two inputs $x_1 = \sin(2\pi t / 25)$, and $x_2 = \cos(2\pi t / 25)$ are given. The responses of the plant and identification model are shown as solid line and dotted line in Fig. 7 respectively.

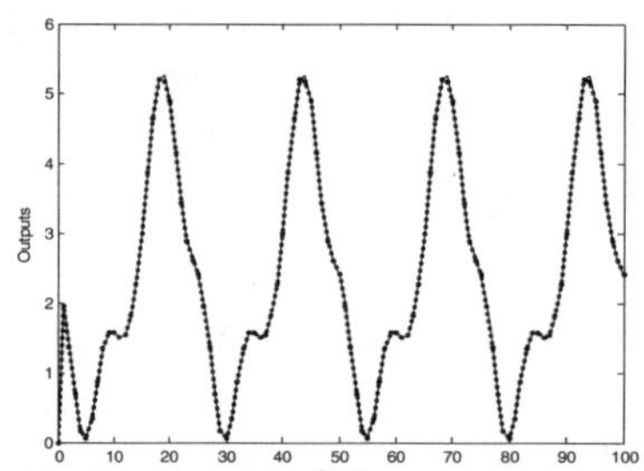

Fig. 7. Outputs of MISO system and identification model

Figure 7 shows that identification error of the fuzzy neural networks for this MISO system is very small.

Example 3 — Identification of the MIMO dynamic system: the plant is described by the equations (35):

$$\begin{bmatrix} y_1(t+1) \\ y_2(t+1) \end{bmatrix} = \begin{bmatrix} \dfrac{y_1(t)}{1+y_2^2(t)} \\ \dfrac{y_1(t)y_2(t)}{1+y_2^2(t)} \end{bmatrix} + \begin{bmatrix} u_1(t) \\ u_2(t) \end{bmatrix} . \tag{35}$$

The identification procedure is based on random inputs $u_1(t)$ and $u_2(t)$ uniformly distributed in the interval $[-1,+1]$. In the verification procedure, for an input

$[\sin(2\pi t/25), \cos(2\pi t/25)]^T$, y_1 responses of the plant and the identification model are presented as dotted curve and solid curve in Fig. 8 respectively, while y_2 responses are shown in Fig. 9. It is demonstrated that our fuzzy neural networks can successfully identify MIMO dynamic systems as well.

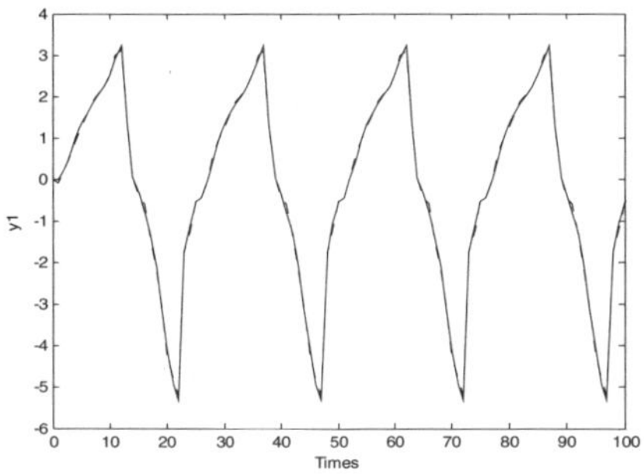

Fig. 8. y_1 response of MIMO system and identification model

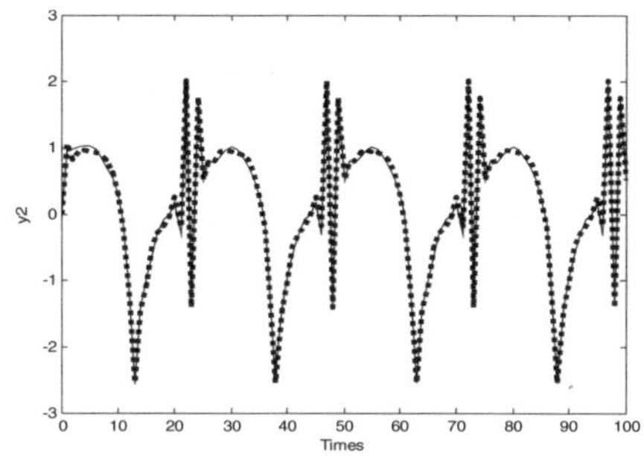

Fig. 9. y_2 response of MIMO system and identification model

6 Conclusions

A new novel clustering algorithm is proposed for the structure learning of fuzzy neural networks. This clustering algorithm can on-line partition the input data and self-organize the fuzzy neural structure. Therefore, no priori knowledge of the distribution of input data is needed for initialization of fuzzy rules. They are automatically generated with the incoming training data. Our fuzzy neural networks can use this on-line training algorithm for the structure and parameter training. The effectiveness of our learning algorithm is verified by the identification of dynamic nonlinear systems.

Acknowledgement

X. Z. Gao's research work was funded by the Academy of Finland under Grant 214144.

References

1. S. Shao: Fuzzy self-organizing controller and its application for dynamic processes. Fuzzy Sets Syst., vol. 26, pp. 151-164, 1998.
2. G. Feng: A novel stable tracking control scheme for robotic manipulators. IEEE Trans. Syst., Man, Cybern., vol. 27, pp. 510-516, June 1997.
3. C. T. Lin and C. S. G. Lee: Neural-network-based fuzzy logic control and decision system. IEEE Trans. Comput., vol. 40, pp. 1320-1336, Dec. 1991.
4. C. J. Lin and C. T. Lin: An ART-based fuzzy adaptive learning control network. IEEE Trans. Fuzzy Syst., vol. 5, pp. 477-496, Aug. 1997.

5. E. Lughofer and E. P. Klement: Premise parameter estimation and adaptation in fuzzy systems with open-loop clustering methods. Proc. of IEEE Int. Conf. Fuzzy Syst., pp. 499-504, July 2004.
6. T. Kohonen: The self-organizing map. Proc. of the IEEE, vol. 78, no. 9, Sep. 1990.
7. G. A. Carpenter, S. Grossberg, N. Markuzon, J. H. Reynolds, and D. B. Rosen: Fuzzy ARTMAP: A neural network architecture for incremental supervised learning of analog multidimensional maps. IEEE Trans. Neural Networks, vol. 3, no. 5, Sep. 1992.
8. K. S. Narendra and K. Pathasarathy: Identification and control of dynamical systems using neural networks. IEEE Trans. Neural Networks, vol. 1, pp. 4-26, Jan. 1990.

Part V

Soft Computing for Modeling, Optimization and Information Processing

A Particle Swarm Approach to Quadratic Assignment Problems

Hongbo Liu[1,3], Ajith Abraham[1,2], and Jianying Zhang[3]

[1] School of Computer Science, Dalian Maritime University, Dalian, 116024, China
[2] School of Computer Science and Engineering, Yonsei University, 134 Shinchon-dong, Sudaemoon-ku, Seoul 120-749, Korea
ajith.abraham@ieee.org
[3] Department of Computer Science, Dalian University of Technology, Dalian, 116023, China
{lhb,zhangjy}@dlut.edu.cn

Summary. Particle Swarm Optimization (PSO) algorithm has exhibited good performance across a wide range of application problems. But research on the Quadratic Assignment Problem (QAP) has not much been investigated. In this paper, we introduce a novel approach based on PSO for QAPs. The representations of the position and velocity of the particles in the conventional PSO is extended from the real vectors to fuzzy matrices. A new mapping is proposed between the particles in the swarm and the problem space in an efficient way. We evaluate the performance of the proposed approach with Ant Colony Optimization (ACO) algorithm. Empirical results illustrate that the approach can be applied for solving quadratic assignment problems and it has outperforms ACO in the completion time.

1 Introduction

Particle Swarm Optimization (PSO) algorithm is inspired by social behavior patterns of organisms that live and interact within large groups. In particular, PSO incorporates swarming behaviors observed in flocks of birds, schools of fish, or swarms of bees, and even human social behavior, from which the Swarm Intelligence(SI) paradigm has emerged [1, 2]. It could be implemented and applied easily to solve various function optimization problems, or the problems that can be transformed to function optimization problems. As an algorithm, the main strength of PSO is its fast convergence, which compares favorably with many global optimization algorithms [3, 4, 5]. PSO has exhibited good performance across a wide range of applications [6, 7, 8]. However, research on discrete problems, especially Quadratic Assignment Problem (QAP), has been done little [9, 10]. In this paper, we design a fuzzy scheme based on discrete particle swarm optimization [11, 12] to solve quadratic assignment problems.

A. Saad et al. (Eds.): Soft Computing in Industrial Applications, ASC 39, pp. 213–222, 2007.
springerlink.com

2 Quadratic Assignment Problem

The quadratic assignment problem (QAP) is a standard problem in location theory. It was introduced by Koopmans and Beckmann in 1957 [13] and is a model for many practical problems [14]. Intuitively, the QAP can be described as the problem of assigning a set of facilities to a set of locations with given distances between the locations and given flows between the facilities. The goal then is to place the facilities on locations in such a way that the sum of the product between flows and distances is minimal. More formally, given n facilities $\{F_1, F_2, \cdots, F_n\}$ and n locations $\{L_1, L_2, \cdots, L_n\}$, two $n \times n$ matrices $FM = [f_{ij}]$ and $DM = [d_{rs}]$, where f_{ij} is the flow between facilities F_i and F_j and d_{rs} is the distance between locations L_r and L_s, the QAP can be stated as follows:

$$\min_{\Pi \in P(n)} Z_\Pi = \sum_{i=1}^{n}\sum_{j=1}^{n} f_{ij} d_{\Pi_i \Pi_j} \tag{1}$$

where $P(n)$ is the set of all permutations (corresponding to the assignment solutions) of the set of integers $\{1, 2, \cdots, n\}$, and Π_i gives the location of facility F_i in the current solution $\Pi \in P(n)$. Here $f_{ij} d_{\Pi_i \Pi_j}$ describes the cost contribution of simultaneously assigning facility F_i to location Π_i and facility F_j to location Π_j. It is to be noted that the number of facilities (n) is assumed to be the same as the number of locations. In the other word, one facility could be assigned to only one location, and one location could be assigned to only one facility in a feasible assignment solution.

The term *quadratic* stems from the formulation of the QAP as an integer optimization problem with a quadratic objective function [14]. Let b_{ij} be a binary variable which takes value 1 if facility F_i is assigned to location L_j and 0 otherwise. Then the problem can be re-formulated as:

$$\min \sum_{i=1}^{n}\sum_{j=1}^{n}\sum_{r=1}^{n}\sum_{s=1}^{n} f_{ij} d_{rs} b_{ir} b_{js} \tag{2}$$

s.t.

$$b_{ij} \in \{0, 1\}, \quad i = 1, 2, \cdots, n, \quad j = 1, 2, \cdots, n; \tag{3}$$

$$\sum_{i=1}^{n} b_{ij} = 1, \quad i = 1, 2, \cdots, n, \quad j = 1, 2, \cdots, n; \tag{4}$$

$$\sum_{j=1}^{n} b_{ij} = 1, \quad i = 1, 2, \cdots, n, \quad j = 1, 2, \cdots, n. \tag{5}$$

The QAP is a NP-hard optimization problem [15]. While some NP-hard combinatorial optimization problems can be solved exactly for relatively large instances, as exemplified by the traveling salesman problem (TSP), QAP instances of size larger than 20 are considered intractable. The QAP is considered as one of the hardest optimization problems, because exact algorithms show a very poor performance on it [16]. Therefore, several heuristics have been proposed for finding near-optimum solutions for large QAP instances, including ant colonies optimization [17, 18, 19].

3 Particle Swarm Model

The classical PSO model consists of a swarm of particles, which are initialized with a population of random candidate solutions. They move iteratively through the d-dimension problem space to search the new solutions, where the fitness, f, can be calculated as the certain qualities measure. Each particle has a position represented by a position-vector $\mathbf{x}_i$ (i is the index of the particle), and a velocity represented by a velocity-vector $\mathbf{v}_i$. Each particle remembers its own best position so far in a vector $\mathbf{x}_i^{\#}$, and its j-th dimensional value is $x_{ij}^{\#}$. The best position-vector among the swarm so far is then stored in a vector $\mathbf{x}^*$, and its j-th dimensional value is x_j^*. During the iteration time t, the update of the velocity from the previous velocity to the new velocity is determined by Eq.(6). The new position is then determined by the sum of the previous position and the new velocity by Eq.(7).

$$v_{ij}(t) = wv_{ij}(t-1)+c_1r_1(x_{ij}^{\#}(t-1)-x_{ij}(t-1))+c_2r_2(x_j^*(t-1)-x_{ij}(t-1)) \quad (6)$$

$$x_{ij}(t) = x_{ij}(t-1) + v_{ij}(t) \quad (7)$$

Where r_1 and r_2 are the random numbers in the interval [0,1]. c_1 is a positive constant, called as coefficient of the self-recognition component, c_2 is a positive constant, called as coefficient of the social component. The variable w is called as the inertia factor, which value is typically setup to vary linearly from 1 to near 0 during the iterated processing. From Eq.(6), a particle decides where to move next, considering its own experience, which is the memory of its best past position, and the experience of its most successful particle in the swarm.

In the PSO model, the particle searches the solutions in the problem space within a range $[-s, s]$ (If the range is not symmetrical, it can be translated to the corresponding symmetrical range.) In order to guide the particles effectively in the search space, the maximum moving distance during one iteration is clamped in between the maximum velocity $[-v_{max}, v_{max}]$ given in Eq.(8), and similarly for its moving range given in Eq.(9):

$$v_{i,j} = sign(v_{i,j})min(|v_{i,j}|, v_{max}) \quad (8)$$

$$x_{i,j} = sign(x_{i,j})min(|x_{i,j}|, x_{max}) \quad (9)$$

The value of v_{max} is $\rho \times s$, with $0.1 \le \rho \le 1.0$ and is usually chosen to be s, i.e. $\rho = 1$. The pseudo-code for particle swarm optimization algorithm is illustrated in Algorithm 1.

Algorithm 1. Particle Swarm Optimization Algorithm

Initialize the size of the particle swarm n, and other parameters.
Initialize the positions and the velocities for all the particles randomly.
While (the end criterion is not met) do
  $t = t + 1$;
  Calculate the fitness value of each particle;
  $\mathbf{x}^* = argmin_{i=1}^{n}(f(\mathbf{x}^*(t-1)), f(\mathbf{x}_1(t)), f(\mathbf{x}_2(t)), \cdots, f(\mathbf{x}_i(t)), \cdots, f(\mathbf{x}_n(t)))$;
  For i= 1 to n
    $\mathbf{x}_i^{\#}(t) = argmin_{i=1}^{n}(f(\mathbf{x}_i^{\#}(t-1)), f(\mathbf{x}_i(t))$;
    For $j = 1$ to d
      Update the j-th dimension value of $\mathbf{x}_i$ and $\mathbf{v}_i$
        according to Eqs.(6),(8),(7),(9);
    Next j
  Next i
End While.

4 A Fuzzy Particle Swarm Approach for QAP

For applying particle swarm algorithm successfully for an objective problem, one of the key issues is how to map the problem solution to the particle space, which directly affects its feasibility and performance. In a "crip" particle swarm model for the assignment problem, it would trend to assign many facilities to the same location or assign many locations to the same facility. This kind of the assignment would be unfeasible. In this section, a fuzzy matrix is introduced to represent the quadratic assignment problem. And then, a new approach to the problem space mapping is depicted for particle swarm optimization for the quadratic assignment problem.

Suppose $F = \{F_1, F_2, \cdots, F_n\}$, $L = \{L_1, L_2, \cdots, L_n\}$, then the fuzzy assignment relation from F to L can be expressed as follows:

$$A = \begin{bmatrix} a_{11} & a_{12} & \cdots & a_{1n} \\ a_{21} & a_{22} & \cdots & a_{2n} \\ \vdots & \vdots & \ddots & \vdots \\ a_{n1} & a_{n2} & \cdots & a_{nn} \end{bmatrix}$$

Here a_{ij} represents the degree of membership of the j-th element F_j in domain F and the i-th element L_i in domain L to relation A. In the fuzzy relation matrix A between F and L, the elements subject to the following constraints:

$$a_{ij} = \mu_R(F_j, L_i), \quad i = 1, 2, \cdots, n, \quad j = 1, 2, \cdots, n. \tag{10}$$

μ_R is the membership function, the value of a_{ij} means the degree of membership that the facility F_j would be assigned to the location L_i in the feasible assignment solution. In the quadratic assignment problem, the elements of the solution must satisfy the following conditions:

$$a_{ij} \in \{0,1\}, \quad i = 1,2,\cdots,n, \quad j = 1,2,\cdots,n; \tag{11}$$

$$\sum_{i=1}^{n} a_{ij} = 1, \quad i = 1,2,\cdots,n, \quad j = 1,2,\cdots,n; \tag{12}$$

$$\sum_{j=1}^{n} a_{ij} = 1, \quad i = 1,2,\cdots,n, \quad j = 1,2,\cdots,n. \tag{13}$$

For applying PSO successfully, one of the key issues is how to map the problem solution to the particle space, which directly affects its feasibility and performance [11]. According to fuzzy matrix representation of the quadratic assignment problem, the position X and velocity V in the particle swarm are re-defined as follows:

$$X = \begin{bmatrix} x_{11} & x_{12} & \cdots & x_{1n} \\ x_{21} & x_{22} & \cdots & x_{2n} \\ \vdots & \vdots & \ddots & \vdots \\ x_{n1} & x_{n2} & \cdots & x_{nn} \end{bmatrix}; \qquad V = \begin{bmatrix} v_{11} & v_{12} & \cdots & v_{1n} \\ v_{21} & v_{22} & \cdots & v_{2n} \\ \vdots & \vdots & \ddots & \vdots \\ v_{n1} & v_{n2} & \cdots & v_{nn} \end{bmatrix}$$

The elements in the matrix X above have the same meaning as Eq.(10). Accordingly, the elements of the matrix X must satisfy the following conditions:

$$x_{ij} \in \{0,1\}, \quad i = 1,2,\cdots,n, \quad j = 1,2,\cdots,n; \tag{14}$$

$$\sum_{i=1}^{n} x_{ij} = 1, \quad i = 1,2,\cdots,n, \quad j = 1,2,\cdots,n; \tag{15}$$

$$\sum_{j=1}^{n} x_{ij} = 1, \quad i = 1,2,\cdots,n, \quad j = 1,2,\cdots,n. \tag{16}$$

Because the position and velocity in the new fuzzy particle swarm model have been transformed to the form of matrices, they are updated by the new Eqs.(17) and (18) with the matrix operations.

$$\begin{aligned} V(t) = w \otimes V(t-1) \oplus (c_1 r_1) \otimes (X^{\#}(t-1) \ominus X(t-1)) \\ \oplus (c_2 r_2) \otimes (X^{*}(t-1) \ominus X(t-1)) \end{aligned} \tag{17}$$

$$X(t+1) = X(t-1) \oplus V(t) \tag{18}$$

The position matrix may violate the constraints (14), (15) and (16) after some iterations, it is necessary to normalize the position matrix. First we make all the negative elements in the matrix to become zero. If all elements in a column of the matrix are zero, they need be re-evaluated using a series of random numbers within the interval [0,1]. And then the matrix undergoes the following transformation:

$$Xnormal = \begin{bmatrix} x_{11}/\sum_{i=1}^{n} x_{i1} & x_{12}/\sum_{i=1}^{n} x_{i2} & \cdots & x_{1n}/\sum_{i=1}^{n} x_{in} \\ x_{21}/\sum_{i=1}^{n} x_{i1} & x_{22}/\sum_{i=1}^{n} x_{i2} & \cdots & x_{2n}/\sum_{i=1}^{n} x_{in} \\ \vdots & \vdots & \ddots & \vdots \\ x_{n1}/\sum_{i=1}^{n} x_{i1} & x_{n2}/\sum_{i=1}^{n} x_{i2} & \cdots & x_{nn}/\sum_{i=1}^{n} x_{in} \end{bmatrix}$$

Since the position matrix indicates the potential assigned solution, the fuzzy matrix can be "decoded" to the feasible solution. We choose the element which has the max value in the column, then tag it as "1", and other numbers in the column and row are set as "0" in the assigning matrix. After all the columns and rows have been processed, we get the assignment solution without violating the constraints (14), (15) and (16), and then calculate the assignment cost of the solution.

5 Experiment Settings, Results and Discussions

In our experiments, Ant Colony Optimization (ACO) was used to compare the performance with PSO. The two algorithms share many similarities. ACO deals with artificial systems that is inspired from the foraging behavior of real ants, which are used to solve discrete optimization problems [21]. The main idea is the indirect communication between the ants by means of chemical pheromone trials, which enables them to find short paths between their nest and food. It is implemented as a team of intelligent agents which simulate the ants behavior, walking around the graph representing the problem to solve using mechanisms of cooperation and adaptation. PSO is a stochastic search technique inspired by social behavior of bird flocking or fish schooling. Both methods are valid and efficient methods in numeric programming and have been employed in various fields due to their strong convergence properties. Specific parameter settings for the algorithms are described in Table 1. We consider the instances from Taillard's datasets[1] and QAPlib[2]. Each experiment (for each algorithm) was repeated 10 times with different random seeds. Each trial had a fixed number of $50 * n * n$ iterations (n is the dimension of the problem). If the value $50 * n * n$ is larger than $2 * 10^4$, the maximum iteration was set to $2 * 10^4$. The average costs (AvgCost) and the standard deviations (std) were calculated from the 10 different trials. The standard deviation indicates the differences in the results during the 10 different trials. Usually the main emphasis will be to generate the assignment solutions at a minimal amount of time. So the completion time for 10 trials were used as one of the criteria to improve their performance.

In order to closely track the performance of our algorithms, first we tested two small scale problems, nug5 and nug8. The nug5 is a simple QAP instance with 5 facilities on 5 locations. Its united matrix DF of the distance and flow is

[1] http://ina2.eivd.ch/collaborateurs/etd/

[2] http://www.opt.math.tu-graz.ac.at/qaplib/

Table 1. Parameter settings for the algorithms

Algorithm	Parameter name	Parameter value
ACO	Number of ants	5
	Weight of pheromone trail α	1
	Weight of heuristic information β	5
	Pheromone evaporation parameter ρ	0.8
	Constant for pheromone updating Q	10
PSO	Swarm size	5
	Self-recognition coefficient c_1	1.49
	Social coefficient c_2	1.49
	Inertia weight w	$0.9 \rightarrow 0.1$

$$\begin{array}{c|ccccc} & 1 & 2 & 3 & 4 & 5 \\ 1 & 0 & 5 & 2 & 4 & 1 \\ 2 & 1 & 0 & 3 & 0 & 2 \\ 3 & 1 & 2 & 0 & 0 & 0 \\ 4 & 2 & 1 & 1 & 0 & 5 \\ 5 & 3 & 2 & 2 & 1 & 0 \end{array}$$

Upper half of the DF matrix is the distance information, and the lower half is the flow information. The two algorithms both search the best value 50 in its 10 runs. The results for 10 ACO runs were all 50, while the results of 10 PSO runs were 50 nine times and 52 once. The optimal result is supposed to be 50 with the best permutation, (4,5,1,2,3). Figure 1 illustrates the performance curves during the search processes. ACO usually searches a better result using a less iteration number than PSO for the smaller scale problem.

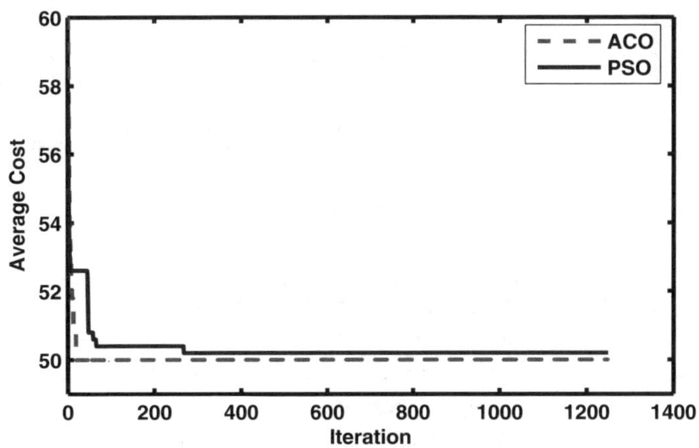

Fig. 1. Performance for nug5

The nug8 is other QAP instance with 8 facilities on 8 locations. The scale is a little larger than nug5. The united matrix DF of the distance and flow is

$$
\begin{array}{c}
\\ 1 \\ 2 \\ 3 \\ 4 \\ 5 \\ 6 \\ 7 \\ 8
\end{array}
\begin{array}{c}
\begin{array}{cccccccc} 1 & 2 & 3 & 4 & 5 & 6 & 7 & 8 \end{array} \\
\begin{pmatrix}
0 & 5 & 2 & 4 & 1 & 0 & 0 & 6 \\
1 & 0 & 3 & 0 & 2 & 2 & 2 & 0 \\
2 & 1 & 0 & 0 & 0 & 0 & 0 & 5 \\
3 & 2 & 1 & 0 & 5 & 2 & 2 & 10 \\
1 & 2 & 3 & 4 & 0 & 10 & 0 & 0 \\
2 & 1 & 2 & 3 & 1 & 0 & 5 & 1 \\
3 & 2 & 1 & 2 & 2 & 1 & 0 & 10 \\
4 & 3 & 2 & 1 & 3 & 2 & 1 & 0
\end{pmatrix}
\end{array}
$$

Figure 2 illustrates the performance of the two algorithms for nug8. The results for 10 ACO runs were {218,224,224,214,224,224,228,224,224,224}, with an average value of 222.8. The results of 10 PSO runs were {214,222,218,214,220,218, 218,218,222,224}, with an average value of 218.8. The optimal result is supposed to be 214 with the best permutation, (2,1,4,5,3,8,7,6). While ACO provided the best result once, PSO provided the best result twice.

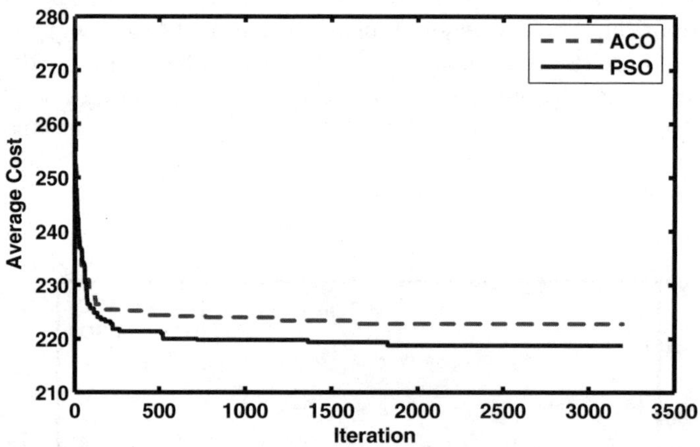

Fig. 2. Performance for nug8

Further, we tested the algorithms for other instances. The average cost (Avg-Cost), the standard deviations (std) and the time for 10 trials were recoded. Empirical results are summarized in Table 2. As the results depict, the ACO is an effective algorithm for the little scale problems, while PSO usually had better averages for a little bigger problem sizes. PSO also had larger standard deviations. The robustness of our algorithm is one of the future works in this study. It is to be noted that PSO usually spent the less time to assign the facilities on the locations.

Table 2. Comparing the results of ACO and PSO for quadratic assignment problems

Problem	ACO			PSO		
	AvgCost	std	time	AvgCost	std	time
nug5	50.0	0	258.4700	50.2	0.6325	103.8130
nug8a	222.8	3.9101	631.1193	218.8	3.2931	322.8750
tai8a	85934	800.4784	872.5578	83294	2698.1	572.0150
chr12a	16557	1661.6	1048.0	13715	2098.0	736.0545
tai12a	256180	3066.5	968.1560	254230	5809.9	653.1410
chr20a	5438.8	261.3909	1142.7	4456.0	389.8974	754.5
dre30	1849.6	82.1998	1514.4	1592.0	118.4736	1040.5
tho40	302840	3603.3	1612.3	286670	5318.3	1233.3
tai50a	5626356	15225	2045.3	5587622	52893	1602.4

6 Conclusions

In this paper, we introduced an approach based on Particle Swarm Optimization (PSO) for quadratic assignment problems. The representations of the position and velocity of the particles in PSO is extended from the real vectors to fuzzy matrices, through which we accomplished the mapping between the quadratic assignment problem and the particle. We evaluated the performance of our proposed approach and compared it with Ant Colony Optimization (ACO). Empirical results illustrated that the proposed approach was an effective approach to solve quadratic assignment problems and it outperformed ACO in the completion time.

Acknowledgements

The first author wishes to thank to Kyriakos Tsourapas, who provided the ACO source code for the compared experiments. This work is supported by NSFC (60373095) and MOE (KP0302).

References

1. Kennedy J, Eberhart R (2001) Swarm Intelligence. Morgan Kaufmann, CA
2. Clerc M (2006) Particle Swarm Optimization. ISTE Publishing Company, London
3. Eberhart R C, Shi Y (1998) Comparison between genetic algorithms and particle swarm optimization. In: Proceedings of IEEE International Conference on Evolutionary Computation, 611–616
4. Settles M, Rodebaugh B, Soule T (2003) Comparison of genetic algorithm and particle swarm optimizer when evolving a recurrent neural network, Cantu-Paz E, et al. (eds.): GECCO 2003, LNCS 2723, 148–149
5. Boeringer D W, Werner D H (2004) Particle swarm optimization versus genetic algorithms for phased array synthesis. IEEE Transactions on Antennas and Propagation, 52(3):771–779

6. Parsopoulos K E, Vrahatis M N (2002) Recent Approaches to Global Optimization Problems through Particle Swarm Optimization, Natural Computing 1:235–306
7. Schute J F, Groenwold A A (2005) A study of global optimization using particle swarms. Journal of Global Optimization, Kluwer Academic Publishers, 31:93–108
8. Ajith Abraham, He Guo and Hongbo Liu, Swarm Intelligence: Foundations, Perspectives and Applications, Swarm Intelligent Systems, Nadia Nedjah and Luiza Mourelle (Eds.), Studies in Computational Intelligence, Springer Verlag, Germany, pp. 3–25, 2006.
9. Kennedy J, Eberhart R C (1997) A discrete binary version of the particle swarm algorithm. in: Proceedings of International Conference on Systems, Man, and Cybernetics, IEEE Computer Society Press, 4104–4108
10. Shi X, Xing X, Wang Q, at el. (2004) A discrete PSO method for generalized TSP problem. in: Proceedings of International Conference on Machine Learning and Cybernetics, IEEE Computer Society Press, 2378–2383
11. Pang W, Wang K, Zhou C, Dong L (2004) Fuzzy discrete particle swarm optimization for solving traveling salesman problem. In: Proceedings of the Fourth International Conference on Computer and Information Technology, IEEE CS Press, 796–800
12. Ajith Abraham, Hongbo Liu, Weishi Zhang and Tae-Gu Chang, Job Scheduling on Computational Grids Using Fuzzy Particle Swarm Algorithm, 10th International Conference on Knowledge-Based & Intelligent Information & Engineering Systems, Springer Verlag, Germany, Lecture Notes in Artificial Intelligence, B. Gabrys, R.J. Howlett, and L.C. Jain (Eds.): Part II, Lecture Notes on Artificial Intelligence 4252, pp. 500-507, 2006.
13. Koopmans T C, Beckman M J (1957) Assignment problems and the location of economic activities. Econometrica, 25:53–76
14. Stützle T (2005) Iterated local search for the quadratic assignment problem. European Journal of Operational Research, In Press, Corrected Proof, ScienceDirect Available online 13 May 2005
15. Garey M R, Johnson D S (1979) Computers and Intractability: a Guide to the Theory of NP-Completeness. Freeman, CA
16. Angel E, Zissimopoulos V (2002) On the hardness of the quadratic assignment problem with metaheuristics. Journal of Heuristics, 8:399–414
17. Misevicivs A (2005) A tabu search algorithm for the quadratic assignment problem. Computational Optimization and Applications, 30:95–111
18. Maniezzo V, Colorni A (1999) The ant system applied to the quadratic assignment problem. IEEE Transactions on Knowlege and Data Engineering, 11(5):769–778
19. Drezner Z, Hann P M, Taillard E D (2005) Recent advances for the quadratic assignment problem with special emphasis on instances that are difficult for metaheuristic methods. Annals of Operations Research, 139:65–94
20. Gambardella L M, Taillard E D, Dorigo M (1999) Ant colonies for the quadratic assignment problem. Journal of the Operational Research Society, 50:167–176
21. Dorigo M, Stützle T (2004) Ant Colony Optimization, MIT Press, MA

Population-Based Incremental Learning for Multiobjective Optimisation

Sujin Bureerat and Krit Sriworamas

Department of Mechanical Engineering, Faculty of Engineering, Khon Kaen University, Thailand, 40002
sujbur@kku.ac.th

Abstract. The work in this paper presents the use of population-based incremental learning (PBIL), one of the classic single-objective population-based optimisation methods, as a tool for multiobjective optimisation. The PBIL method with two different updating schemes of its probability vectors is presented. The performance of the two proposed multiobjective optimisers are measured and compared with four other established multiobjective evolutionary algorithms i.e. niched Pareto genetic algorithm, version 2 of non-dominated sorting genetic algorithm, version 2 of strength Pareto evolutionary algorithm, and Pareto archived evolution strategy. The optimisation methods are implemented to solve 8 bi-objective test problems where design variables are encoded as a binary string. The Pareto optimal solutions obtained from the various methods are compared and discussed. It can be concluded that, with the assigned test problems, the multiobjective PBIL methods are comparable to the previously developed algorithms in terms of convergence rate. The clear advantage in using PBILs is that they can provide considerably better population diversity.

Keywords: Multiobjective Evolutionary Optimisation, Population-Based Incremental Learning, Non-dominated Solutions, Pareto Archive, Performance Comparison.

1 Introduction

In the past, the well-established optimisation methods were mostly developed for single-objective optimisation. They can, however, be employed to solve a multiobjective problem by using numerical strategies such as the weighted-sum technique [14], global criteria method [9] and normalised normal constraint technique [15]. The principle of such numerical strategies is to convert an optimisation problem with multiple design objectives to be a single-objective problem to suit the available optimisers. This means that the optimisation methods need to be performed for as many simulation runs as the number of Pareto optimal points required.

Optimisation methods can be classified as methods with and without using function derivatives. In comparison of EAs, the methods without using derivatives, to the derivative-based methods, the latter are far superior in terms of convergence rate and consistency. Nevertheless, using EAs are advantageous in that they can deal with all kinds of design variables [2], and most importantly the multiobjective evolutionary methods can search for a set of Pareto optimum points within one attempt whereas the

A. Saad et al. (Eds.): Soft Computing in Industrial Applications, ASC 39, pp. 223–232, 2007.
springerlink.com

gradient-based techniques need many operations. Although having such benefits, the evolutionary methods are still not as powerful as expected due to a complete lack of consistency and low convergence rate. Since there is no convergence guaranteed, the results obtained from using an EA are mostly classified as an approximated Pareto front.

In recent years, a number of evolutionary algorithms have been developed as multiobjective optimisers and they are termed multiobjective evolutionary algorithms (MOEAs). In the early days, the well-known methods were, for example, vector evaluation genetic algorithm (VEGA) [16], multiobjective genetic algorithm (MOGA) [7] and none-dominated sorting genetic algorithm (NSGA) [18]. Later, there have been numerous new algorithms developed. Some work on comparing their performance has been made e.g. in [20] and [22]. The development of new approaches, improvement of the existing algorithms, and implementation of the methods on real world applications are still a great challenge.

The work in this paper is aimed at developing population-based incremental learning (PBIL), one of the classical single-objective EAs, as a tool for multiobjective optimisation. The proposed algorithm is said to be a mix of some advantages of the predecessors. Two PBIL algorithms with different probability vector updating procedures are presented: one is modified from that presented in [12] while the other one deals with the weighted-sum technique. The two proposed optimisers along with four recently developed MOEAs including NPGA [4] [10] and [11], NSGAII [6], SPEA2 [21] and PAES [13] are implemented to solve 8 bi-objective test problems. Design variables are encoded as a series of binary strings. The performances of the optimisers are measured, compared and discussed. It can be concluded that, with the given design conditions, the PBILs are said to be as good as some of the existing MOEAs in terms of convergence rate, and superior to them when considering population diversity.

2 Multiobjective Optimisation

A particular multiobjective design problem can be posed as:

Find **x** such that

$$\begin{aligned}
&\text{Min: } \mathbf{f} = \{f_1(\mathbf{x}),\ldots,f_m(\mathbf{x})\} \\
&\text{Subject to} \\
&g_i(\mathbf{x}) \leq 0 \\
&h_i(\mathbf{x}) = 0
\end{aligned}$$

MOEAs are normally created to deal with unconstrained optimisation; however, they can be applied to constrained problems by using a penalty function technique. Moreover, the non-dominated scheme for constrained optimisation given in reference [5] is found to be greatly efficient and effective. All of the evolutionary methods mentioned in this paper are categorised as Pareto-based methods. The basic concept of exploring Pareto optimum points via such an algorithm is that, on each generation while a new population is created, non-dominated solutions are classified and carried on to the next generation. The term, non-dominated solutions, defines the local Pareto solutions among the members of the current population during evolutionary search.

3 PBIL for Multiobjective Design

The original PBIL algorithm is based upon binary searching space similar to GAs [1]. Later, there has been further development of PBIL for continuous and discrete design spaces [8], [17] and [19]. The method is classified as an estimation of distribution algorithm (EDA) that achieves its search through probability estimation and sampling techniques. The application of PBIL to multiobjective design has been presented in [12] where the approach is named MOSA. The principle of PBIL search can be thought of as iteratively limiting the design space depending on current best design variables and random process. The design space is iteratively narrowed until approaching the optimum. Rather than keeping all binary genes or a population as with GA, the population in PBIL is represented by the probability vector of having '1' at each bit position in the binary strings. For more details, see [1].

Initially, for single objective optimisation, the search procedure starts with the initial probability vector whose elements are full of '0.5'. An initial population corresponding to the probability vector is created. The binary population is decoded and objective values are computed. The best gene, whether it is a minimum or maximum, is chosen to update the next probability vector P_i^{new} using the relation

$$P_i^{new} = P_i^{old}(1-LR) + b_i LR \tag{1}$$

where $LR \in (0,1)$ is called the learning rate to be defined and b_i is the ith bit of the best binary solution. It is also useful to apply mutation to the probability vector at some predefined probability such that

$$P_i^{new} = P_i^{old}(1-ms) + \text{rand}(0 \text{ or } 1).ms \tag{2}$$

where ms is the amount of shift used in the mutation. The best solutions are carried over into the next generation ensuring that the best solution during the search is not lost. The probability vector is updated iteratively until convergence is achieved.

When employed as a multiobjective optimiser, more probability vectors should be used in order to obtain a more diverse population; therefore, it is called a probability matrix instead. The search starts with an (empty) external Pareto set and initial probability matrix whose elements are full of '0.5'. Each row of the probability matrix is a probability vector that will be used to create a sub-population. Let N be the number of design solutions in a population, l be the number of probability vectors and n_b be the number of binary bits. The probability matrix, therefore, has the size of $l \times n_b$ where each row of the matrix results in approximately N/l design solutions as one sub-population. Having generated the population and evaluated their corresponding objective values, the non-dominated members sorted from the union set of the current population and the old external Pareto set are taken as a new external Pareto set. If the external Pareto set is full (the number of non-dominated members exceeds the archive size), some solutions are removed from the external Pareto set using the adaptive grid algorithm [13]. The probability matrix and the non-dominated solutions are improved iteratively until some termination criterion is met.

In this paper, two updating schemes for the probability matrix are proposed. The first scheme is quite similar to that presented in [12]. To update a row vector of the probability matrix, $n_0 < N$ binary solutions from the current Pareto archive are selected at random. The mean value of each ith bit position of the selected binary solutions is computed and used as b_i in equation (1). The mutation (2) is also performed with the given probability.

The second scheme uses the weighted-sum technique. In updating each row of the probability matrix, m weighting factors are generated randomly while the condition $\sum w_i = 1$ is preserved. A binary solution from the union set of the current population and the external Pareto set, which gives the minimum value of the weighted-sum function (3), is chosen to update the row probability vector. The mutation is also allowed to occur by the predefined probability.

$$f_w(\mathbf{b}) = \sum_{i=1}^{m} w_i f_i \tag{3}$$

Selection is performed if the number of non-dominated solutions obtained exceeds the predefined size of the external Pareto set. By using the adaptive grid algorithm, one of the members in the most crowded region is removed from the archive. The crowded regions are updated and the member in the most crowded region is removed iteratively until the number of non-dominated solutions is equal to the size of the archive.

4 Performance Testing

Numerical experiments were made to measure the performances of the proposed PBIL. Six unconstrained bi-objective problems $\mathbf{F}_1$, $\mathbf{F}_2$, $\mathbf{F}_3$, $\mathbf{F}_4$, $\mathbf{F}_5$ and $\mathbf{F}_6$ taken from reference [20] are used to benchmark the presented approaches and the other MOEAs. The optimisation problems are said to cover all aspects of difficulty in approximating a Pareto front via MOEAs [20]. The number of design variables is 30. Two more unconstrained bi-objective test problems, apart from the six test problems previously mentioned, are: $\mathbf{F}_7$ the FON problem in [6], and $\mathbf{F}_8$ the SPH-m problem in [21]. For $\mathbf{F}_8$, the bounds are set to be $x_i \in [-5, 5]$.

The multiobjective evolutionary algorithms i.e. NPGA, NSGAII, SPEA2 and PAES along with PBIL1 and PBIL2 are implemented to solve the bi-objective minimisation problems. The two PBIL strategies are PBIL1 the multiobjective PBIL using the first probability matrix updating scheme, and PBIL2 the multiobjective PBIL using the second probability matrix updating scheme. The algorithms can be categorised as using the crossover-based method (NPGA, NSGA and SPEA), the mutation-based method (PAES) and the estimation of distribution method (PBIL). All of the methods employ a population size of 100 and an iteration number of 100 for every design problem except for the $\mathbf{F}_7$ test-case where the population is sized 30 and the number of iterations is 30. The methods that perform elitism have the archive size of 100. Each design variable is encoded with 30 binary strings unless otherwise specified. Design conditions set for each optimiser are detailed as followed.

NPGA the number of randomly selected individuals for tournament selection is 30, the next generation consists of 50 (15 for $\mathbf{F}_7$) non-dominated solutions and 50 (15 for $\mathbf{F}_7$) members from tournament selection, crossover probability is 1.0 and mutation probability is 0.1.

NSGAII crossover probability is 1.0 and mutation probability is 0.1.
SPEA2 crossover probability is 1.0 and mutation probability is 0.1.
PAES uses (1+1)-PAES and adaptive grid archiving technique.

PBIL1 uses the first probability matrix updating scheme similar to [12], learning rate LR = 0.5 (constant), the number of probability vectors l = 20, mutation shift ms = 0.2 and mutation probability is 0.02.

PBIL2 uses the second probability matrix updating scheme as in equation (3), learning rate LR = 0.5 (constant), the number of probability vectors l = 20, mutation shift ms = 0.2 and mutation probability is 0.02.

Each method is employed to solve each problem over 30 runs while on each operation the non-dominated solutions of the final iteration are taken as the optimal front. The performance assessment is somewhat the same as presented in [20]. Two performance parameters are as follows.

The first indicator is the C value defined as:

$$C(A,B) := \frac{\left|\{\mathbf{b} \in B ; \exists\, \mathbf{a} \in A : \mathbf{a} \preceq \mathbf{b}\}\right|}{|B|} \tag{4}$$

From the definition, if $C(A, B)$ = 1, all the solutions in B are dominated by or equal to solutions in A whereas $C(A,B)$ = 0 implies that none of the solutions in A cover B.

The second criterion is the combination of M_1, M_2 and M_3 whose definitions were given in [20] as

$$M = \frac{M_1}{M_2 + M_3} \tag{5}$$

where M_1 is used to measure the average distance of a set of non-dominated solutions to the true optimal front (lower is better), M_2 is used to measure the front distribution (higher is better), and M_3 is used to measure the extent of the front (higher is better).

Form the relation, a lower value of M means a better Pareto front. Note that the parameter M is proposed as an attempt to have an additional prospective view in evaluating a Pareto front.

The value of the C indicator is used to compare a pair of evolutionary methods. Therefore, there are 6×5 comparisons for a test problem. When solving a test problem, there are 30 M values for each method. The mean value of M is used to interpret the convergence performance.

5 Comparison Results

Fig. 1 (a) displays plots of approximate Pareto fronts of $\mathbf{F}_1$ obtained from the various optimisers. Note that the approximate Pareto front from each method is the best of 30 runs sorted by using C values. Dashed lines are the true optimal Pareto front of $\mathbf{F}_1$. From the figure, the Pareto fronts obtained from PBIL1 and PBIL2 are more evenly distributed than the rest. The Pareto fronts of $\mathbf{F}_2$ from the various methods are shown in Fig. 1 (b). For this design case, the fronts from PBIL1 and PBIL2 are more evenly distributed than those from the other methods. Fig. 2 (a) shows the Pareto fronts of $\mathbf{F}_3$ from the various optimisers. The true optimal front for this problem is non-contiguous. It is shown that PBIL1 and PBIL2 can explore all the Pareto sub-fronts whereas the fronts obtained from other optimisers cannot cover all sub-fronts.

The Pareto fronts of $\mathbf{F}_4$, multimodal problem, are displayed in Fig. 2 (b). For this test problem, SPEA2 totally outperforms the rest while the second best is PBIL1 and the third best is PBIL2. The fronts from PAES, PBIL1 and PBIL2 are the most evenly distributed. Fig. 3 (a) illustrates the Pareto fronts of the deceptive problem explored by the various optimisers. From the figure, it can be said that using SPEA2 and NSGAII results in better fonts. The approximate Pareto fronts from SPEA2, NSGAII and PAES are the most evenly distributed. Fig. 3 (b) displays the Pareto fronts of $\mathbf{F}_6$ test problem explored by the six algorithms. In this figure, the best Pareto front is that obtained from PBIL1 while the second best is from SPEA2. Fig. 4 (a) displays the Pareto front of $\mathbf{F}_7$ obtained from using the various algorithms. All of the optimisation methods give poor front distribution with PBIL1 being the best. This is caused by the number of iterations and population size being used. The non-dominated fronts of $\mathbf{F}_8$ from using the various optimisers are given in Fig. 4 (b). It can be seen that PBIL2 gives the best front considering the front distribution and the distance from the true optimal front.

The comparison made in the previous two paragraphs is based on observation. The boxplots of C values comparing 6×5 pairs of optimisers are shown in Fig. 5 and 6. From the results, NPGA is less efficient than the others. According to the boxplots, for the first three test problems, all of the methods with the exception of NPGA are said to be equally good. In the $\mathbf{F}_4$ test problem, PBIL1 gives the best results whereas NSGAII and SPEA2 give the best results in the $\mathbf{F}_5$ test problem. NSGAII, SPEA2, PAES and PBIL1 are said to be equally good for the $\mathbf{F}_6$ test problem. SPEA2 is the best method for the case of $\mathbf{F}_7$ while PBIL1 is the close second best. For the $\mathbf{F}_8$ problem, PAES the mutation-based method give the best results whereas PBIL2 is the close second best.

The bar-chart illustration of the M criterion is shown in Fig. 7. A bar graph represents the average of 30 M values obtained from using a particular optimiser for solving a test problem. Based upon this measure, in cases of $\mathbf{F}_1$, $\mathbf{F}_2$ and $\mathbf{F}_3$ the methods that give the best Pareto sets are PBIL1 and PBIL2 whereas PAES is the third best and NPGA has the worst results. In the $\mathbf{F}_4$ case, the best method is PBIL1 but from Fig. 2 the obviously best method is SPEA2. This shows that the population diversity has more weight in the M value. The best results of the problem $\mathbf{F}_5$ are from SPEA2 and NSGAII while the best results of $\mathbf{F}_6$ are from SPEA2, NSGAII, PBIL1 and PBIL2. PBIL2 gives the best M value in the case of $\mathbf{F}_7$ whereas PAES, PBIL1 and PBIL2 are equally good and considered the best for the $\mathbf{F}_8$ problem.

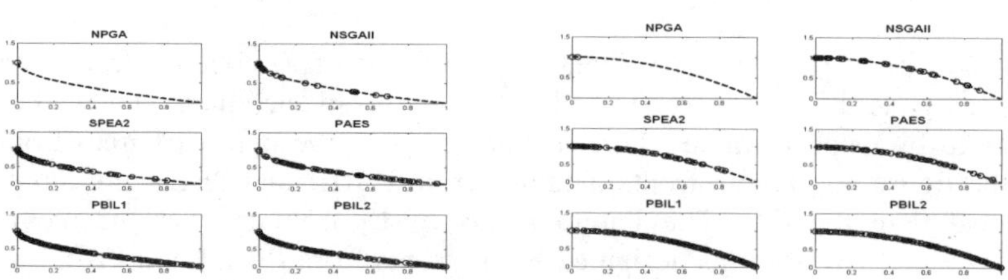

Fig. 1. Pareto front of (a) $\mathbf{F}_1$ and (b) $\mathbf{F}_2$ from the various methods

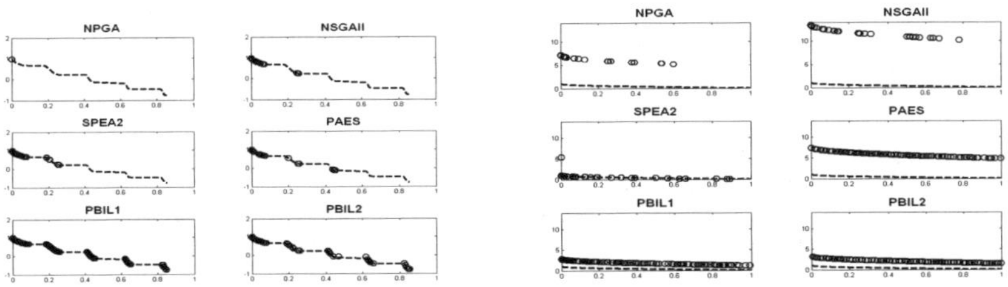

Fig. 2. Pareto front of (a) $\mathbf{F}_3$ and (b) $\mathbf{F}_4$ from the various methods

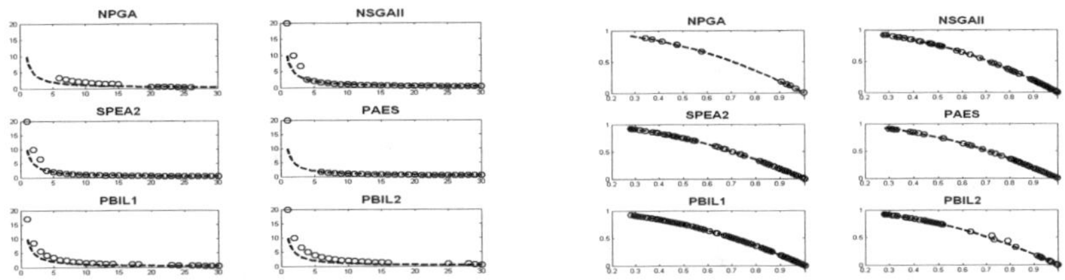

Fig. 3. Pareto front of (a) $\mathbf{F}_5$ and (b) $\mathbf{F}_6$ from the various methods

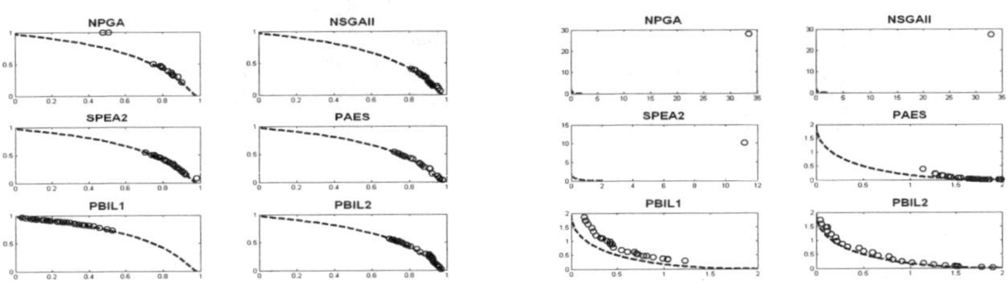

Fig. 4. Pareto front of (a) $\mathbf{F}_7$ and (b) $\mathbf{F}_8$ from the various methods

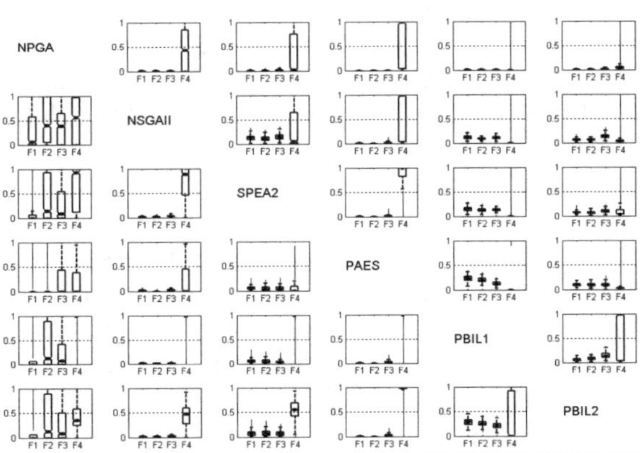

Fig. 5. Boxplot of C values $\mathbf{F}_1$ $\mathbf{F}_2$ $\mathbf{F}_3$ & $\mathbf{F}_4$

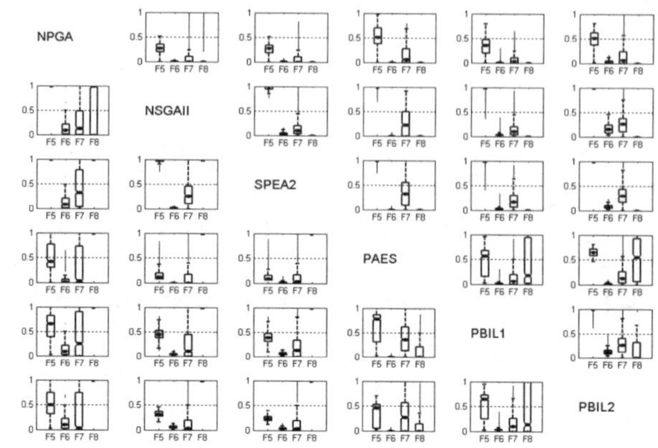

Fig. 6. Boxplot of C values $\mathbf{F}_5$ $\mathbf{F}_6$ $\mathbf{F}_7$ & $\mathbf{F}_8$

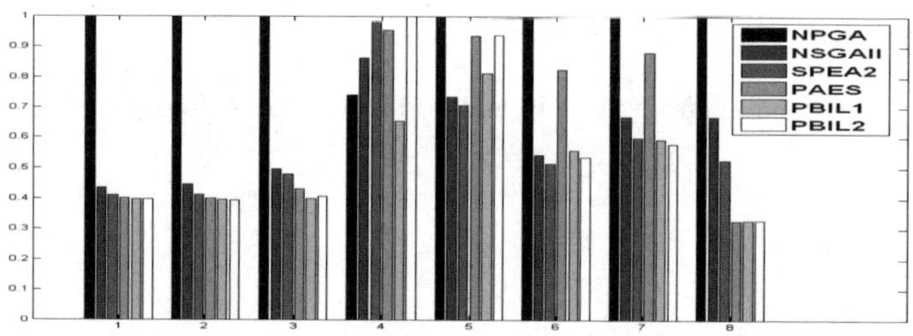

Fig. 7. Comparison of M values

6 Conclusions and Discussion

According to the numerical experiment results and several comparative criteria, it can be said that PBIL is one of the most powerful tools for multiobjective optimisation. The method is overall superior to or as good as the other established multiobjective optimisers in most of the test-cases except for the $\mathbf{F}_4$ and $\mathbf{F}_5$ test problems. It can be concluded that PBIL1 is as good as PBIL2 or vice versa although PBIL1 is better than PBIL2 in terms of front distribution. The most outstanding capability of PBIL is its unmatched ability in providing population diversity.

The conclusions drawn in this paper cannot, nonetheless, be applied to all kinds of optimisation problems as the performance of evolutionary search is rather dependent on the type of design problem. For examples, a crossover-based method is the best for global optimisation whereas a mutation-based method is far superior to others when dealing with some large-scale problem e.g. in references [14] and [3].

The multiobjective PBIL is said to be the simplest form of the estimation of distribution algorithm for multiobjective design. The effect of parameters e.g. the number of probability vectors n_b on the searching performance needs to be investigated. It has not yet been compared to the more advanced EDAs like the Bayesian network. Furthermore, the test of the presented technique in solving real world problems needs to be proven before being accepted as well-established.

Acknowledgement. The corresponding author is grateful of the support from the Thailand Research Fund (TRF), SIRDC and The Faculty of Engineering, Khon Kaen University.

References

1. Baluja S (1994) Population-based incremental learning: a method for integrating genetic search based function optimization and competitive learning. Technical Report CMU_CS_95_163, Carnegie Mellon University
2. Bureerat S, Cooper JE (1998) Evolutionary methods for the optimisation of engineering systems. In: IEE Colloquium Optimisation in Control: Methods and Applications, IEE, London, Uk, pp 1/1-1/10
3. Bureerat S, Limtragool J (2006) Performance enhancement of evolutionary search for structural topology optimization, Finite Element in Analysis and Design, 42:547-566
4. Coello CC, Romero CEM (2002) Evolutionary algorithms and multiple objective optimization. multicriteria optimization. Ehrgott, Matthias; Gandibleux, Xavier (Eds.), pp 277-331
5. Deb K, Pratap A, Meyarivan T: Constrained test problems for multi-objective evolutionary optimization. KanGAL Report No. 200002, Kanpur Genetic Algorithms Laboraotry (KanGAL), Indian Institute of Technology, Kanpur, India.
6. Deb K, Pratap A, Agarwal S, Meyarivan T (2002) A fast and elitist multiobjective genetic algorithm: NSGAII. IEEE Trans. On Evolutionary Computation 6(2):182-197
7. Fonseca CM, Fleming PJ (1993) Genetic algorithms for multiobjective optimization: formulation, discussion and generalization. In: Proc. of the 5th Inter. Conf. on Gas, pp 416-423
8. Fyfe C (1999) Structured population-based incremental learning. Soft Computing 2(4): 91 – 198
9. Grandhi RV, Bharatram, G (1993) Multiobjective optimization of large-scale structures. AIAA 31(7):1329-1337
10. Horn J, Nafpliotis N (1993) Multiobjective optimization using niched Pareto genetic algorithm. Tech. Report I11iGA1 Report 93005, UIUC
11. Horn J, Nafpliotis N, Goldberg DE (1994) A niched pareto genetic algorithm for multiobjective optimization. In: The 1st IEEE Conf. on Evolutionary Computation, pp 82-87
12. Ivvan S, Pena V, Rionda SB, Aguirre AH (2005) Multiobjective shape optimization using estimation distribution algorithms and correlated information. In: EMO2005, pp 664-676
13. Knowles JD, Corne DW (2000) Approximating the non-dominated front using the Pareto archive evolution strategy. Evolutionary Computation 8(2):149-172.
14. Kunakote T (2004) Topology optimization using evolutionary algorithms: comparison of the evolutionary methods and checkerboard suppression technique. Master thesis, Khon Kaen University
15. Messac A, Ismail-Yahaya A, Mattson CA (2003) The normalized normal constraint method for generating the Pareto frontier. Structural and Multidisciplinary Optimization 25(2):86-98

16. Schaffer JD (1985) Multiobjective optimization with vector evaluated genetic algorithms. In: GAs and their Application: Proc. of 1st Inter Conf. on Gas, pp 93-100
17. Sebag M, Ducoulombier A (1998) Extending population-based incremental learning to continuous search spaces. Lecture Notes in Computer Science 1498: 418
18. Srinivas N, Deb K (1994) Multiobjective optimization using non-dominated genetic algorithms. Evolutionary Computation 2(3):221-248
19. Yuan B, Gallagher M (2003) Playing in continuous spaces: some analysis and extension of population-based incremental learning. In: Proceedings of the 2003 Congress on Evolutionary Computation, IEEE, Canberra, Australia, pp 443-450
20. Zitzler E, Deb K, Thiele L (2000) Comparison of multiobjective evolutionary algorithms: empirical results. Evolutionary Computation 8(2):173-195
21. Zitzler E, Laumanns M, Thiele L (2002) SPEA2: improving the strength Pareto evolutionary algorithm for multiobjective optimization. In: Evolutionary Methods for Design, Optimization and Control, Barcelona, Spain
22. Zitzler E, Thiele L, Laumanns M, Fonseca CM, Fonseca VG (2003) Performance assessment of multiobjective optimizers: an analysis and review. IEEE Trans. on Evolutionary Computation 7(2):117-132

Combining of Differential Evolution and Implicit Filtering Algorithm Applied to Electromagnetic Design Optimization

Leandro dos Santos Coelho[1] and Viviana Cocco Mariani[2]

[1] Production and Systems Engineering Graduate Program, PPGEPS
[2] Mechanical Engineering Graduate Program, PPGEM
Pontifical Catholic University of Parana, PUCPR
Imaculada Conceição, 1155, Zip code 80215-901, Curitiba, Parana, Brazil

Abstract. Differential evolution (DE) is a population-based and stochastic search algorithm of evolutionary computation that offers three major advantages: it finds the global minimum regardless of the initial parameter values, it involves fast convergence, and it uses few control parameters. This work presents a global optimization algorithm based on DE approaches combined with local search using the implicit filtering algorithm. The implicit filtering algorithm is a projected quasi-Newton method that uses finite difference gradients. The difference increment is reduced as the optimization progresses, thereby avoiding some local minima, discontinuities, or nonsmooth regions that would trap a conventional gradient-based method. Problems involving optimization procedures of complex mathematical functions are widespread in electromagnetics. Many problems in this area can be described by nonlinear relationships, which introduce the possibility of multiple local minima. In this paper, the shape design of Loney's solenoid benchmark problem is carried out by DE approaches. The results of DE approaches are also investigated and their performance compared with those reported in the literature.

Keywords: evolutionary computation, electromagnetic optimization, differential evolution.

1 Introduction

The literature about electromagnetics contains several optimization algorithms for solving Loney's solenoid design [1]-[3]. With the development of artificial intelligence, some alternative methods for the nonlinear, nonconvex, and discontinuous optimization problems in electromagnetics are revealed. Recent attempts to overcome the limitations of the traditional mathematical programming approaches include the application of evolutionary algorithms [4], taboo search [5], and particle swarm optimization [6].

In this context, Evolutionary Algorithms (EAs) are a class of stochastic search and optimization methods. These algorithms, based on the principles of natural biological evolution, have received considerable and increasing interest over the past decade. EAs operate on a population of potential solutions, applying the principle of survival of the fittest to produce successively better approximations to a solution.

A. Saad et al. (Eds.): Soft Computing in Industrial Applications, ASC 39, pp. 233–240, 2007.
springerlink.com

EAs encompass a range of different 'nature-inspired' methods, including genetic algorithms, evolution strategies, evolutionary programming, genetic programming, and their variants.

DE is one of the best EAs, and has proven to be a promising candidate to solve real valued optimization problems. DE is a population-based algorithm introduced by Storn and Price [7], [8] is a very simple but very powerful stochastic global optimizer. The crucial idea behind DE is a scheme for generating trial parameter vectors. DE combines simple arithmetic operators with the classical operators of crossover, mutation and selection to evolve from a randomly generated starting population to a final solution.

This paper contributes by presenting hybrid approaches of DE combined with implicit filtering algorithm for the optimization of Loney's solenoid design. The Loney's solenoid design problem is an ill conditioned objective function with a very narrow valley that makes it very difficult to handle.

2 Differential Evolution

The fundamental idea behind DE is a scheme whereby it generates the trial parameter vectors. In each time step, DE mutates vectors by adding weighted random vector differentials to them. If the cost of the trial vector is better than that of the target, the target vector is replaced by trial vector in the next generation.

A number of alternative versions of basic configurations of DE have been proposed by [9], [10]. However, there is a variant of DE that has been reported, DE/*best*/1/*exp*. The different variants are classified according to the following notation: DE/α/β/δ, where α indicates the method for selecting the parent chromosome that will form the base of the mutated vector, β indicates the number of difference vectors used to perturb the base chromosome, and δ indicates the recombination mechanism used to create the offspring population. The *bin* acronym indicates that the recombination is controlled by a series of independent binomial experiments. The *exp* indicates a series of independent exponential experiments. The variant implemented here was the DE/*best*/1/*exp*, which is given by the following steps:

A. Initialization of the parameter setup

The user must choose the key parameters that control the DE, i.e., population size, boundary constraints of optimization variables, mutation factor (f_m), recombination rate (CR), and the stop criterion (G_{max}).

B. Initialization of individual populations

Set generation t=0. Initialize a population of i=1,..,M individuals (real-valued n-dimensional solution vectors) with random values generated according to a uniform probability distribution in the n dimensional problem space. Initialize the entire solution vector population in the given upper and lower limits of the search space.

C. Evaluation of individual populations

Evaluate the fitness value of each individual (in this work, the objective of the fitness function is to minimize the cost function).

D. Mutation operation (or differential operation)

Mutation is an operation that adds a vector differential to a population vector of individuals, according to equation (1):

$$z_i(t+1) = x_{best}(t) + f_m(t)[x_{i,r_2}(t) - x_{i,r_3}(t)] \tag{1}$$

where i=1,2,...,M is the individual's population index; j=1,2,...,n is the position in n of the dimensional individual; t is the time (generation); $x_i(t) = \left[x_{i_1}(t), x_{i_2}(t), ..., x_{i_n}(t)\right]^{\mathrm{T}}$ stands for the position of the i-th individual of a population of N real-valued n-dimensional vectors; $z_i(t) = \left[z_{i_1}(t), z_{i_2}(t), ..., z_{i_n}(t)\right]^{\mathrm{T}}$ stands for the position of the i-th individual of a *mutant vector*; $x_{best}(t)$ is the best individual of the population in generation t; r_2 and r_3 are mutually different integers and are also different from the running index, i, randomly selected with uniform distribution from the set $\{1, 2, \cdots, i-1, i+1, \cdots, N\}$; $f_m(t) > 0$ is a real parameter called a *mutation factor*, which controls the amplification of the difference between two individuals so as to avoid search stagnation and is usually a constant value taken from the range [0.1, 1].

The mutation operation using the difference between two selected randomly individuals may cause the mutant individual to escape from the search domain. If an optimized variable for the mutant individual is outside of the domain search, then this variable is replaced by its lower bound or its upper bound so that each individual should be restricted within the search domain.

E. Recombination operation

Following the mutation operation, recombination is applied to the population. Recombination is employed to generate a trial vector by replacing certain parameters of the target vector by the corresponding parameters of a randomly generated donor vector.

For each vector, $z_i(t+1)$, an index $rnbr(i) \in \{1, 2, \cdots, n\}$ is randomly chosen using uniform distribution, and a *trial vector*, $u_i(t+1) = \left[u_{i_1}(t+1), u_{i_2}(t+1), ..., u_{i_n}(t+1)\right]^{\mathrm{T}}$, is generated with

$$u_{i_j}(t+1) = \begin{cases} z_{i_j}(t+1), & \text{if } (randb(j) \le CR) \text{ or } (j = rnbr(i)), \\ x_{i_j}(t), & \text{if } (randb(j) > CR) \text{ or } (j \ne rnbr(i)) \end{cases} \tag{2}$$

where $randb(j)$ is the j-th evaluation of a uniform random number generation with [0, 1]; CR is a *crossover* or *recombination rate* in the range [0, 1]; and F is the evaluation of the cost function. Usually, the performance of a DE algorithm depends on three variables: the population size N, the mutation factor $f_m(t)$ and the CR.

F. Selection operation

Selection is the procedure whereby better offspring are produced. To decide whether or not the vector $u_i(t+1)$ should be a member of the population comprising the next generation, it is compared with the corresponding vector $x_i(t)$. Thus, if f denotes the objective function under minimization, then

$$x_i(t+1) = \begin{cases} u_i(t+1), & \text{if} \quad f(u(t+1)) < f(x_i(t)), \\ x_i(t), & \text{otherwise} \end{cases} \tag{3}$$

In this case, the cost of each trial vector $u_i(t+1)$ is compared with that of its parent target vector $x_i(t)$. If the cost, f, of the target vector $x_i(t)$ is lower than that of the trial vector, the target is allowed to advance to the next generation. Otherwise, the target vector is replaced by a trial vector in the next generation.

G. Verification of the stopping criterion

Set the generation number for $t = t + 1$. Proceed to Step *C* until a stopping criterion is met, usually a maximum number of iterations (generations), G_{max}. The stopping criterion depends on the type of problem.

3 Combining of DE and Implicit Filtering Method

A fundamental and practical remark is that after a certain amount of time, the population in DE is quite uniform and the cost function of the population is no longer decreasing, the odds to produce fitter solutions (individuals) being very low. That is, the process has fallen into a basin of attraction from which it has a (very) low probability to escape [11].

It is experimentally clear that the exploitation of the basin of attraction that has been found may be more efficiently performed by another algorithm than by a DE. Hence, it is much more efficient to use a local search algorithm such as a hill-climbing. This schema of algorithm is qualified hybrid. DE and IF methods have potentialities that complement each other. The proposed combination of DE with Implicit Filtering (IF) for local search consists of a form of sequential hybridization.

Basically, in this combined method, the DE is applied to the optimization problem and the best solution obtained by DE is used as starting point for the IF method. In this work, this hybrid approach is called DE-IF method.

The quasi-Newton implicit filtering algorithms differ from other methods in the literature that use either inaccurate gradient information, only samples of the function, or difference or interpolatory approximations to gradients and/or Hessians.

Implicit filtering, originally proposed in the context of computer aided design of semiconductors [12] is a generalization of the gradient projection algorithm of [13] in which derivatives are computed with difference quotients. The step sizes (called scales) in the difference quotients are changed as the iteration progresses with the goal of avoiding local minima that are caused by high-frequency, low amplitude oscillations. Real filtering could be performed, but this requires sampling and filtering the entire solution space and thus, is computationally quite expensive. Implicit filtering is very similar to adaptive meshing schemes used by the computational fluid mechanics community to avoid unwanted harmonics. The algorithm is fully described in [14] and [15].

4 Case Study: Optimization of Loney's Solenoid Design

Loney's solenoid problem presents an outstanding potential as a benchmark in the field of magnetostatic inverse problems [1]-[3], [16]. The design problem of Loney's solenoid is to determine the distribution of the current density J which generates a uniform magnetic flux density B in a given interval on the axis of the solenoid. The upper half plane of the axial cross-section of the system is presented in figure 1. The interval of the axis, where the magnetic flux density must be uniform, is ($-z_o$, z_o). The current density J_0 in the coils is assumed to be constant [6]. The position s and the length l of the correcting coils are unknown. The other dimensions are presented in figure 1.

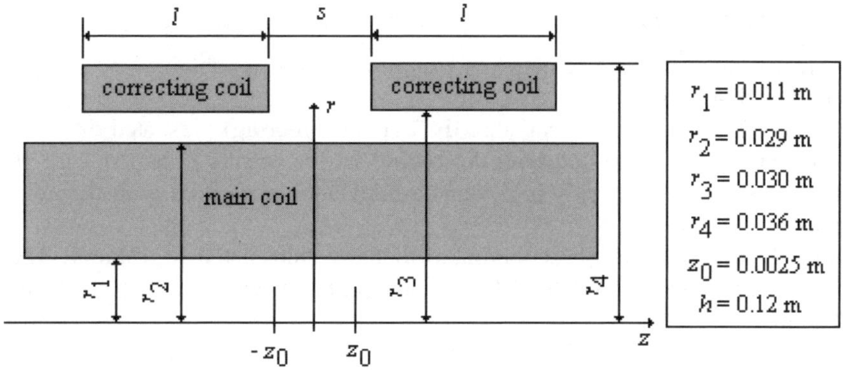

Fig. 1. Axial cross-section of Loney's solenoids (upper plane)

The variables s and l are optimized in this work by DE approaches. In this case, the optimization problem to be solved is the following minimization problem:

$$\min F(s,l) \tag{4}$$

where the objective function F of the minimization problem is given by:

$$F = \frac{B_{max} - B_{min}}{B_0} \tag{5}$$

where B_{max} and B_{min} are the maximum and minimum values of the magnetic flux density in the interval ($-z_o$, z_o) and are functions of s and l. In particular, three different basins of attraction of local minima are placed in the domain of F with values of $F > 4{\cdot}10^{-8}$ (high level region), $3{\cdot}10^{-8} < F < 4{\cdot}10^{-8}$ (low level region), and $F < 3{\cdot}10^{-8}$ (global minimum region). Further details of Loney's solenoid problem are discussed in [1]-[6].

5 Optimization Results

The DE and DE-IF approaches were calculated using MathWorks' Matlab© software. For each of the above described optimization algorithm, a total of 50 independent runs (using different seeds to generate random numbers in each run) were made, using the aforementioned parameters and different initial random seeds. All the runs were terminated after G_{max} = 50 generations. A total of 1,000 cost function evaluations were made by each DE and DE-IF approach in each run.

The variables were allowed to span (space search) within the range of $0 \leq s \leq 20$ and $0 \leq l \leq 20$. The DE/*best*/1/*exp* were selected, along with the values of *CR*=0.8 and a population size *N* of 20 individuals was selected.

Other specific parameters and design procedures used in the standard optimization methods, which were set empirically, were:

- DE(1): classical DE with a constant mutation factor given by $f_m(t) = 0.4$;
- DE(2): $f_m(t)$ was generated with uniform distribution in the range of [0.1; 0.7];
- DE(3): $f_m(t)$ was generated with uniform distribution in the range of [0.3; 0.8].

Table 1 summarizes the statistics of experimental results obtained by applying the DE and DE-IF approaches.

An analysis of Table 1 reveals that IF (Implicit Filtering) presented difficulty and does not provide a good solution. According to the results obtained with IF, the solution presented a relatively high standard deviation compared with the other two methods.

DE(2)-IF presented the best minimum of the results listed in Table 1. The best solution for the Loney's solenoid benchmark problem was *s*= 11.3108 cm and *l* = 1.3484 cm.

However, DE(3)-IF found mean values inferior to those resulting from the other methods. All DE(1-3)-IF methods found a solution very close to the best solution of DE (1-3) methods. However, the DE(1-3)-IF improve all the solutions obtained by DE(1-3) methods.

The mean time of each run in a PC-compatible with Pentium IV 3.2 GHz processor and 2 GB RAM using Matlab 6.5 was 0.02 seconds for IF, 0.05 seconds for DE, and 0.08 seconds for DE-IF approaches.

Table 1. Results (50 runs) for objective function minimization of Loney's solenoid using DE and DE-IF approaches

Optimization Method	Objective Function in 50 Runs, $F(s, l)\cdot 10^{-8}$			
Tested Approaches	Minimum	Mean	Standard Deviation	Maximum
IF	28254.3503	51371.1260	49774.1623	153427.5173
DE(1)	3.4223	3.8729	**0.1703**	4.4179
DE(1)-IF	3.2245	3.8428	0.2052	4.4179
DE(2)	2.7010	3.8732	0.2926	4.7663
DE(2)-IF	**2.0593**	3.8336	0.3376	4.7663
DE(3)	3.1521	3.8586	0.1774	**4.1812**
DE(3)-IF	3.0564	**3.8214**	0.2287	**4.1812**

6 Conclusion and Future Research

In this work, Loney's solenoid design was optimized using DE, IF and DE-IF approaches. DE algorithms offer potential advantages: they find the true global minimum regardless of the initial parameter values, they display a fast convergence, and use few control parameters. In this paper, the efficiency of the DE(2)-IF approach was validated, as shown in Loney's solenoid benchmark (see Table 1). In comparison with the latest results reported in the literature [5], [6], the result listed in Table 1 for DE(2)-IF is better for Loney's solenoid design.

In future research, more detailed studies relating to $f_m(t)$ setup using DE approaches will be carried out to optimize Loney's solenoid design.

References

[1] Di Barba, P.; Savini, A. (1995). Global optimization of Loney's solenoid by means of a deterministic approach, *International Journal of Applied Electromagnetics and Mechanics*, Vol. 6, No. 4, pp. 247-254.

[2] Di Barba, P.; Gottvald, A.; Savini, A. (1995). Global optimization of Loney's solenoid: a benchmark problem, *International Journal of Applied Electromagnetics and Mechanics*, Vol. 6, No. 4, pp. 273-276.

[3] Di Barba, P.; Dughiero, F.; Trevisan, F. (1997). Optimization of the Loney's solenoid through quasi-analytical strategies: a benchmark problem reconsidered, *IEEE Transactions on Magnetics*, Vol. 33, No. 2, pp. 1864-1867.

[4] Borghi, C. A.; Fabbri, M. (1999). Loney's solenoid multi-objective optimization problem, *IEEE Transactions on Magnetics*, Vol. 35, No. 3, pp. 1706-1709.

[5] Cogotti, E.; Fanni, A.; Pilo, F. (2000). A comparison of optimization techniques for Loney's solenoids design: an alternative tabu search algorithm, *IEEE Transactions on Magnetics*, Vol. 36, No. 4, pp. 1153-1157.

[6] Ciuprina, G.; Ioan, D.; Munteanu, I. (2002). Use of intelligent-particle swarm optimization in electromagnetics, *IEEE Transactions on Magnetics*, Vol. 38, No. 2, pp. 1037-1040.

[7] Storn, R.; Price, K. (1995). Differential evolution: a simple and efficient adaptive scheme for global optimization over continuous spaces, *Technical Report TR-95-012*, International Computer Science Institute, Berkeley, USA.

[8] Storn, R. (1997). Differential evolution — a simple and efficient heuristic for global optimization over continuous spaces, *Journal of Global Optimization*, Vol. 11, No. 4, pp. 341-359.

[9] Price, K. (1999). An introduction to differential evolution, in Corne, D., Dorigo, M. and F. Glover (editors), *New ideas in optimization*, London, McGraw-Hill, UK, pp. 79-108.

[10] Storn, R. (1999). Designing digital filters with differential evolution, in Corne, D., Dorigo, M. and F. Glover (editors), *New ideas in optimization*, London, McGraw-Hill, UK, pp. 109-125.

[11] Preux, Ph.; Talbi, E. -G. (1999). Towards hybrid evolutionary algorithms, *International Transactions in Operational Research*, Vol. 6, pp. 557-570.

[12] Stoneking, D.; Bilbro, G.; Trew, R.; Gilmore, P.; Kelley, C.T. (1991). Yield optimization using a GaAs process simulator coupled to a physical device model, *Proceedings IEEE/Cornell Conference on Advanced Concepts in High Speed Devices and Circuits*, Ithaca, NY, USA, pp. 374-383.

[13] Bertsekas, D. B. (1976). On the Goldstein-Levitin-Polyak gradient projection method, *IEEE Transactions on Automatic Control*, Vol. 21, pp. 174-184.
[14] Gilmore, P.; Kelley, C. T. (1995). An implicit filtering algorithm for optimization of functions with many local minima, *SIAM Journal of Optimization*, Vol. 5, No. 2, pp. 269-275.
[15] Gilmore, P.; Kelley, C. T. (1994). An implicit filtering algorithm for optimization of functions with many local minima, Tech. Report CRSC-TR94-23, North Carolina State University, Center for Research in Scientific Computation.
[16] Coelho, L. S.; Mariani, V. C. (2006). Use of cultural particle swarm optimization for Loney's solenoids design, *Proc. 20th Biennial IEEE Conference on Electromagnetic Field Computation*, Miami, FL, USA.

A Layered Matrix Cascade Genetic Algorithm and Particle Swarm Optimization Approach to Thermal Power Generation Scheduling

Siew Chin Neoh, Norhashimah Morad, Chee Peng Lim*, and Zalina Abdul Aziz

School of Electrical and Electronic Engineering
University of Science Malaysia, Engineering Campus
14300 Nibong Tebal, Penang, Malaysia
cplim@eng.usm.my

Abstract. A layered matrix encoding cascade genetic algorithm and particle swarm optimization approach (GA-PSO) for unit commitment and economic load dispatch problem in a thermal power system is presented in this paper. The tasks of determining and allocating power generation to different thermal units in a way that the total power production cost is at the minimum subject to equality and inequality constraints makes unit commitment and economic load dispatch challenging. A case study, based on the thermal power generation problem presented in [1], is used to demonstrate the effectiveness of the proposed method in generating a cost-effective power generation schedule. The schedule obtained is compared with that of Linear Programming (LP) as reported in [1]. The results show that the proposed GA-PSO approach outperforms LP in solving the unit commitment and economic load dispatch problem for thermal power generation system in this case study.

Keywords: Genetic Algorithms, Particle Swarm Optimization, Thermal Power Scheduling.

1 Introduction

Thermal power generator scheduling represents a critical activity in power system operations. It includes a number of production constraints such as types and number of generation units available, minimum and maximum generation level, hourly cost of running at different generation level, generator start up cost, and also expected reserved capacity.

According to [2], the thermal generator scheduling is a combinatorial problem that consists of two sub problems; the first is to develop a daily generator schedule whereas the second is to evaluate the total production cost associated with the generated schedule. The power generation scheduling problem can also be viewed as the unit commitment (number of power generation units dedicated to serve the load demand) and economic dispatch problem (the allocation of power generation to different generator units). In [3], the commitment and dispatch problems are decoupled, while the first attempt to consider the coupled problem is presented in [4].

*Corresponding author.

A. Saad et al. (Eds.): Soft Computing in Industrial Applications, ASC 39, pp. 241–250, 2007.
springerlink.com

Various numerical optimization techniques have been employed to optimize power generation scheduling, e.g. priority list methods, dynamic programming, branch-and-bound methods, and Lagrangian relaxation. According to [3], the priority list method gives a solution that is far from the optimum, whereas dynamic programming is computationally prohibitive due to its increasing search space. A simulated-annealing approach that is able to find the global optimum schedule for unit commitment is developed in [5]; however, the approach does not always satisfy the constraints. Branch-and-bound method has a drawback on the exponential growth in the execution time with the size of unit commitment problem. Lagrangian relaxation may suffer from numerical convergence even though it could provide fast solutions [6].

Besides the methods mentioned above, Genetic Algorithms (GAs) have been identified to be able to find solutions for problems involving multiple constraints and objectives, especially in various scheduling problems [7-9]. Similar to GA, Particle Swarm Optimization (PSO) is a population based optimization method that has been shown to be very effective when applied to diverse set of optimization problems [6, 10]. In this research, we investigate the effectiveness of a GA-PSO hybrid approach, using a layered matrix encoding representation structure, in finding solutions to the thermal generator scheduling problem.

2 Problem Description

Unit commitment and economic load dispatch are important in power generation scheduling. In this study, the problem and data are taken from [1]. The daily electricity load demand that need to be fulfilled according to five different time periods is shown in Table 1.

Three types of thermal power generating units are available in this study: twelve of type 1, ten of type 2, and five of type 3. For each type of generator, the minimum and maximum operating levels, hourly costs of running for each generator at the minimum level or above the minimum level, and the start up cost need to be considered in generating the power generator schedule. The details of each type of thermal power generator are given in Table 2.

Table 1. Daily Electricity load demand according to time period

Time Period	Electricity Load Demand (Megawatts)
12 p.m. to 6 a.m.	15,000
6 a.m. to 9 a.m.	30,000
9 a.m. to 3 p.m.	25,000
3 p.m. to 6 p.m.	40,000
6 p.m. to 12 p.m.	27,000

Besides meeting the electricity load demand as shown in Table 1, there must be sufficient generators to support an increase of load of up to 15%. In other words, the extra load requirement must be supported by the chosen generators, without starting up any more generators. Generators that are already operating should be adjusted

Table 2. Information of thermal power generator

Generator Type	Units Available	Minimum Level (MW)	Maximum Level (MW)	Cost per hour at minimum (£)	Cost per hour per megawatt above minimum (£)	Start-up Cost (£)
Type 1	12	850	2000	1000	2	2000
Type 2	10	1250	1750	2600	1.30	1000
Type 3	5	1500	4000	3000	3	500

within their permitted limits to accomplish the increase of 15% electricity load demand. By fulfilling the demand constraint and generator operation level constraint, the main objective of this case study is to decide the best combination of generators that should be working in each particular period over a day to generate the best power dispatch in order to minimize the total production cost.

2.1 Constraints and Mathematical Formulation

The main variables in this problem are as follows:

n_{ik} = number of generating units of type k working in period i (where i =1,2,3,4,5 are the five period of the day, and k =1,2,3 are the generator type).
z_{ik} = number of generator of type k started up in period i.
x_{ik} = output power from generator of type k in period i.

In the proposed model, the power scheduling problem is represented in two layers: (1) randomly generate the unit commitment schedule, (2) randomly allocate power to

Layer 1: Unit Commitment

Period (P_i)	Generator (G_k) G_1	G_2	G_3	Electricity Demand (D_i)
P_1	n_{11}	n_{12}	n_{13}	D_1
P_2	n_{21}	n_{22}	n_{23}	D_2
P_3	n_{31}	n_{32}	n_{33}	D_3
P_4	n_{41}	n_{42}	n_{43}	D_4
P_5	n_{51}	n_{52}	n_{53}	D_5

⇩ ⇩ ⇩

Layer 2: Power Dispatch

Period (P_i)	Generator (G_k) G_1	G_2	G_3	Electricity Demand (D_i)
P_1	x_{11}	x_{12}	x_{13}	D_1
P_2	x_{21}	x_{22}	x_{23}	D_2
P_3	x_{31}	x_{32}	x_{33}	D_3
P_4	x_{41}	x_{42}	x_{43}	D_4
P_5	x_{51}	x_{52}	x_{53}	D_5

Fig. 1. The layered matrix encoding structure for power generation schedule in the proposed model

each generator type based on the generated schedule for unit commitment. These two layers are decision outputs for the power generation scheduling and the overall representation structure is given in Fig.1. In layer 1, $n_{ik,}$ represents the number of generator unit to be running for generator type k in period i; whereas in layer 2, x_{ik} refers to the amount of power to be allocated to generator type k in period i based on layer 1.

There are three important constraints in the proposed model:

- Demand constraints
- Generator's operation constraint
- Start-up generator constraint

2.1.1 Demand Constraint

The electricity load demand, as shown in Table 1, must be met in each period. Therefore, the sum of power to be generated by each generator type k in period i should fulfill the demand for period i, as indicated in Equation (1), where D_i is the demand in period i.

$$\sum_{k=1}^{k=3} x_{ik} \geq D_i \tag{1}$$

The extra guaranteed load requirement must be able to be met without starting up any more generators, as shown in Equation (2) for all period i.

$$\sum_{k=1}^{k=3} M_k n_{ik} \geq \frac{115}{100} D_i \tag{2}$$

where M_k is the maximum output levels for generators of type k and n_{ik} is the committed units of generator k at period i.

2.1.2 Generator's Operation Constraint

The output power of each generator must lie within the limits of the minimum and maximum operation levels. Equations (3) and (4) show the operation constraints of the power generator.

$$x_{ik} \geq m_k n_{ik} \tag{3}$$

$$x_{ik} \leq M_k n_{ik} \tag{4}$$

where m_k and M_k are the given minimum and maximum operating levels for generator type k, x_{ik} is the generated output power of generator k in period i, and n_{ik} is the committed units of generator k at period i.

2.1.3 Start-Up Generator Constraint

The number of generator started up in period i, z_{ik}, can be obtained by Equation (5).

$$z_{ik} = \begin{cases} n_{ik} - n_{(i-1)k} & \text{if generator } n_{ik} \geq n_{(i-1)k} \\ 0 & \text{if otherwise} \end{cases} \tag{5}$$

where n_{ik} refers to the committed units of generator k at period i. Note that when $i=1$, period i-1 is taken as the 5th period.

2.1.4 Evaluation Function

The weighted-sum approach is applied to minimize the total power generation cost. Equation (6) shows the objective function used.

Objective Function

$$= \text{minimize} \left\{ \sum_{i=1}^{i=5} \left[\sum_{k=1}^{k=3} \left(C_k (x_{ik} - m_k n_{ik}) + E_k n_{ik} + F_k z_{ik} \right) \right] \right. \quad (6)$$

where C_k is the cost per hour per megawatt above the minimum level of generator k multiplied by the number of hours in period i, E_k is the cost per hour for operating at the minimum level of generator k multiplied by the number of hours in period i, and F_k is the start up cost of generator k.

3 The GA-PSO Optimization

3.1 A Layered Matrix Structure

Many real world problems involve multi-decision and multi-constraint. As a result, multi-dimensional encoding structure may be necessary in order to incorporate all required constraints and decisions into one single solution representation. However, it is clear that increase in dimension for a multi-dimensional encoding structure (e.g. 4 or 5 dimensions) makes the solution representation and evaluation process tedious. This paper proposed a layered matrix encoding structure to solve multi-dimensional encoding problems.

The proposed layered matrix encoding structure is different from the existing multi-dimensional encoding approaches in which it separates different decision outputs into different layers so as to simplify the problem representation. With the layer structure, constraints and decision outputs can be analyzed more distinctively and effectively, and at the same time makes the evaluation process much easier. One of the main advantages of the layered encoding structure is that it could allow multi-stage cascade optimization where different optimizers can be used to optimize different decision based on the characteristic of the decision search space.

In the proposed model for power generation scheduling, a two layers 2D matrix is used to represent the power generation schedule for each period throughout a day which is similar to the structure shown in Fig.1. Each layer represents different decision outputs, layer 1 is used to decide unit commitment (number of generator units to be running), and layer 2 is used to decide economic power dispatch (determination of the amount of power to be generated from the selected generator to fulfill demand).

3.2 Cascade GA-PSO Optimization

For every single schedule of unit commitment, there could be thousands of possible schedules of power dispatch that could fulfill the similar unit commitment. Therefore, it is easy to imagine how large the searching space of stochastic search can be. This paper proposed a layered encoding GA-PSO cascade optimization to narrow down the searching space and fasten the process of finding both global and local optimums. The flow of cascade GA-PSO optimization model is depicted in Fig.2.

Layers in the layered matrix structure can be viewed as the optimization stages in cascade optimization. In the proposed cascade optimization for power generation scheduling, a population of possible unit commitment schedules is generated based on the power generation constraints in the first stage. For each schedule of unit commitment, a population of possible power dispatch is generated in the second stage. The GA

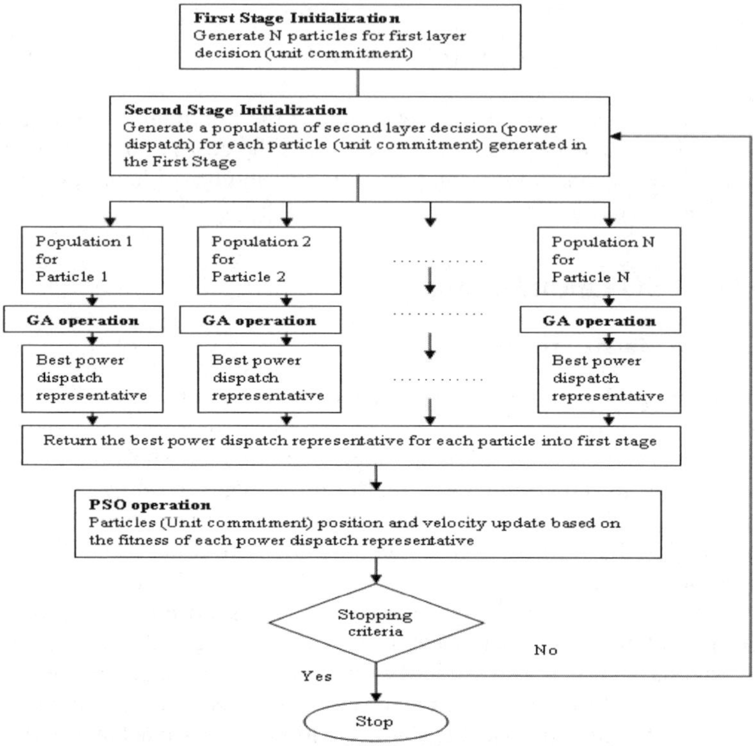

Fig. 2. A cascade GA-PSO optimization model

is used to find the optimal power dispatch for each schedule of unit commitment. The best power dispatch for each unit commitment schedule will be returned back to the first stage. Subsequently, based on all the best combinations of unit commitment and power dispatch in the first generation, PSO is used to fine-tune the possible unit commitment schedules and thereafter generate the second generation of unit commitments. The evolution loop of GA-PSO proceeds until the best solution converges for 20 iteration.

The GA is a stochastic search method that mimics the metaphor of natural biological evolution [11], whereas PSO is an optimization tool driven by the social behavior of organisms [12]. The GA is used to search for the best power dispatch to represent a particular schedule of unit commitment. After each schedule of unit commitment has its best representative of economic dispatch, PSO will be used to learn from the behavior of the pool of representatives and adjust the search for better unit commitment. The combination of GA and PSO provides good balance between exploration and exploitation which will also helps to balance the individuality and sociality of the search.

The cascade GA-PSO optimization is as follow:

Step 1: First Stage Initialization
The schedules of unit commitment (positions of particles, n_{ikq}) and the initial velocity of each position of the particle, V_{ikq}, are randomly initialized. Check the feasibility of each particle. Regenerate the particle if it does not satisfy the constraints.

Step 2: Second Stage Initialization
Based on each schedule of unit commitment, generate a population of appropriate and feasible power dispatch. Integrate GA to find the best representative of power dispatch for each unit commitment schedule. Return the best representatives into first stage.

Step 3: Particles Updating
Adjust the position and velocity of each particle with Equation (7) and (8) to generate a new generation of unit commitment schedules.

Step 4: Repeat
Repeat step 2 to step 3 until the number of PSO iteration reaches *T*, a maximum PSO iteration which is specified before termination.

3.3 PSO in Unit Commitment

PSO is a stochastic population-based approach in which each potential solution (also called as particle) searches through the problem space, refining its knowledge, adjust its velocity based on the social information that it gathered and update its position. The best found position for a particle is given as pbest whereas the best found position for all particles is denoted as gbest. Equation (7) and (8) calculate the velocity (V_{ikq}) and position (n_{ikq}) of particle q in the dimensional search space of $i \times k$ respectively. Position n_{ikq} refers to the unit commitment, n_{ik}, of particle q in the population of particle swarm.

$$V_{ikq} = w \times V_{ikq} + c_1 \times rand\,(\ ) \times (pbest - n_{ikq}) + c_2 \times rand\,(\ ) \times (gbest - n_{ikq}) \tag{7}$$

$$n_{ikq} = n_{ikq} + V_{ikq} \tag{8}$$

where w is the inertia weight, c_1 and c_2 are two constant factors, and rand( ) is a randomly generated value between 0 and 1.

A decreasing w from the range of 0.9 to 0.4 is applied in this case study as it has been shown to perform well in [13, 14]. The values of c_1 and c_2 are both set to 2 in this research as mentioned in [12] where the recommended choice for constant c_1 and c_2 is integer 2 since it on average makes the weight for "social" and "cognition" parts to be 1. The PSO loop is terminated after convergence for 20 iterations.

3.4 GA in Power Dispatch

In the proposed model, layer 2 of the layered matrix structure is used to represent chromosomes in the GA population pool. A probabilistic selection is performed based upon the individual's fitness such that the better individuals have an increased chance of being selected. Stochastic Universal Sampling which is a single phase sampling algorithm with minimum spread and zero bias is used as the selection function in the proposed model. The basic strategy underlying this approach is to keep the expected number of copies of each chromosome in the next generation [15].

3.4.1 GA Operators

For this case study, *Randomly Selected Crossover* is used as the crossover operator whereas *Randomly Selected Mutation* is used as the mutation operator.

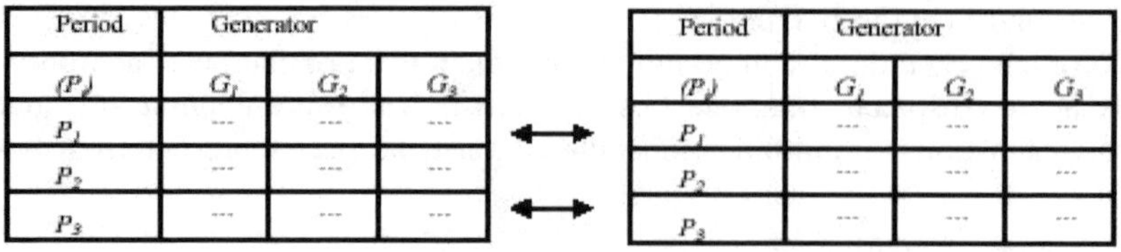

Fig. 3. Randomly Selected Crossover

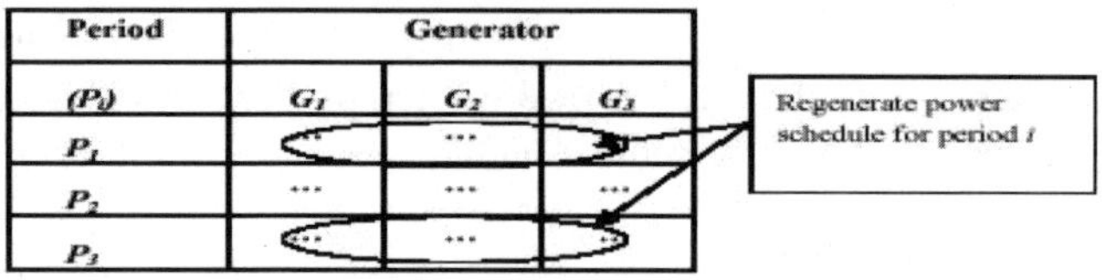

Fig. 4. Randomly Selected Mutation

The principle of *Randomly Selected Crossover* for the power generation scheduling problem is depicted in Fig. 3. As the chromosomes are represented by 2D matrices, *Randomly Selected Crossover* is applied to the selected matrix rows. The crossover probability is given as *Pxovr*, in which each individual has *Pxovr* chances to be selected for a crossover. For each crossover, the number of rows that are randomly selected from a 2D matrix will be inter-exchanged with another individual to produce offsprings (new individuals). The purpose of having *Randomly Selected Crossover* is to allow inter-exchange of power scheduling distribution among individuals in the population and, at the same time, to ensure that the load demand for a particular period, D_i, is always met.

Randomly Selected Mutation is applied to randomly regenerate output power for a selected period over a day. The probability for an individual to be selected to undergo mutation is given as *Pm*. In *Randomly Selected Mutation*, the randomly selected rows from the 2D matrix (selected period of power schedule) will be regenerated (Fig. 4).

3.4.2 Genetic Parameter Selection

The techniques developed in this model were implemented on a 4.0GHz PC using MATLAB language. According to experience, the GA population size and the maximum number of generation convergence for GA termination are both selected as 20; whereas the crossover probability, *Pxovr*, and mutation probability, *Pm*, are set as 0.7 and 0.3 respectively.

4 Results

The results for power dispatch and unit commitment based on the layered matrix cascade GA-PSO approach and LP (reported in [1]) are depicted in Tables 3 and 4, respectively. From the tables, the layered matrix cascade GA-PSO approach gives lower daily power production cost (£ 954,260) compared to that of LP (£ 988,540). A total cost of £ 34,280 has been saved using the proposed approach. It is clearly showed that layered matrix GA-PSO cascade optimization outperforms LP in this case study. As a result, it can be used as an alternative way of solving unit commitment and power dispatch problem in thermal power generation scheduling. In the point of view of soft computing, the layered matrix encoding structure of GA-PSO cascade optimization helps to narrow down the search space and fasten the process of finding both local and global minimum. Searching by layers in a GA-PSO cascade flow is easier then searching randomly in a big pool of possible solutions without a proper direction.

Table 3. Power dispatch and unit commitment using the layered matrix cascade GA-PSO model

Period	Units commitment for generator type	Power dispatch (MW)	Total daily cost
Period 1	12 of type 1	10450	£ 954,260
	3 of type 3	4550	
Period 2	5 of type 1	5300	
	4 of type 2	16800	
	5 of type 3	7900	
Period 3	12 of type 1	10300	
	4 of type 2	14700	
Period 4	11 of type 1	15850	
	3 of type 2	16450	
	5 of type 3	7700	
Period 5	12 of type 1	10500	
	7 of type 2	16500	

Table 4. Power dispatch and unit commitment using LP (as reported in [1])

Period	Units commitment for generator type	Power dispatch (MW)	Total daily cost
Period 1	12 of type 1	10200	£ 988,540
	3 of type 2	4800	
Period 2	12 of type 1	16000	
	8 of type 2	14000	
Period 3	12 of type 1	11000	
	8 of type 2	14000	
Period 4	12 of type 1	21250	
	9 of type 2	15750	
	2 of type 3	3000	
Period 5	12 of type 1	11250	
	9 of type 2	15750	

5 Conclusions

A layered matrix cascade GA-PSO approach has been developed and applied to solve unit commitment and power dispatch problem in thermal power generation scheduling.

Constraints of the power generation scheduling have been formulated and explained in this paper and the advantages of the layered matrix structure and the cascade GA-PSO approach is mentioned. The detail procedure of cascade GA-PSO has been given and the performance between the layered matrix cascade GA-PSO model and LP in saving total daily power generation cost has been compared. The intrinsic nature of particles position and velocity update in PSO and the genetic evolution character in GA makes the cascade GA-PSO a superior approach for solving unit commitment and power dispatch problem. Similarly, the layered matrix structure that is able to reduce the complexity of multi-dimensional representation and to allow combination of different optimizers enhanced the construction of the cascade GA-PSO model.

References

1. Williams HP (1999) Model Building in Mathematical Programming. 4th edn John Wiley and Sons Inc
2. Wong KP and Wong YW (1995) Thermal generator scheduling using hybrid genetic / simulated -annealing approach. In: IEE Proceedings Part C –Generation Transmission, and Distribution, vol 142, pp 372–380
3. Aldridge CJ, McKee S, McDonald JR, Galloway SJ, Dahal KP, Bradley ME, and Macqueen JF (2001) A knowledge-based genetic algorithm for unit commitment. In: IEE Proceedings on Generation, Transmission and Distribution, vol 148, pp 146–152
4. Garver LL (1963) Power scheduling by integer programming. IEEE Trans. Power Apparatus and Systems 81: 730–735
5. Zhuang F and Galiana FD (1990) Unit commitment by simulated annealing. IEEE Transactions on Power Systems 5: 311–317
6. Ting TO, Rao MVC, and Loo CK (2006) A novel approach for unit commitment problem via an effective hybrid particle swarm optimization. IEEE Transactions on Power Systems 21: 411–418
7. Sarker R and Newton C (2002) A genetic algorithm for solving economic lot size scheduling problem. Computers and Industrial Engineering 42: 189–198
8. Ip WH, Li Y, Man KF., and Tang KS (2000) Multi-product planning and scheduling using genetic algorithm approach. Computer & Industrial Engineering 38: 283–296
9. Ono I, Yamamura M, and Kobayashi S (1996) A genetic algorithm for job-shop scheduling problems using job-based order crossover. In: Proceedings of 1996 IEEE International Conference on Evolutionary Computation, pp 547–552
10. Eberhart RC and Shi Y (2001) Particle swarm optimization: developments, applications, and resources. In: Proceedings of the IEEE Congress on Evolutionary Computation, Seoul, Korea, pp 81–86
11. Holland J (1975) Adaptation in Natural and Artificial Systems. The University of Michigan Press, Ann Arbor
12. Kennedy J, Eberhart R (1995) Particle Swarm Optimization. In: Proceedings of IEEE International Conference on Neural Networks, pp 1942–1948
13. Naka S, Genji T, Yura T, and Fukuyama Y (2003) A hybrid particle swarm optimization for distribution state estimation. IEEE Transactions on Power Systems 18: 60–68
14. Yoshida H, Kawata K, Fukuyama Y, Takayama S, and Nakanishi Y (2000) A particle swarm optimization for reactive power and voltage control considering voltage security assessment. IEEE Transactions on Power Systems 15: 1232–1239
15. Gen M and Cheng R (2000) Genetic Algorithms and Engineering Optimization. John Wiley & Sons Inc, New York

Differential Evolution for Binary Encoding

Tao Gong and Andrew L. Tuson

Department of Computing, City University, London, EC1V 0HB
t.gong@city.ac.uk,
a.tuson@city.ac.uk

Summary. Differential Evolution (DE) is a competitive optimization technique for numerical optimization problems with real-parameter representation. This paper aims to investigate how DE can be adapted with binary encoding and to study its behaviors on the binary level.

1 Introduction

Although DE [1] has been competitive in numerical optimization, only a few works [1] exploit its usage for discrete optimization problems. This limitation is mainly caused by the working mechanism of DE (which is based on real vectors) and the lack of a principled generalization of DE.

This paper addresses this issue in a rigorous manner by using *forma analysis* to derive discrete DE operators, that are also empirically studied to justify the success of our formal approach.

2 Differential Evolution: The Basics

Differential Evolution (DE) was introduced by Storn and Price [2] as an efficient evolutionary algorithm initially designed for global optimization problems over continuous spaces.

DE [2] uses a population of real vectors which evolve simultaneously along with the increasing generation number g. The vector can be represented as:

$$X(i, g) = (x_{ig}(1), x_{ig}(2), \ldots, x_{ig}(D)),$$

where i stands for the index of the vector in the population, D represents the total number of parameters.

A crucial component of DE is the *mutant vector*. After randomly initializing the population, for each target vector $X(i, g)$, DE generates a mutant vector by "*adding the weighted difference vector between two population members to a third member*" [2], formally as (1):

$$M(i, g+1) = X(r_1, g) + F(X(r_2, g) - X(r_3, g)), \tag{1}$$

A. Saad et al. (Eds.): Soft Computing in Industrial Applications, ASC 39, pp. 251–262, 2007.
springerlink.com

where r_1, r_2, r_3 are 3 mutually exclusive members of the population, F stands for the amplification of the random deviation $(X(r_2, g) - X(r_3, g))$ and $M(i, g+1)$ represents the mutant vector generated based on the base vector $X(r_1, g)$ for target vector $X(i, g)$.

Then, the mutant vector undergoes a crossover with the *target vector* $X(i, g)$ to generate the *trial vector* $U(i, g+1)$. There are two types of crossover operator commonly used for DE: *Exponential* Crossover and *Binomial* Crossover [1]. The exponential crossover simulates a 2-point crossover (when cycling of chromosome is considered) with an *exponentially* determined segment length L (parameterized by CR) and a random starting point n, while the binomial crossover can be regarded as a uniform crossover adopting a binomial distribution $B(D, CR)$.

After the crossover, a "binary" selection process chooses the better candidate from the trial vector $U(i, g+1)$ and the target vector $X(i, g)$ to be the candidate for the next generation $X(i, g+1)$.

3 Forma Analysis

Forma analysis [3] is a formal but practical method that allows the problem representation and its operators to be structured in a formal manner by using *equivalence relations*. Each equivalence relation Ψ divides the search space into disjoint equivalence classes Ξ_ψ (depending which value the solutions match), with individual equivalence classes being denoted by ξ, which gathers solutions that are equivalent under a certain equivalence relation.

The initial aim of forma analysis [3] was to codify knowledge of the problem domain using a set of *equivalence classes* (or *formae*) which is assumed to be able to cluster solutions with related performance in order to guide the search process more effectively, e.g. edges if we are considering the travelling salesman problem. Since equivalence relations/classes have the ability to capture the properties of solutions, the operators can thus be mathematically derived with regards to these equivalence relations to manipulate these properties in a formal way.

Some of the characteristics and operator templates related to forma analysis [3, 4, 5] are given below to facilitate our further discussion.

3.1 Describing the Search Space

The key concept is that of a basis: a set of equivalence relations that allows us to properly describe the search space S.

Definition 1. ***(Basis)*** *A subset* Ψ *of a set of equivalence relations is termed as a* ***basis*** *for the set of equivalence relations, if* Ψ *spans the set and* Ψ *is independent.*

An encoding can thus be derived by taking the image of the basis equivalence classes corresponding to a particular solution in the search space.

3.2 Operator Templates: Domain Independent Operators

Forma analysis can derive operators that explicitly manipulate the given equivalence relations. This is achieved by combining the basis with domain independent operators for specifying operator behavior in terms of basis. Two of these templates are key to the work presented in this paper.

One such template corresponds to the (*strict*) k-change operator template [5], formally as:

$$O_k(x, k, \Psi) = \{y \in S \mid D_\Psi(x, y) = k\}, \quad (2)$$

where $D_\Psi(x, y)$ represents the *forma distance* [6] between solutions x and y under basis Ψ.

Given a basis Ψ for a set of equivalence relations over a search space S, Random Transmitting Recombination (RTR) [3] is defined to select a child solution z out of the *dynastic potential* of the parent solutions x and y, formally as:

$$\mathrm{RTR}(x, y, \Psi) = \{z \in S \mid \forall \psi \in \Psi : \psi(x) = \psi(z) \vee \psi(y) = \psi(z)\}, \quad (3)$$

where the actual child solution z is chosen from the set above uniformly at random.

4 DE Operator Templates: Generalization of DE

In this section, operator templates are formalized for both DE mutation and DE crossovers.

4.1 DE Mutation Operator Template

The generation process of a mutant vector is in fact applying to a randomly selected vector a k-change operator, where k is decided by the scaled distance of another two sampled vectors on each dimension.

Thus, based on the (strict) k-change operator and the *forma distance* which is defined as the weighted sum of the number of equivalence relations where two individuals are different, the DE mutation operator is defined as:

$$M_{de}(x_1, x_2, x_3, F, \Psi_i) = \{m \in S | \ D_{\Psi_i}(x_1, m) = k \ \wedge \ k = F \times D_{\Psi_i}(x_2, x_3)\}, \quad (4)$$

where x_1 represents the base vector, x_2 and x_3 represent the vectors to produce the difference, m stands for the mutant vector, and Ψ_i represents the basis constructed for the i-th dimension.

4.2 DE Crossover Operator Templates

Since the exponential crossover simulates a 2-point crossover with an *exponentially* [1] determined segment length L and a random starting point n, the *Generalized Exponential Crossover* (GEX) template can be defined by specifying the random starting point n and the segment length L for a 2-point crossover. Thus, we claim that GEX can be formalized by two steps of specification of the RTR operator template.

1. By adding the restriction of continuous blocks to RTR, we obtain the *Generalized N-point Crossover* (GNX) [4];
2. By further specifying the generation of the crossover segment in GNX with (n, L), we can get GEX operator template.

Effectively, GEX represents a subset of both GNX and RTR, by restricting the mappings as:

$$Link(\Psi) : \text{RTR} \rightarrow \text{GNX} \rightarrow \text{GEX}, \tag{5}$$

while the linkage specification can be formalized as follows:

$$block(m, t, u, n, L) = \begin{cases} \psi_j(m, u) = 1, \text{ if } (n \le j < n + L), \\ \psi_j(t, u) = 1, \quad \text{otherwise.} \end{cases} \tag{6}$$

This formulation means that, the equivalence classes of the mutant vector m in the linked equivalence relations ψ_j $(j \in (n, n + L))$ are passed to the trial vector u, while the target vector t contributes its equivalence classes to the other equivalence classes of the trial vector u.

Thus, the GEX operator template can be formally defined as following:

$$\text{GEX}(m, t, n, L, \Psi) = \{u \in S \mid u = \text{RTR}(m, t, \Psi) \ \wedge \ block(m, t, u, n, L)\}. \tag{7}$$

For DE binomial crossover [1], since CR decides whether the corresponding formae in the mutant vector m should replace the target vector t to be copied to the trial vector u, the *Generalized Binomial Crossover* (GBX) can be defined by a traditional uniform crossover (RTR) with a crossover rate CR, formally as:

$$\text{GBX}(m, t, CR, \Psi) = \{u \in S \mid u = \text{RTR}(m, t, \Psi) \ \wedge \ P^{CR}_{\psi(u,m)=1}\}, \tag{8}$$

where $P^{CR}_{\psi(u,m)=1}$ represents the probability that the trial vector u takes the same value as the mutant vector m for some equivalence class ψ is CR.

5 Differential Evolution Operators for Binary Encoding

In this section, discrete DE mutation and crossover operators are formally derived respectively, based on the formal description of binary encoding.

To represent the binary-string solution $(x_1, x_2, \ldots, x_D)$ in a way that can be manipulated and utilized by forma analysis, the basis is defined as:

$$\psi_j(X, Y) = \begin{cases} 1, \text{ if } x_j = y_j; \\ 0, \text{ otherwise.} \end{cases} \tag{9}$$

According to the binary values valid for a variable x_j, we define the equivalence classes (formae) Ξ_{ψ_j} induced by the corresponding equivalence relation ψ_j to be $\Xi_{\psi_j} = \{\xi^0_j, \xi^1_j\}$. The derivation of DE operators will be based on the above formal description of binary basis.

5.1 DE Mutation Operator

According to the DE mutation template, the distance of two random solutions in each dimension is applied to another solution in the corresponding dimension. Since the issue of dimensionality is actually inherited from the continuous domain, how can we interpret the dimensionality in the binary case appropriately?

There are two possible answers. The first would be to regard each binary decision variable as one dimension, while the other would be to consider all decision variables being in a single dimension.

Each Decision Variable as a Single Dimension

Given the basis for a binary string as $\Psi = \{\psi_1, \psi_2, \ldots, \psi_D\}$, if we consider each decision variable as a single dimension, the distance of two solutions for each dimension is either 1 or 0.

In Table 1 as an example, the distance between $X(r_2, g)$ and $X(r_3, g)$ for each dimension is decided by the values they take for this dimension, while the forma distance [6] is a *hamming distance* of 3.

Then, we apply a scaled difference to another solution $X(r_1, g)$. Assuming F is 0.6, the scaled difference for each dimension will be either 0.6 or 0. Although the distance being a floating number is not a issue for real vectors, for binary string certain interpretation is required.

To interpret the scaled difference, we randomly decide whether the difference should be rounded to 1 or 0 as:

$$D_j(Mutant, X(r_1, g)) = \begin{cases} 1, \text{ if } rand_j < F \ \wedge \ D_j(X(r_2, g), X(r_3, g)) = 1, \\ 0, \text{ otherwise.} \end{cases} \tag{10}$$

In (10), for each position j, $F * D_j(X(r_2, g), X(r_3, g))$ is used to decide the probability that a change is applied to $X(r_1, g)$ in the corresponding position to produce the mutant solution. As shown in Table 1, the distance for ψ_1 is rounded to 0 because $rand_1 > F$, while for ψ_3 and ψ_5 the distances are successfully rounded to 1 with probability F.

In the case of considering each binary decision variable as a single dimension, we derive a mutation operator that applies the rounded scaled-distance to each position of the base solution. We call this type of DE mutation the "*restricted-change DE mutation*" (M_{res}), where only the decision variables for which $X(r_2, g)$ and $X(r_3, g)$ are different have the chance to be mutated.

All Decision Variables as a Single Dimension

If we consider all decision variables as a single dimension, the distance of two solutions effectively reduces to the hamming distance, which is the same as the forma distance.

Using the same example shown in Table 1, the distance between $X(r_2, g)$ and $X(r_3, g)$ for the single dimension becomes 3. Assuming the scaling factor F is

Table 1. Illustration of restricted-change and any-change DE mutations

	ψ_1	ψ_2	ψ_3	ψ_4	ψ_5
$X(r_2, g)$	1	1	0	1	1
$X(r_3, g)$	0	1	1	1	0
Distance	1	0	1	0	1
$X(r_1, g)$	0	1	0	1	1
Mask$_{res}$	$rnd_1 > F$	$*$	$rnd_3 < F$	$*$	$rnd_5 < F$
M$_{res}$	0	1	1	1	0
Mask$_{any}$	$*$	$\diamond$	$*$	$\diamond$	$*$
M$_{any}$	0	0	0	0	1

0.6, the scaled difference is 1.8. Again, we need to interpret the floating number distance to a valid distance for binary solution.

Given the distance $D(X(r_2, g), X(r_3, g))$ is d, the scaled distance is

$$d' = F * d.$$

In the case of d' being a floating number, there are always two options to round d' to its nearest integers: $(int)d'$ and $(int)d' + 1$. By calculating the distance between d' and $(int)d'$ (or d' and $(int)d' + 1$), we can produce the probability of rounding d' to each option, as:

$$P(d' \rightarrow (int)d' + 1) = d' - (int)d',$$

$$P(d' \rightarrow (int)d') = 1 - (d' - (int)d').$$

Thus, the mutant solution can be generated as shown in (11).

$$D(Mutant, X(r_1, g)) = \begin{cases} (int)d' + 1, \text{ if } rand < d' - (int)d', \\ (int)d', \qquad \text{otherwise.} \end{cases} \tag{11}$$

In the example, the scaled distance 1.8 means that it has a probability of 0.8 to be rounded to 2 and a probability of 0.2 to be rounded to 1. Assuming that $rand$ is smaller than 0.8, the distance applied to the base solution to produce the mutant solution should be 2.

Applying any 2 *mutually exclusive* (in accord with strict k-change) changes in the base solution produces a possible mutant solution. As shown in Table 1, the positions with $\diamond$ are selected to change to produce the mutant solution.

In the case of considering all decision variables in the solution as a single dimension, we derive a mutation operator that applies the rounded scaled-distance to the base solution by flipping the values for some randomly selected decision variables. We call this type of DE mutation the "*any-change DE mutation*" (M$_{any}$), where changes can be made to any decision variables.

5.2 DE Crossover Operator

Since DE's exponential and binomial crossovers already manipulate solutions in a discrete form, they can be directly regarded as the results of instantiating the GEX and GBX templates respectively with binary formal description.

6 Experimental Study

Based on the derived binary-adapted DE, we carry out experimental study with some artificial binary optimization problems to understand how these operators work for discrete domain and their behaviors. The problems we consider here include MAX-ONE problems [7], Order-3 Deceptive problems [8], Royal Road problems [9], and Long Path problems [10], which are very useful for revealing the behaviors and search patterns of optimization techniques.

6.1 Test Problem Descriptions

MAX-ONE Problems

The aim of a MAX-ONE problem [7] is simply to maximize the ones in a binary string. The fitness of a string is the number of ones it contains. The string length 500 is used for this study, with the optimum 500.

Order-3 Deceptive Problems

Deceptive problems are a class of problems where some low order Building Blocks do not combine to form higher order Building Blocks. Instead, deceptive Building Blocks are formed to produce sub-optimal or near sub-optimal solutions. These problems were devised to challenge GAs [8].

Goldberg [8] defined an order-3 fully deceptive problem as shown in the following Table 2. The deceptive problem used as the test problems consists of the

Table 2. The Order-3 Deceptive Problem

f(0,0,0) = 28	f(0,0,1) = 26
f(0,1,0) = 22	f(0,1,1) = 0
f(1,0,0) = 14	f(1,0,1) = 0
f(1,1,0) = 0	f(1,1,1) = 30

concatenation of subproblems of length 3. In this study, we consider deceptive problem containing 100 copies of the above basic deceptive subproblem, which has the optimum 3000.

Royal Road Problems

The motivation for designing the Royal Road problems [9] was to devise a class of non-deceptive problems for GAs.

The Royal Road problem considered in this study ($R1$) is computed by separating a bit string with length 80 into 10 continuous Building Blocks s_i of 8 bits, each of which contributes $c_i = 8$ to the total fitness if the solution x is an instance of s_i. This problem has an optimal fitness of 80.

Long Path Problems

The intention of devising the Long Path problems [11, 12] was to clarify the issue about modality and difficulty of landscape–without local optima, the problem can still be very hard for hill-climbers. Although always leading upwards, the path to optimum is artificially made extremely long.

The particular problem used in this study is the *Root2path* long path problem, where the path length increases exponentially according to the bit string length. For the construction and decoding of *Root2path*, readers are redirected to the original work [10]. The fitness of a bit string with length l can be calculated as $3 * 2^{\lfloor (l-1)/2 \rfloor} + l - 2$. We consider length 29 in this study, with its optimum being $49,179$.

6.2 Experimental Settings

In this study, we consider the following 4 strategies with different combinations of the binary-DE operators:

1. any-change mutation (M_{any}) and exponential crossover–`DE/any/exp`;
2. any-change mutation (M_{any}) and binomial crossover–`DE/any/bin`;
3. restricted-change mutation (M_{res}) and exponential crossover–`DE/res/exp`;
4. restricted-change mutation (M_{res}) and binomial crossover–`DE/res/bin`.

For each of the test problems, tuning is carried out for each of the strategy to reach its best performance among different combinations of parameter settings with equal effort. The parameter setting with the best performance[1] over 20 independent runs for each strategy will be used for the experimental study. The free parameters for each problem are F and CR, while NP (size of population) is fixed to be an appropriate size for each problem through observations. The tuning results of the 4 strategies for all the problems are shown in Table 3.

For MAX-ONE problems, it is obvious that relatively lower F is beneficial to get a better performance for all the strategies, while larger CR is required for exponential crossover to generate crossover segment at a suitable size and lower CR is useful for binomial crossover to perform robust convergence. These tuned parameters consequentially reflect the nature of MAX-ONE problems that gentle jumps are more likely to produce fast and effective convergence for the binary-adapted DEs.

[1] The performance is evaluated by considering both its average best solution found and its average number of generations to reach its best solution. The number of generations to reach its best solution is only considered when two parameter settings have the same average best solution.

Table 3. Tuning results of binary-DE

MAXONE	POPSIZE	LENGTH	GEN
	50	500	500
DE/any/exp	F=0.1, CR=0.9		
DE/any/bin	F=0.1, CR=0.2		
DE/res/exp	F=0.1, CR=0.9		
DE/res/bin	F=0.2, CR=0.3		

ORDER3	POPSIZE	LENGTH	GEN
	100	300	2000
DE/any/exp	F=0.1, CR=0.8		
DE/any/bin	F=0.3, CR=1.0		
DE/res/exp	F=0.1, CR=0.8		
DE/res/bin	F=0.2, CR=0.9		

RoyalRoad	POPSIZE	LENGTH	GEN
	30	80	500
DE/any/exp	F=0.5, CR=0.9		
DE/any/bin	F=0.4, CR=0.2		
DE/res/exp	F=0.6, CR=0.9		
DE/res/bin	F=0.5, CR=0.3		

LongPath	POPSIZE	LENGTH	GEN
	30	29	300
DE/any/exp	F=1.0, CR=0.8		
DE/any/bin	F=0.9, CR=0.5		
DE/res/exp	F=1.0, CR=0.3		
DE/res/bin	F=0.9, CR=0.2		

However, it is quite reasonable that F for Royal Road problems is relatively larger. For Royal Road problems, it is notorious that some building blocks might dominate the whole population which in turn heavily reduces the diversity. A larger F (thus a longer *annealing schedule*) in DE mutation can avoid the risk of premature convergence by exploring more building blocks. It is even more obvious that Long Path problems, where larger jump is rewarding, require an even larger F to quickly reach the optimum.

6.3 Experimental Results and Explanations

According to the tuned parameter settings, 50 independent runs were executed for each of these strategies for each problem, with the average best solutions against generation recorded, as shown in Fig. 1. (For more experimental results, readers are suggested to contact the author.)

MAX-ONE Problems

For MAX-ONE problems, the results indicate that both `DE/any/bin` and `DE/res/bin` outperform `DE/any/exp` and `DE/res/exp` with fast and constant convergence. This is explained by the nature of MAX-ONE problems. Because all the decision variables in MAX-ONE problems are independent, the binomial crossover can optimize each variable separately with a tuned bias towards a certain parent, thus performing better than the exponential crossover whose concept of linked *block* is not useful. The advantage of the any-change mutation is also reflected, as it has the probability to make changes at any positions, it can explore the search space more quickly than the restricted-change mutation where the positions to change are restricted.

Order-3 Deceptive Problems

For Order-3 deceptive problems, `DE/any/exp` and `DE/res/exp` with the exponential crossover always perform better. The explanation is that the Order-3 deceptive problems are constructed by a set of basic Order-3 subproblems (blocks), where the exponential crossover with linkage specification is very efficient by combining optimized subproblems. In contrast, the binomial crossover is more destructive to these subproblems. Again, any-change mutation gives more exploration power for `DE/any/bin` to outperform `DE/res/bin` with restricted-change mutation. However, this is not necessarily the case for `DE/any/exp` and `DE/res/exp`, because any-change mutation might become destructive when the exponential crossover is combining blocks of sub-solutions.

Royal Road Problems

For Royal Road problems, strategies with any-change mutation perform relatively better than those with restricted-change mutation. This is because the increase in the problem size may enhance the "hitchhiking" effect [9], which

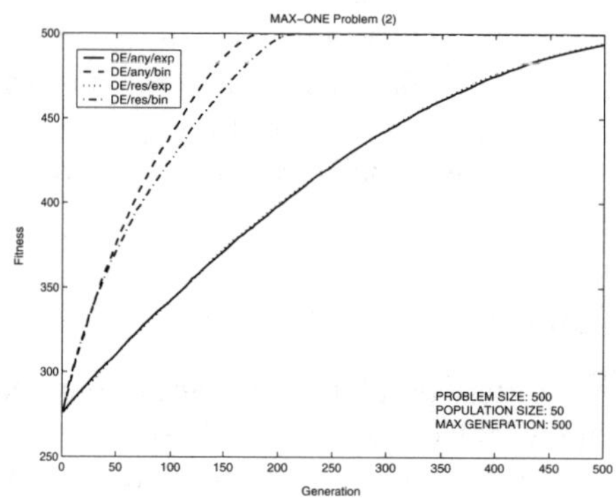

(a) Max-One

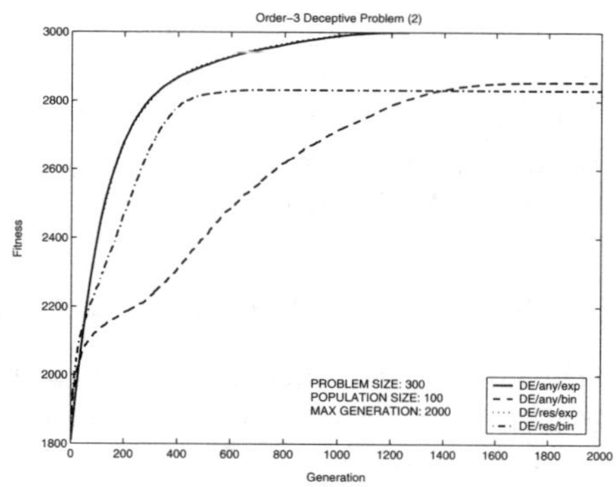

(b) Order-3

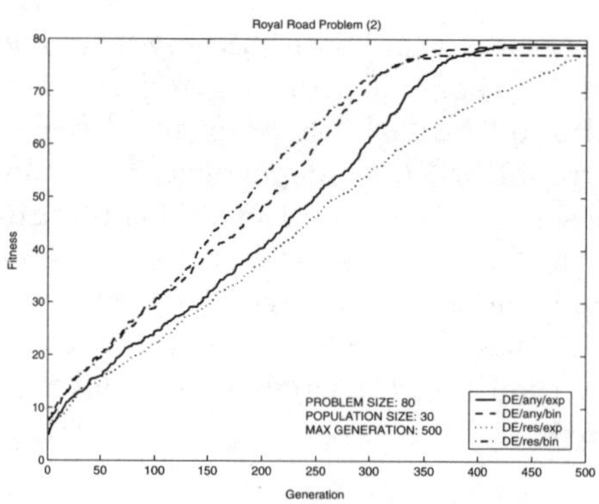

(c) Royal Road

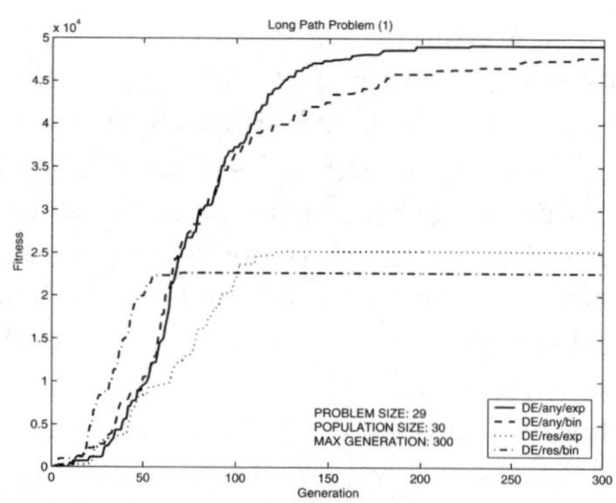

(d) Long Path

Fig. 1. The average-best against generation for the test problems seriously decreases the diversity. Adopting any-change mutation can explore the search space better to discover more building blocks without dominated by only a few of them.

Long Path Problems

For Long Path problems, it is observed with significance that `DE/any/exp` and `DE/any/bin` perform much better than `DE/res/exp` and `DE/res/bin`, due to any-change mutation's better exploration ability. Restricted-change mutation also brings the problem of *premature convergence*. The decoding algorithm in [10] implies these decision variables are highly linked to decide the fitness; thus, the exponential crossover with linkage specification almost guarantees a better performance over the binomial crossover in the study.

6.4 Brief Discussions

From the above experimental study, we get an in-depth understanding of how DE works for binary/discrete domain, with the behaviors fully justified. The assumptions made for the operator templates are also reflected from the behaviors of the derived operators.

7 Conclusions

In this paper, DE operator templates are formally defined using forma analysis to derive binary-adapted DE operators. An empirical study has been carried out to study these operators with results comparable to those in the EC literature. Through both the tuning results and the execution results, we obtained insights to how DE works for binary optimization problems, with the behaviors fully justified by our analysis.

Due to the fact that the application of DE for discrete optimization problems is very limited, this preliminary work with binary encoding can be regarded as a good starting point for our future research on this issue. Generalizing DE to some difficult real-world discrete optimization problems (e.g. TSP and JSSP) is an obvious next step.

References

1. Price, K.V., Storn, R., Lampinen, J.: Differential Evolution: A Practical Approach to Global Optimization. Springer-Verlag, London, UK (2005)
2. Storn, R., Price, K.: Differential evolution–a simple and efficient adaptive scheme for global optimization over continuous spaces. Technical Report TR–95–012, ICSI (1995)
3. Radcliffe, N.: The algebra of genetic algorithms. Annals of Maths and Artificial Intelligence **10** (1994) 339–384

4. Surry, P.D.: A Prescriptive Formalism for Constructing Domain-specific Evolutionary Algorithm. PhD thesis, University of Edinburgh, Edinburgh, Scotland, UK (1998)
5. Tuson, A.L.: No optimization without representation: a knowledge based systems view of evolutionary/neighbourhood search optimization. PhD thesis, University of Edinburgh, Edinburgh (1999)
6. Gong, T., Tuson, A.L.: Formal descriptions of real parameter optimisation. In: Proceedings of IEEE Congress on Evolutionary Computation, July 16-21, 2006, Vancouver, BC, Canada, IEEE Press (2006) 2119–2126
7. Ackley, D.: A Connectionist Machine for Genetic Hillclimbing. Kluwer Academic Publishers, Boston, MA (1987)
8. Goldberg, D.: Genetic algorithms and walsh functions: Part I, a gentle introduction. Complex Systems **3** (1989) 129–152
9. Mitchell, M., Forrest, S., Holland, J.: The royal road for genetic algorithms: Fitness landscapes and GA performance. In Varela, F., Bourgine, P., eds.: Proceeding of the 1st European Conference on Artificial Life, MIT Press (1992) 245–254
10. Horn, J., Goldberg, D.E., Deb, K.: Long path problems. In Davidor, Y., Schwefel, H.P., Männer, R., eds.: PPSN. Volume 866 of Lecture Notes in Computer Science., Springer (1994) 149–158
11. Horn, J., Goldberg, D.E.: Genetic algorithm difficulty and the modality of fitness landscapes. In Whitley, L.D., Vose, M.D., eds.: FOGA, Morgan Kaufmann (1994) 243–269
12. Rudolph, G.: How mutation and selection solve long path problems in polynomial expected time. Evolutionary Computation **4** (1996) 195–205

Part VI

Soft Computing in Civil Engineering and Other Applications

Prioritization of Pavement Stretches Using Fuzzy MCDM Approach – A Case Study

A.K. Sandra*, V.R. Vinayaka Rao, K.S. Raju, and A.K. Sarkar

Civil Engineering Group,
Birla Institute of Technology and Science (BITS) – Pilani, Rajasthan, India
aksandra@bits-pilani.ac.in

Abstract. Effective pavement management requires the prioritization of the road stretches for logical disbursement of the funds available towards maintenance of the pavement. Several methods have been developed and implemented towards this goal. However, the uncertainty involved with some of the parameters has not been addressed adequately in most of the works. One such parameter has been identified as the severity of distress which is difficult to assess accurately. Hence a Fuzzy Multi Criteria Decision Making (FMCDM) approach has been proposed in this paper. For demonstration of the approach, pavement distresses with respect to their extent and severity have been collected over a number of stretches. In addition, an expert opinion survey has been carried out to quantify the influence of these parameters on the functional condition of the pavement. Priority Index (PI) has been worked out, based on which the ranking of the stretches has been arrived at.

Keywords: Pavement Stretches Prioritization, Pavement Distress Parameters, Priority Index, FMCDM, Fuzzy Logic.

1 Background

Proper upkeep of the existing pavements is essential in addition to the development of new highway and road links for the economical growth of any country. Inadequate and inappropriate maintenance policies adopted by several countries is resulting in heavy financial loss in the form of ever increasing Road User Costs (RUC) and causes discomfort to the road users. This problem multiplies itself several folds with the increasing road length.

In a developing country like India, the funds being limited for the maintenance of the existing pavement, it is important to utilize the money in the most appropriate manner. A wholesome pavement management approach at both network and project levels will definitely result in optimal solution for road maintenance.

Pavement Management System (PMS) is an ideal tool in this kind of situation and offers a methodical way of upkeeping the road network in its best possible serviceability level. In any PMS, prioritization plays a major role especially when the funds available for road maintenance are limited [Ramadhan et al. (1999)]. Several researchers have worked in this direction to develop priority ranks for the available

* Corresponding author.

A. Saad et al. (Eds.): Soft Computing in Industrial Applications, ASC 39, pp. 265–278, 2007.
springerlink.com

network of roads. Hass et al. (1994), FHWA (1994) and NCHRP (1995) have discussed in detail, different ranking and optimization methods. Reddy & Veeraragavan (2002) have proposed a Priority Index (PI) as a function of Pavement Distress Index (PDI) and prioritization factor for ranking the road network. Chen et al. (1993) and Sharaf (1993) have demonstrated the use of Composite Index (CI) method for prioritization and indicated that this technique has yielded optimal solutions for the case studies considered. However, in these studies, the extent and severity of failures are considered simultaneously in quantifiable terms inspite of the fact that the magnitude of severity of distress has unavoidable uncertainty. Golabi & Pereira (2003) have used optimization process to work out the solutions for prioritization of pavement sections based on road condition, traffic and environmental characteristics. NCHRP (1995), Broten (1996), Golabi & Pereira (2003) have discussed the applicability of various optimization tools such as Linear Programming, Integer Programming, Markov Decision Analysis, Dynamic Programming etc. for network prioritization. Broten (1996) has indicated that the use of optimization tool alone will not be able to solve the problem in a wholesome manner since these tools tend to achieve the network goals while failing from the point of view of ensuring optimal treatment at a given point of time.

In such uncertain situations, Fuzzy Multi-Criteria Decision Making (FMCDM) approach provides an ideal option and as such it has been tried and tested by number of researchers for raking alternatives for different situations [Bandara & Gunaratne (2001), Chen-Tung Chen (2001), Sanja & Radivoj (2002), Mei-Fang Chen et al.(2003), Prakash (2003), Wang & Fenton]. Bandara & Gunaratne (2001) have used fuzzy approach for prioritization and have included both severity and extent of pavement failure in the overall process. However, fuzzy approach has to be utilized only when uncertainty is predominant. In other words, when particular parameter is quantifiable with fair degree of accuracy, this approach need not be used. It is in this direction that the present work has been carried out with fuzzy logic being applied to only those parameters which are predominantly uncertain in nature.

Most usually, the policies regarding the maintenance of roads are taken with respect to the functional condition of road as it directly affects the RUC. Hence, Different functional distress parameters are collected with respect to both severity and extent of failure. Since it is possible to measure the extent of failure with high accuracy, this parameter has been proposed as a direct parameter in the prioritization process. As such, there is an inbuilt ambiguity while assessing the severity of the distress. Hence the Fuzzy approach has been suggested to assess the severity. The broad objectives of this work are mentioned below.

- To conduct expert opinion survey for assessing the level of influence of selected pavement deterioration parameters on the functional condition of the pavement.
- To collect the different distresses influencing the functional condition of pavement on selected road stretches.
- Prioritizing the road pavement sections using Fuzzy Multi Criteria Decision Making FMCDM approach.

2 Data Collection

Primary data has been collected through field investigations as well as expert opinion surveys. The opinion of selected experts from all over India has been sought to ascertain the influence of different distress parameters on the functional condition of the pavement. The distresses considered are cracking, potholes, rutting, patching, raveling and edge failure with respect to three severity levels namely low, medium and high. A questionnaire was prepared and sent to selected experts all over India for this purpose. In addition, they were also given photographic cues and clues with a view to reduce the possibility of variability and bias amongst themselves. Further. they were asked to indicate their preferences regarding the influence of severity of various distress parameters in terms of linguistic variables such as Negligible (N), Low (L), Moderate (M), High (H) and Very High (VH) as it would be difficult to express the weights in quantifiable terms. The responses given by a group of 15 experts have been summarized and presented in *Table 1.*

Table 1. Summery of Experts Opinions

	Experts														
Criteria	**E1**	**E2**	**E3**	**E4**	**E5**	**E6**	**E7**	**E8**	**E9**	**E10**	**E11**	**E12**	**E13**	**E14**	**E15**
CL	L	N	N	L	N	N	N	N	N	N	L	L	N	L	N
CM	M	M	L	M	N	L	L	L	L	L	M	M	L	M	L
CH	H	H	M	H	L	M	M	VH	L	M	H	H	M	H	M
PL	M	L	L	M	N	M	M	M	M	M	M	L	L	L	M
PM	H	H	M	H	L	H	H	H	H	H	H	M	M	M	H
PH	VH	VH	H	VH	H	VH	VH	VH	VH	VH	VH	VH	H	H	VH
RL	M	L	L	N	N	M	L	L	L	L	N	N	N	N	N
RM	H	M	M	L	N	H	M	M	L	M	L	L	L	L	N
RH	VH	H	H	M	M	VH	VH	H	M	H	M	H	M	H	M
PAL	L	N	L	N	N	M	N	N	M	M	N	N	L	L	L
PAM	M	M	M	L	L	H	L	M	M	H	L	L	M	M	M
PAH	H	H	H	M	M	VH	M	VH	H	VH	M	M	M	H	H
RUL	M	N	L	N	L	L	N	N	L	L	L	N	L	N	L
RUM	H	M	L	L	M	M	L	L	M	M	M	L	M	M	M
RUH	VH	VH	M	M	VH	H	M	M	H	H	H	H	M	H	H
EL	M	L	L	L	N	N	L	L	L	N	N	N	N	L	N
EM	H	H	M	M	L	L	M	H	M	L	L	L	L	M	N
EH	VH	VH	H	H	VH	M	H	VH	H	M	M	H	H	H	N

LEGEND

CL=Low level cracking
CM=Medium level cracking
CH =High level cracking
PL=Low level potholes
PM=Medium level potholes
PH =High level potholes
RL=Low level raveling
RM=Medium level raveling
RH=High level raveling
PAL=Low level patching
PAM=Medium level patching
PAH=High level patching
RUL=Low level rutting
RUM=Medium level rutting
RUH=High level rutting
EL=Low level edge failure
EM=Medium level edge failure
EH=High level edge failure
N=Negligible
L= Low
M=Medium
H=High
VH=Very High

Note: Definitions regarding the severity of different distresses have been detailed in *APPENDIX 1.*

For collecting pavement condition data, four different stretches dominated by different types of functional failures have been chosen in the state of Rajasthan, India for a total length of 25km. Care has been taken to see that these stretches are representing different functional classes of highways *viz.* National Highways (NH), State Highways (SH) and Major District Roads (MDR).

A team of enumerators have been engaged in collecting the severity and extent of all the failures mentioned in the Expert Opinion Survey proforma. Each 1km length of road was divided into 20 stretches of 50m long. For maintaining uniformity in the data, all the enumerators were trained in the field and also were supplied with photographic cues and clues. Distress parameters such as cracking, potholes, raveling, patching, edge failure were measured in %of total area of pavement stretch and rutting was measured in millimeters. To demonstrate the proposed methodology of prioritization a typical stretch of 1km length has been considered for further consideration.

3 Pavement Prioritization Through Fuzzy Logic

3.1 Fuzzy Numbers

The fuzzy set theory was proposed by Zadeh, L. A. in 1965, to represent the uncertainty involved in any situation in linguistic terms. A fuzzy number $\widetilde{A}$ is a fuzzy set, and its membership function is $\mu_{\tilde{A}}(x): R \rightarrow [0, 1]$ [Dubois & Prade (1978); Yeou-Geng Hsu et al. (2003); Mei-Fang Chen et al.(2003)], where *'x'* represents the criteria. A linear membership function is the widely used and the corresponding fuzzy numbers are called Triangular Fuzzy Numbers (TFNs). TFNs are the special class of

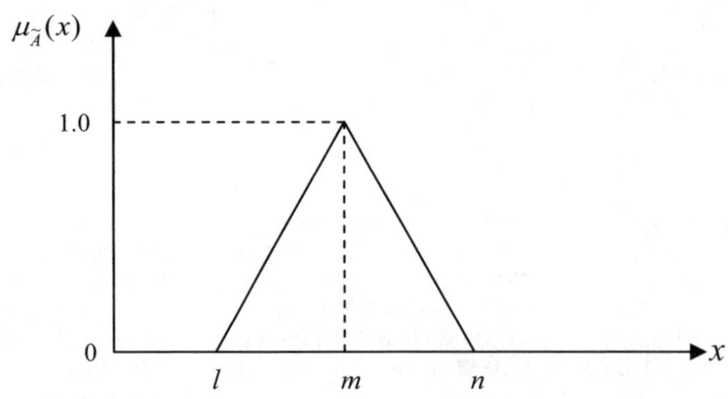

Fig. 1. Membership Function for the Triangular Fuzzy Numbers

fuzzy numbers whose membership is defined by three real numbers *(l, m, n)* i.e. $\mu_{\tilde{A}}(x)$ = *(l, m, n))*, which is pictorially shown in *Fig. 1.* The TFNs can be expressed as follows.

$$\mu_{\tilde{A}}(x) = \begin{cases} \dfrac{(x-l)}{(m-l)}; & l \le x \le m; \\ \dfrac{(n-x)}{(n-m)}; & m \le x \le n; \\ 0; & \text{Otherwise}; \end{cases} \quad (1)$$

3.2 Operations on Fuzzy Numbers

When $\widetilde{A} = (l,m,n)$ and $\widetilde{B} = (p,q,r)$ two TFNs, the general operations are as follows [Mei-Fang Chen et al. (2003), Prakash (2003)].

- Addition of two fuzzy number
  $(l,m,n) \oplus (p,q,r) = (l+p, m+q, n+r)$
- Subtraction of two fuzzy numbers
  $(l,m,n) \Theta (p,q,r) = (l-r, m-q, n-p)$
- Multiplication of any real number "*k*" and a fuzzy number
  $k \otimes (l,m,n) = (kl, km, kn)$

3.3 Prioritization Process

In the present study, pavement section alternatives have been prioritized based on the methods proposed by various researchers [Bandara & Gunaratne (2001), Chen-Tung Chen (2001), Huang (1989)]. Prioritization process is explained in the following stages.

Stage 1: Data collected in the filed is being normalized in the scale of 0 to 100 with respect to the maximum value in the series through a simple normalization as shown below.

Normalized Data Point = (Data Point) x 100 / (Mode of the Data Series) (2)

A summary of normalization data is presented in *Table 2.* Further, these values are being arranged into 10 groups with a uniform interval of 10 and ratings have been given, which is presented in *Table 3.*

These rating values are being arranged in a matrix form, named as Rating Matrix $(R_{ij})_{N \times M}$ with each row representing alternative (A_1, A_2......., A_N) and each column representing criteria. The Rating Matrix has been presented in *Table 4.*

Table 2. Normalized Pavement Condition data on Selected Stretch

Stretch No.	Criteria																	
	CL	CM	CH	PL	PM	PH	RL	RM	RH	PAL	PAM	PAH	RUL	RUM	RUH	EL	EM	EH
A_1	0	0	0	6	3	0	3	17	84	16	100	67	0	0	0	46	0	0
A_2	0	0	0	3	3	0	38	54	7	56	58	0	0	0	0	37	0	0
A_3	7	0	0	14	0	0	56	34	3	0	13	11	0	73	0	0	0	0
A_4	100	0	0	3	0	0	52	17	0	0	0	0	0	0	0	17	0	0
A_5	8	14	0	7	2	3	58	17	48	0	6	0	0	0	100	33	0	0
A_6	69	100	0	5	0	0	50	31	0	0	0	0	0	82	0	10	0	0
A_7	0	0	0	3	7	4	21	86	33	0	0	0	100	0	0	0	0	0
A_8	0	0	0	1	1	16	0	35	53	0	42	0	0	55	0	100	0	0
A_9	0	0	0	0	0	24	0	83	48	0	67	0	0	55	0	12	0	0
A_{10}	0	0	0	3	3	23	0	42	100	0	0	27	0	0	0	0	0	0
A_{11}	0	0	0	100	100	100	0	74	0	0	11	0	0	0	0	0	0	0
A_{12}	0	0	0	3	5	13	2	0	70	30	92	100	0	0	0	0	0	0
A_{13}	0	0	0	25	39	20	0	21	59	0	3	0	0	73	0	0	73	73
A_{14}	0	0	0	8	9	39	0	100	15	0	15	0	0	0	0	25	30	0
A_{15}	0	0	0	2	4	62	15	24	69	100	0	0	0	100	0	27	0	100
A_{16}	0	0	0	3	3	45	31	38	26	90	0	0	0	0	0	33	100	0
A_{17}	8	0	0	4	0	0	100	0	0	0	0	0	0	0	0	0	0	0
A_{18}	0	0	0	1	2	22	73	28	11	0	0	0	0	91	0	0	0	0
A_{19}	15	0	0	3	0	0	90	0	0	0	22	0	0	0	0	0	0	0
A_{20}	15	0	0	6	2	0	71	16	0	1	53	0	0	0	0	0	0	0

Table 3. Ratings for the Normalized Values

Normalized Value	0-10	11-20	21-30	31-40	41-50	51-60	61-70	71-80	81-90	91-100
Rating	1	2	3	4	5	6	7	8	9	10

Table 4. Rating Matrix

$[R_{ij}]_{N \times M} =$

CL	CM	CH	PL	PM	PH	RL	RM	RH	PAL	PAM	PAH	RUL	RUM	RUH	EL	EM	EH	
1	1	1	1	1	1	1	2	9	2	10	7	1	1	1	5	1	1	A_1
1	1	1	1	1	1	4	6	1	6	6	1	1	1	1	4	1	1	A_2
1	1	1	2	1	1	6	4	1	1	2	2	1	8	1	1	1	1	A_3
10	1	1	1	1	1	6	2	1	1	1	1	1	1	1	2	1	1	A_4
1	2	1	1	1	1	6	2	5	1	1	1	1	1	10	4	1	1	A_5
7	10	1	1	1	1	6	4	1	1	1	1	1	9	1	2	1	1	A_6
1	1	1	1	1	1	3	9	4	1	1	1	10	1	1	1	1	1	A_7
1	1	1	1	1	2	1	4	6	1	5	1	1	6	1	10	1	1	A_8
1	1	1	1	1	3	1	9	5	1	7	1	1	6	1	2	1	1	A_9
1	1	1	1	1	3	1	5	10	1	1	3	1	1	1	1	1	1	A_{10}
1	1	1	10	10	10	1	8	1	1	2	1	1	1	1	1	1	1	A_{11}
1	1	1	1	1	2	1	1	8	4	10	10	1	1	1	1	1	1	A_{12}
1	1	1	3	4	3	1	3	6	1	1	1	1	8	1	1	8	8	A_{13}
1	1	1	1	1	4	1	10	2	1	2	1	1	1	1	3	4	1	A_{14}
1	1	1	1	1	7	2	3	7	10	1	1	1	10	1	3	1	10	A_{15}
1	1	1	1	1	5	4	4	3	10	1	1	1	1	1	4	10	1	A_{16}
1	1	1	1	1	1	10	1	1	1	1	1	1	1	1	1	1	1	A_{17}
1	1	1	1	1	3	8	3	2	1	1	1	1	10	1	1	1	1	A_{18}
2	1	1	1	1	1	9	1	1	1	3	1	1	1	1	1	1	1	A_{19}
2	1	1	1	1	1	8	2	1	1	6	1	1	1	1	1	1	1	A_{20}

Stage 2: The linguistic variables utilized for expressing the severity of different distresses have been expressed as TFNs. TFNs assigned for various linguistic variables are shown in *Table 5.*

Table 5. Triangular Fuzzy Numbers (TFNs) for Linguistic Variables

Linguistic Variable	TFN
Negligible	(0, 0, 0.1)
Low	(0, 0.1, 0.3)
Medium	(0.3, 0.5, 0.7)
High	(0.7, 0.9, 1)
Very High	(0.9, 1, 1)

Stage 3: Experts opinion available for the various distress severities in the form of linguistic variable as presented in *Table 1* are being converted into fuzzy numbers. To normalize differences existing in expert opinion, simple average of fuzzy numbers for all the linguistic variables has been calculated and the corresponding weights are being worked out and presented in the *Table 6.* Fuzzy weights for all criteria can be expressed in the form of following row matrix.

$$\widetilde{W} = [\widetilde{w}_1, \widetilde{w}_2 ...., \widetilde{w}_M] \tag{3}$$

Where, $\widetilde{w}_1, \widetilde{w}_2 ...., \widetilde{w}_M$ are the fuzzy weights for all criteria expressed in Triangular Fuzzy Numbers i.e $\widetilde{w}_j = (\widetilde{w}_{j1}, \widetilde{w}_{j2}, \widetilde{w}_{j3})$ $\forall$ *j= 1, 2, 3.......M*

Table 6. Fuzzy weights for various distress parameters

Criteria	Fuzzy Weight
CL	(0.000, 0.033, 0.167)
CM	(0.120, 0.253, 0.447)
CH	(0.460, 0.640, 0.787)
PL	(0.180, 0.333, 0.527)
PM	(0.547, 0.740, 0.873)
PH	(0.847, 0.973, 1.000)
RL	(0.040, 0.107, 0.260)
RM	(0.193, 0.327, 0.500)
RH	(0.580, 0.760, 0.880)
PAL	(0.060, 0.133, 0.287)
PAM	(0.253, 0.420, 0.607)
PAH	(0.580, 0.760, 0.880)
RUL	(0.020, 0.087, 0.247)
RUM	(0.227, 0.393, 0.587)
RUH	(0.607, 0.787, 0.900)
EL	(0.020, 0.080, 0.233)
EM	(0.240, 0.387, 0.560)
EH	(0.627, 0.787, 0.880)

Stage 4: Fuzzy evaluation value $(\tilde{p}_i)$ is then calculated by multiplying the rating matrix with the weight matrix and summed up for all the stretches, which are presented in *Table 7*. This process is mathematically expressed as follows.

$$\tilde{P}_i = \sum_{j=1}^{M} R_{ij} \otimes \tilde{w}_j, \quad \forall\ i=1,2,.......N \ and \ \forall\ j=1,2,3.......M \tag{4}$$

Table 7. Fuzzy Evaluation Values for all the stretches

Stretch No.	Fuzzy Evaluation value $(\tilde{p}_i)$
A_1	(16.33, 23.20, 30.12)
A_2	(8.31, 12.96, 19.07)
A_3	(8.98, 13.78, 19.54)
A_4	(6.01, 9.24, 14.15)
A_5	(13.95, 19.47, 25.19)
A_6	(9.29, 15.22, 23.37)
A_7	(9.15, 13.89, 20.00)
A_8	(12.25, 18.12, 24.98)
A_9	(13.83, 20.17, 26.95)
A_{10}	(14.45, 19.61, 24.30)
A_{11}	(21.37, 29.13, 36.33)
A_{12}	(18.19, 25.31, 32.02)
A_{13}	(20.23, 28.25, 35.88)
A_{14}	(11.47, 16.36, 21.75)
A_{15}	(22.85, 31.14, 39.41)
A_{16}	(13.61, 19.63, 26.98)
A_{17}	(5.96, 8.96, 12.96)
A_{18}	(10.58, 15.65, 21.60)
A_{19}	(6.43, 9.73, 14.08)
A_{20}	(7.34, 11.21, 16.14)

Stage 5: To establish the relative preference of all the stretches, difference between all combinations of the fuzzy values has been computed. This is mathematically expressed as

$$(\tilde{F}_{ij}) = (\tilde{p}_i - \tilde{p}_j) \qquad \forall\, i=1 \ to \ N \qquad \forall\, j=1 \ to \ N \quad \text{and} \quad i \neq j \tag{5}$$

It is to be noted that $\tilde{p}_i$ and $\tilde{p}_j$ are the Triangular Fuzzy Numbers hence $(\tilde{p}_i - \tilde{p}_j)$ is also Triangular Fuzzy Number. A sample of these values are summarized and presented in *Table 8*.

Table 8. Difference between two fuzzy evaluation values (Sample data)

$\tilde{p}_1 - \tilde{p}_2$	(-2.73, 10.24, 21.81)
$\tilde{p}_1 - \tilde{p}_3$	(-3.21, 9.42, 21.14)
:	:
$\tilde{p}_{19} - \tilde{p}_{20}$	(-6.74, 1.48, 9.71)

Stage 6: The fuzzy preference relation matrix *[E]* has been developed, to know the degree of preference stretch A_i over stretch A_j.

$$E = \begin{bmatrix} e_{11} & e_{12} & \cdots & e_{1N} \\ e_{21} & e_{22} & \cdots & e_{2N} \\ \cdots & \cdots & \cdots & \cdots \\ e_{N1} & e_{N2} & \cdots & e_{NN} \end{bmatrix} \tag{6}$$

Where, e_{ij} is the real number indicates the degree of preference between the respective i^{th} and j^{th} pavement stretches. It has been calculated using positive (S_{ij}^{+}) and negatives areas (S_{ij}^{-}) of difference between two fuzzy values ($\tilde{p}_i - \tilde{p}_j$).

$$e_{ij} = \frac{S_{ij}^{+}}{S_{ij}^{+} + \left|S_{ij}^{-}\right|} \tag{7}$$

Where, $(S_{ij}^{+} + \left|S_{ij}^{-}\right|)$ = Total area of ($\tilde{p}_i - \tilde{p}_j$)

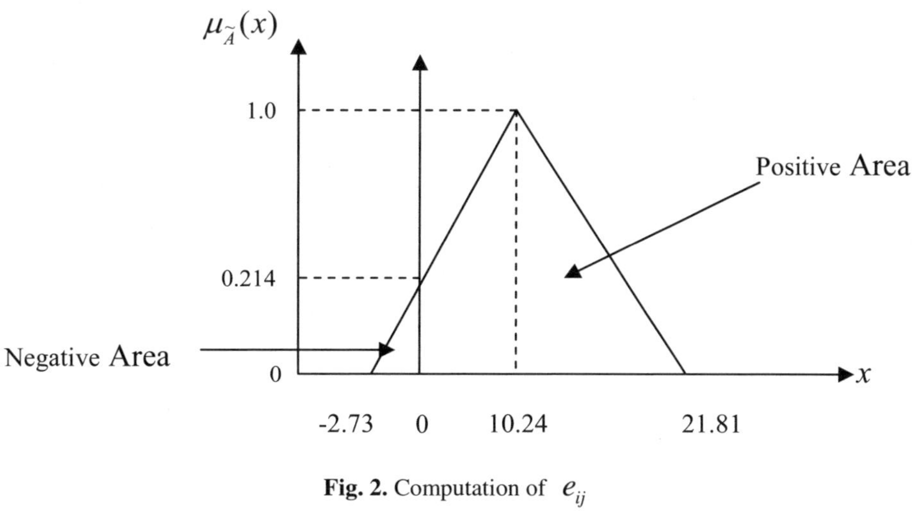

Fig. 2. Computation of e_{ij}

Positive and negative areas have been computed using the membership function $[\mu_{\tilde{F}_{ij}}(x)]$ of the $(\tilde{p}_i - \tilde{p}_j)$. An example of computation of e_{ij} is shown in *Fig. 2.*

For example, if the $\tilde{F}_{12} = (\tilde{p}_1 - \tilde{p}_2) = (2.73, 10.24, 21.81)$.

Total area form the Fig 5 = 12.27 Positive area = 11.97 Negative area= 0.292

$$e_{12} = (11.97)/12.27 = 0.98$$

Here $e_{ii} = 0.5$ and $e_{ij} + e_{ji} = 1.0$. If $e_{ij} > 0.5$ the stretch A_i is to be given priority over stretch A_j and vice versa. Computed values of e_{ij} for all the combinations are summarized and presented in *Table 9.*

Stage 7: Priority Index (PI) for all the pavement stretches are computed from the fuzzy preference relation matrix using the following mathematical form.

$$(PI)_i = \sum_{j=1}^{n} (e_{ij} - 0.5) \qquad \forall\ i = 1 \text{ to } N \tag{8}$$

Based on the PI, all the stretches have been ranked and presented in *Table 10.* The prioritization process, as explained in the above stages is quite complex and cumbersome due to a large number of stretches and criterion. Hence, a code has been developed in MATLAB [(www.mathworks.com)] and being used in the present study.

From *Table 10,* it can be noted that section no. 15 with priority index of 7.35 is to be given the first priority and section no. 17 with priority index of -6.95 is to be given last priority.

Table 9. Fuzzy Preference Relation Matrix

$[E_{ij}]_{N \times N} =$

0.50	0.98	0.97	0.98	0.75	0.88	0.96	0.79	0.69	0.77	0.18	0.37	0.22	0.90	0.11	0.71	0.95	0.91	0.98	1.00
0.02	0.50	0.44	0.81	0.10	0.32	0.42	0.17	0.09	0.09	0.01	0.00	0.00	0.25	0.04	0.10	0.86	0.30	0.80	0.67
0.03	0.56	0.50	0.85	0.13	0.36	0.48	0.20	0.11	0.12	0.01	0.01	0.00	0.29	0.03	0.12	0.90	0.34	0.84	0.73
0.02	0.19	0.15	0.50	0.00	0.10	0.14	0.02	0.00	0.00	0.18	0.06	0.12	0.04	0.23	0.00	0.56	0.07	0.46	0.31
0.25	0.90	0.87	1.00	0.50	0.74	0.86	0.59	0.44	0.50	0.04	0.15	0.07	0.74	0.01	0.46	0.99	0.77	1.00	0.98
0.12	0.68	0.64	0.90	0.26	0.50	0.62	0.33	0.23	0.25	0.01	0.06	0.02	0.45	0.00	0.24	0.94	0.49	0.90	0.81
0.04	0.58	0.52	0.86	0.14	0.38	0.50	0.21	0.13	0.13	0.01	0.01	0.00	0.31	0.02	0.14	0.91	0.36	0.86	0.74
0.21	0.83	0.80	0.98	0.41	0.67	0.79	0.50	0.36	0.41	0.03	0.12	0.05	0.65	0.01	0.38	1.00	0.69	0.98	0.94
0.31	0.91	0.89	1.00	0.56	0.77	0.87	0.64	0.50	0.57	0.08	0.20	0.11	0.77	0.04	0.52	1.00	0.80	1.00	0.98
0.23	0.91	0.88	1.00	0.50	0.75	0.87	0.59	0.43	0.50	0.03	0.13	0.05	0.75	0.01	0.46	0.99	0.78	1.00	0.98
0.82	0.99	0.99	0.82	0.96	0.99	0.99	0.97	0.92	0.97	0.50	0.73	0.55	1.00	0.37	0.93	0.73	1.00	0.81	0.91
0.63	1.00	0.99	0.94	0.85	0.94	0.99	0.88	0.80	0.87	0.27	0.50	0.32	0.96	0.18	0.80	0.88	0.96	0.93	0.98
0.78	1.00	1.00	0.88	0.93	0.98	1.00	0.95	0.85	0.95	0.45	0.68	0.50	0.99	0.33	0.90	0.81	0.99	0.87	0.95
0.10	0.75	0.71	0.96	0.26	0.55	0.69	0.35	0.23	0.25	0.00	0.04	0.01	0.50	0.00	0.25	0.99	0.56	0.96	0.88
0.89	0.96	0.97	0.77	0.99	1.00	0.98	0.99	0.96	0.99	0.63	0.82	0.67	1.00	0.50	0.96	0.66	1.00	0.75	0.87
0.29	0.90	0.88	1.00	0.54	0.76	0.86	0.62	0.48	0.54	0.07	0.20	0.10	0.75	0.04	0.50	1.00	0.78	1.00	0.97
0.05	0.14	0.10	0.44	0.01	0.06	0.09	0.00	0.00	0.01	0.27	0.12	0.19	0.01	0.34	0.00	0.50	0.03	0.40	0.25
0.09	0.70	0.66	0.93	0.23	0.51	0.64	0.31	0.20	0.22	0.00	0.04	0.01	0.44	0.00	0.22	0.97	0.50	0.93	0.84
0.02	0.20	0.16	0.54	0.00	0.10	0.14	0.02	0.00	0.00	0.19	0.07	0.13	0.04	0.25	0.00	0.60	0.07	0.50	0.34
0.00	0.33	0.27	0.69	0.02	0.19	0.26	0.06	0.02	0.02	0.09	0.02	0.05	0.12	0.13	0.03	0.75	0.16	0.66	0.50

Table 10. Ranking of the Pavement Stretches

Stretch No.	Priority Index	Rank
A_1	4.60	5
A_2	-4.00	16
A_3	-3.38	15
A_4	-6.86	19
A_5	1.87	8
A_6	-1.56	13
A_7	-3.14	14
A_8	0.83	10
A_9	2.50	6
A_{10}	1.84	9
A_{11}	6.94	2
A_{12}	5.68	4
A_{13}	6.82	3
A_{14}	-0.97	11
A_{15}	7.35	1
A_{16}	2.27	7
A_{17}	-6.95	20
A_{18}	-1.56	12
A_{19}	-6.62	18
A_{20}	-5.64	17

4 Conclusions

The following conclusions have been drawn from the present work.

- The proposed Fuzzy Multi Criteria Decision Making approach is demonstrated with the data collected from the field and expert opinion and this approach can be extended for prioritization of any given road network.
- The developed software interface is expected to help in establishing the priorities with ease and there is no limitation as far as the number of roads in the given network is concerned.
- The road link which has the highest Priority Index (PI) will be given top priority and vice versa.
- The work can be extended by including more number of variables and the same philosophy can be extended for the additional variables considered.
- A more detailed Distress identification manual including clues and photographic cues, exclusively developed for Indian conditions, will improve the quality of the data as it helps in reducing the variability / bias seen amongst different experts.

References

1. Bandara, N & Gunaratne, M; "Current and Future Pavement Maintenance Prioritization Based on Rapid Visual Condition Evaluation", Journal of Transportation Engineering, Vol. 127, No.2, March/April (2001).
2. Broten, M; "Local agency pavement Management application guide", The Northwest Technology Transfer Center, Washington State Department of Transportation, Olympia, WA, (1996).
3. Chen, X., Weissmann, J., Dossey, T. and Hudson, W. R.; "URMS: a graphical urban roadway management system at network level", Transportation Research Record 1397, Transportation Research Board, Washington, DC, pp. 103-1, (1993).
4. Chen-Tung Chen; "A fuzzy approach to select the location of the distribution center", Fuzzy Sets and Systems 118, pp. 65-73, (2001).
5. Dubois, D & Prade, H; "Operations on Fuzzy Numbers", International Journal of system science, 9 (3), 613-626, (1978).
6. FHWA; "Pavement and Road Surface Management for Local Agencies" Course Notebook, prepared by Texas Transportation Institute Texas A&M University for the Pavement Division of the Federal Highway Administration, Washington, D.C (1994).
7. Golabi, K & Pereira, P; "Innovative Pavement Management and Planning System for Road Network of Portugal", journal of Infrastructure Systems, ASCE, June 2003, pp. 75-80 (2003).
8. Haas, R; Hudson, W. R.; and Zaniewski, J.; Modern Pavement Management, Krieger Publishing Co., Malablar, Florida, (1994).
9. Huang, C. C; "A study on the fuzzy ranking and its application on the decision support systems," Ph.D. dissertation, Tamkang Univ., Taiwan, R.O.C., 1989.
10. Mei-Fang Chen; Gwo-Hshiung Tzeng; Cherng G. Ding; "Fuzzy MCDM Approach Select Service Provider" The IEEE International Conference on Fuzzy Systems, (2003). http://ieeexplore.ieee.org/iel5/8573/27217/01209427.pdf?arnumber=1209427
11. NCHRP Synthesis 222; "Pavement Management Methodologies to Select Projects and Recommend Preservation Treatments", National Cooperative Highway Research Program, Transportation Research Board, Washington, D.C (1995).
12. Prakash, T.N; "Land Suitability Analysis for Agricultural Crops: A Fuzzy Multicrtiteria Decision Making Approach", Master of Science thesis submitted to International Institute for Geo-Information Science and Earth observation, Enschede, The Netherlands, December, (2003).
13. Ramadhan, R.H; Al-Abdul Wahhab & Duffuaa, S.O ; "The use of an analytical hierarchy process in pavement maintenance priority ranking" Journal of Quality in Maintenance Engineering, Vol. 5, No. 1, pp 25-39 (1999).
14. Reddy, B.B & Veeraragavan (2002); "Priority Ranking model for managing flexible pavements at network level" Technical Paper Published in 62nd annual session of Indian Roads Congress ·(IRC), 8th-11th January , Kochi, India, (2002).
15. Sanja, P & Radivoj, P; "A new fuzzy multi-criteria methodology for ranking alternatives", International Transactions in Operational Research, Res. 9, pp. 73-84, (2002).
16. Sharaf, E.; "Ranking versus simple optimization in setting pavement maintenance priorities: a case study from Egypt", Transportation Research Record 1397, Transportation Research Board, Washington DC, pp. 34-8, (1993).
17. Wang. W &Fenton. N; " Risk and Confidence Analysis for Fuzzy Multicriteria Decision Making", Risk Analysis and Decision Research Group, Department of Computer Science, Queen Mary University of London, Mile End Road, London. http://www.dcs.qmul.ac.uk/~norman/papers/Wang_Fenton_Risk_and_Confidence.pdf

18. www.mathworks.com
19. Yeou-Geng Hsu; Gwo-Hshing Tzeng & Joseph Z.shyu; " Fuzzy Multiple Criteria Selection of Government-sponsored frontier technology R&D projects", R&D Management, Blackwell Publishing Ltd. 33, 5, (2003).
20. Zadeh, L.A.; "Fuzzy Sets" Information and Control, Vol. 8,pp.338-353, (1965).

Appendix 1. Description of Pavement Distress Severity Levels

S.No	Type of Distress	Severity	Description
1	Cracking	Low	Width of the cracking is less than 3mm
		Medium	Width of the cracking is greater than 3mm and less than 6mm
		High	Width of the cracking is greater than 6mm
2	Potholes	Low	Depth of the pothole is less than 25mm
		Medium	Depth of the pothole is more than 25 mm and less than 50 mm
		High	Depth of the pothole is more than 50 mm
3	Ravelling	Low	The aggregate or binder has started to wear away but has not progressed significantly. The pavement appears only slightly aged and s lightly rough.
		Medium	The aggregate or binder has worn away and the surface texture is moderately rough and pitted. Loose particles may be present and fine aggregate is partially missing.
		High	The aggregate and/or binder have worn away significantly, and the surface texture is deeply pitted and very rough. Fine aggregate is essentially missing from the surface, and pitting extends to a depth approaching one half (or more) of the coarse aggregate size.
4	Patching	Low	Patch has low severity distress of any type including rutting < 6 mm; pumping is not evident
		Medium	Patch has moderate severity distress of any type or rutting from 6 mm to 12 mm; pumping is not evident.
		High	Patch has high severity distress of any type including rutting > 12 mm, or the patch has additional different patch material within it; pumping may be evident.
5	Rutting	Low	Barely noticeable, depth less than 6 mm
		Medium	Readily noticeable, depth more than 6 mm less than 25 mm
		High	Definite effect upon vehicle control, depth greater than 25 mm

Appendix 1. (*Continued*)

6	Edge Failure	Low	Appearance of edge step with a few initial cracks on the bituminous surface along the edge portion of the carriageway
		Medium	Appearance of edge step with a number of inter-connected high intensity cracks on the bituminous surface along the edge portion of the carriageway
		High	Permanent loss of part of carriageway and pothole formation along the edge portion

A Memetic Algorithm for Water Distribution Network Design

R. Baños[1], C. Gil[1], J.I. Agulleiro[1], and J. Reca[2]

[1] Departamento de Arquitectura de Computadores y Electrónica, Universidad de Almería, La Cañada de San Urbano s/n, 04120 Almería, Spain
{rbanos,cgil,jiagulleiro}@ace.ual.es

[2] Departamento de Ingeniería Rural, Universidad de Almería, La Cañada de San Urbano s/n, 04120 Almería, Spain
jreca@ual.es

Summary. The majority of real optimization problems cannot be solved exactly because they have very large and highly complex search spaces. One of these complex problems is the design of looped water distribution networks, which consists of determining the best way of conveying water from the sources to the users, satisfying their requirements. This paper is to present a new memetic algorithm and evaluate its performance in this problem. With the aim to establish an accurate conclusion, other four heuristic approaches have also been adapted, including simulated annealing, mixed simulated annealing and tabu search, iterated local search, and scatter search. Results obtained in two water distribution networks demonstrate that the memetic algorithm works better when the size of the problem increases.

Keywords: meta-heuristics, memetic algorithms, water distribution systems, optimization.

1 Introduction

Recently, the interest in the design of memetic algorithms (MA) [1] for solving high complexity optimization problems has increased remarkably. Like (GA) [2], MA are population-based methods, although in MA the population is not formed by individuals, but agents. These methods are inspired by models of natural systems that combine the evolutionary adaptation of a population with individual learning within the lifetimes of its members. In other words, MA combine evolutionary computation and local search aspects. Additionally, MAs are inspired by concept of a meme, which represents a unit of cultural evolution that can exhibit local refinement [3]. In the case of MAs memes refer to the strategies (e.g. local refinement, perturbation or constructive methods, etc) that are employed to improve individuals. The quality of MA has been proved in many real optimization problems [4].

One important real life optimization problem is the design of water distribution networks, i.e. the best way of conveying water from the sources to the

A. Saad et al. (Eds.): Soft Computing in Industrial Applications, ASC 39, pp. 279–289, 2007.
springerlink.com

users, satisfying their requirements. These systems have been usually branched networks due to their lower investment costs. However, other alternatives, like looped water distribution networks seem to be a suitable alternative to branched ones as their greater reliability can trade-off the slight increase in the cost of the network when closing some loops. When designing this class of systems, many decisions have to be taken: the network layout, the network connectivity, pipe diameters, capacity of the tanks, power of the pumps, etc.. Optimization of looped water distribution systems is a much more complex problem than optimization of branched ones, as the pipe flows are unknown variables. It can be regarded as a NP-hard [5] problem. Particularly, it is non-linear, constrained, non-smooth and non-convex, and hence multi-modal problem. In last decades, a large number of methods have been applied, including linear programming techniques [6, 7], non-linear optimization models [8], global optimization methods [9], heuristic algorithms [10, 11, 12].

The aim of this work is to present a new memetic algorithm for solving high complexity optimization problems. The behavior of this new method is evaluated in the design of looped water distribution networks. In our implementation, we consider it as a least-cost optimization problem with pipe diameters being decision variables, while pipe layout, connectivity and demands are imposed. Other meta-heuristic approaches previously proposed, including Simulated Annealing (SA) [13], the Mixed Simulated Annealing and Tabu Search (MSATS) [14], Iterated Local Search [15] using SA in as local searcher (ILSSA), and Scatter Search (SS) [16] have also been adapted to this problem. While SA, MSATS, ILSSA are local search based methods that use a single solution in the search process, SS and MA are evolutionary based methods.

The remainder of the paper is organized as follows. Section 2 offers a description of the new memetic algorithm here presented. Section 3 defines the looped water distribution network problem, including a brief description of two test networks used in the experimental analysis. Section 4 details the results obtained by all the methods when solving the test networks described in Section 4. Conclusions of this paper are provided in Section 5.

2 Description of the Memetic Algorithm

It is now well established that pure evolutionary algorithms are not the alternative solve high complex combinatorial problems, and that hybridization with other techniques can improve the efficiency of search process [17]. Memetic algorithms (MAs) [1] are extensions of evolutionary algorithms (EA) that apply local search processes in the agents, trying to improve their fitness by using, for example, hill-climbing. EAs and MAs have been applied in a number of different areas, for example operational research and optimization, automatic programming, machine and robot learning, economic models, immune systems, social systems, etc. [4]. Despite the good results obtained by some MAs, the process of designing efficient MAs often depends of the problem-specific details. In general, the basic structure of MA is formed by four main methods: a Initialization Method to generate a set of diverse initial solutions; a Reproduction Method to create a set of children from the agents of the main population (parents); a Combination Method to create the population of the next iteration with the best solutions of the parents and children; and a Local Optimization Method to improve the current agents.

Algorithm 1. Memetic algorithm pseudo-code

```
Input: PopSize, ChildRate, EntropyRate, StopCondition;
Population ← Initialization();
Population ← LocalOptimizer();
Repeat
   Children ← Reproduction(Population,ChildRate);
   Children ← LocalOptimizer(Children);
   Population ← UpdatePopulation(Population,Children);
   convergence ← EvaluateEntropy(Population,EntropyRate);
   If (convergence = TRUE) then
       Initialization();
While (StopCondition is not fulfilled);
```

Algorithm 1 describes the memetic method here implemented. Input parameters are the population size (PopSize), the percentage of parents used to obtain children (ChildRate), the entropy rate (EntropyRate) which determines the degree of similarity required to the individuals of the population to be re-initialized, and the number of iterations/evaluations which determine the stop condition. The population is initialized by generating random agents, and optimized by applying an hill climbing optimizer. While the stop condition is not fulfilled, the agents are improved using an iterative process. A reproduction method is the applied. It consist of taking a percentage of agents of the main population, modify them by applying a variation operator, and generate the children. They are optimized with the local optimization method, and the population is updated by combining the previous agents and the new ones generated in the reproduction phase. In this algorithm we include a tester of premature convergence based on the concept of entropy [18]. Given *EntropyRate*, the population is re-initialized if it is considered to have converged, i.e. if the fitness of the individuals is below this rate.

3 Problem Formulation and Test Networks

The mathematical formulation of the optimal design of a looped water distribution network has been set up in many previous works ([19, 20]). The objective is to minimize the network investment cost. Further, the problem is constrained by the physical laws of mass and energy conservation. Also, minimum pressure requirements for users, minimum and maximum flow velocities and pipe size restrictions are imposed. For this reason, the fitness of a certain solution (F)

is calculated as the sum of the cost of the pipes making up the network plus a penalty function applied to take into account nodal pressure head deficits (see Equation 1).

$$\mathrm{F}{=}\sum_{i=1}^{n_d} c_i L_i + c_p \sum_{j=1}^{n} \max[(\mathrm{hr_j} - \mathrm{h_j}), 0] \tag{1}$$

where: n_d is the number of possible diameter sizes, c_i the cost of the pipe of diameter i per unit length, L_i the total length of pipe i in the network, c_p the penalty multiplier, n the number of nodes, hr_j the required pressure head in node j, and h_j the actual pressure head computed by the hydraulic solver for node j. In these experiments, a very large penalty value (c_p=100000) was used in order to discard solutions with pressures below the requirements.

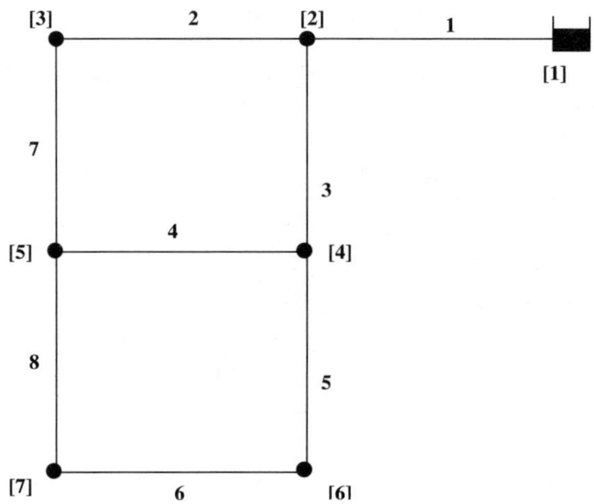

Fig. 1. Layout of Alperovits and Shamir's network

The first test problem is the network designed by Alperovits and Shamir [6]. It is the simple two-loop network, seven nodes and eight pipes arranged in two loops and is fed by gravity from a reservoir of 210 m fixed head. The minimum pressure limitation is 30 m above ground level for each node. There are 14 commercial diameters to be selected. Under this context, there exist $14^8 = 1,4758\ 10^9$ possible configurations. To be consistent with previous papers and to compare the results obtained, the pipe head losses were calculated using the Hazen-Williams equation [21]. The friction coefficient (C) value is 130. The values of the other parameters of the Hazen-Williams equation were the defaults of the EPANET 2.0 network analysis software (α=4.277, a=1.852, b=4.5871). The layout of this network is shown by Figure 1.

The second benchmark network is the pipeline network for water supply in Hanoi (Vietnam) proposed by Fujiwara and Khang [7]. This network, whose

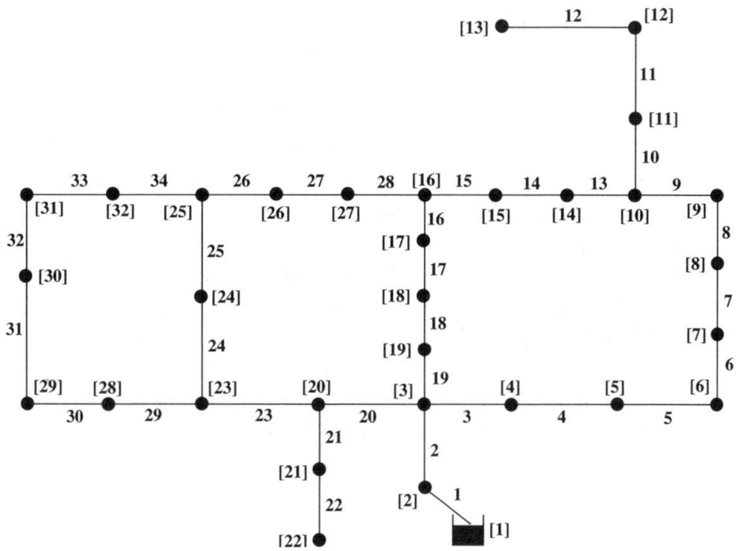

Fig. 2. Layout of Hanoi network

layout is displayed in Figure 2. consists of 32 nodes, 34 pipes, and 3 loops. No pumping station is considered since a single fixed-head source at an elevation of 100 m is available. The minimum pressure-head requirement at all notes is fixed at 30m. A set of 6 available commercial-diameter pipes is used. Under this context, there exist $6^{34} = 2,8651\ 10^{26}$ possible configurations. This network has been evaluated using the following parameters in Hazen-Williams equation: roughness coefficient C=130, α=4.277, a=1.852, b=4.5871.

4 Empirical Analysis

4.1 Programming Environment

The experimental analysis has been performed by using a recently proposed computer model, called MENOME (MEta-Heuristic pipe Network Optimization ModEl). MENOME integrates several meta-heuristic optimizers, a hydraulic network solver, a graphical user interface programmed in the Visual-Basic programming language, and database management module that uses ActiveX Data Objects (ADO) model. The model uses the well known, robust and tested network solver EPANET (Version 2.00.07) [22]. It performs extended period simulation of hydraulic and water quality behavior within pressurized pipe network. It employs the gradient method proposed by Todini and Pilati [23] for solving the mass and energy conservation equations. The EPANET Programmer's Toolkit is an extension of the EPANET simulation package that provides several functions that allow programmers to customize the use of the hydraulic and water quality solution engine provided by EPANET to their own applications. EPANET toolkit DLL is called from the main program written in Visual-Basic language. Two input data files are needed to run the model: the network configuration and the pipeline database. Network configuration is described in a standard EPANET file format, and the pipeline database is given in a relational database that must include two fields: pipe inner diameter and cost per unit length. MENOME include not only the MA described above, but also other four heuristic methods adapted to this problem: Simulated Annealing (SA) [13], Mixed Simulated Annealing and Tabu Search (MSATS) [14], Iterated Local Search (ILS) [15], and Scatter Search (SS) [16]. Some of these methods have been compared in other optimization problems [24].

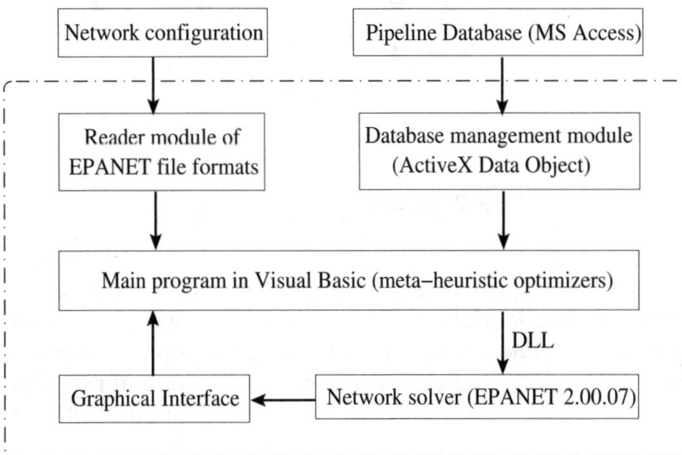

Fig. 3. MENOME flowchart

4.2 Parameter Settings

When comparing different meta-heuristics, it is very important to adjust the parameters according to their particular characteristics. The stop criterion in the experiments cannot be fixed to a number of iterations, since population-based methods (MA, SS) would require more runtime than non-population-based ones (SA, MSATS, ILSSA). Given this circumstance, the best way to guarantee the equality of conditions is that all the methods perform the same number of evaluations of the fitness function. That number of evaluations, n_e, should depend on the complexity of the network. The search space is a function of the number of links n_l and the number of possible pipe diameters n_d. Equation 2 has been proposed to establish the number of evaluations. Considering a multiplication coefficient K_m=1000, the resulting fitness function evaluations are 9169, and 26457 for the Alperovits and Shamir, and Hanoi networks, respectively.

$$n_e = K_m * n_l * \log_{10}(n_d) \quad (2)$$

With the aim of increasing the information about the performance of the metaheuristics, we execute several experiments before giving conclusions about the robustness of a method. In particular, we have executed the MA and the other heuristic adaptations the same number of executions using different parameters in both networks (see Table 1). Each execution of MA uses different percentages of agents to reproduce to obtain children (recombination phase) and several population sizes. In the Simulated Annealing based methods, each execution uses a different initial temperatures, Ti={250,150,100,75,50,25,10,5,2}, which allows analyze the impact of using several annealing scheduling. Cooling rates are adjusted in order to obtain a value of n_e.

Table 1. Parameters used in the experiments

SA	Ti	{500,250,150,100,75,50,25,10,5,2}
	Cooling Rate	f(Ti,n_e)
	Stop condition	temperature ¡ 0.01
MSATS	Ti	{500,250,150,100,75,50,25,10,5,2}
	Cooling Rate	f(Ti,n_e)
	Stop condition	temperature ¡ 0.01
	Tabu list length	5
ILSSA	Ti	{500,250,150,100,75,50,25,10,5,2}
	Cooling Rate	f(Ti,n_e)
	Stop condition	temperature ¡ 0.01
	Perturbation Rate	0.25
SS	Population/Reference Set Size	100/10
	Stop condition	n_e
MA	Population Size	50/200/500
	EntropyRate	5%
	Local Optimizer iterations	100
	% parents used in recombination	25%, 50%, 75%

4.3 Results and Discussion

Table 2 shows the average and minimum cost when solving the Alperovits and Shamir problem. Using their best parametric configuration, all them obtain a minimum cost of 419000 monetary units. However, if we calculate an average result of the nine executions, only the MA is able to obtain the best average result (419000), i.e. always obtains this result, while the other methods are slightly worse, less than 1.5% of difference. Figure 4 displays the evolution of the cost when the search advances, using their best parametric configuration. ILSSA has not been included due to each run of ILSSA depends on the previous one, and the best separate run of the other methods are included in this graphic. As we see, all the methods converge to 419000 monetary units.

Table 2. Results obtained in Alperovits and Shamir

	avg. cost	Δ	min. cost	Δ	best config.
SA	420100	1.00262	419000	1	Ti=50
ILSSA	424900	1.01408	419000	1	Ti=50
MSATS	424300	1.01265	419000	1	Ti=50
SS	426100	1.01695	419000	1	Ti=50
MA	419000	1	419000	1	INDIFFERENT

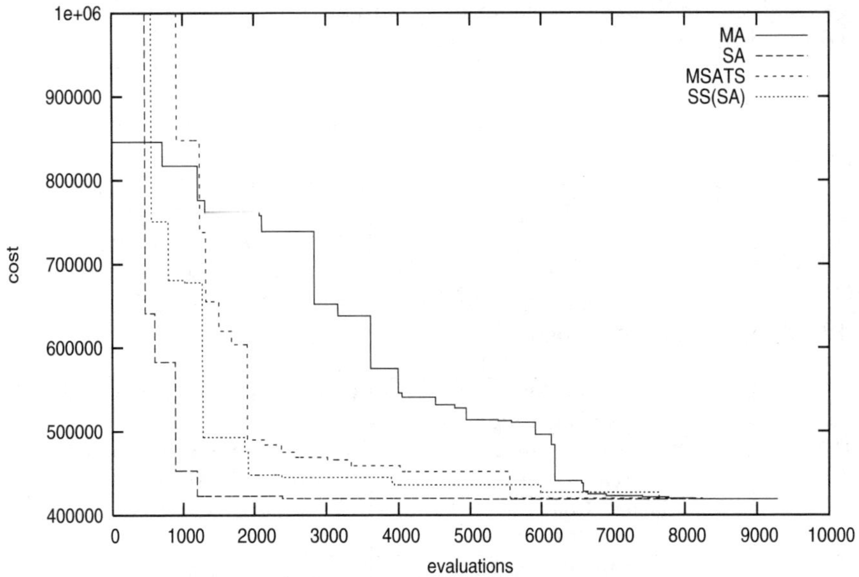

Fig. 4. Comparing MA, SA, MSATS, and SS at Alperovits network

Table 3 shows the average and minimum cost obtained by all them when solving the Hanoi problem. The best result is obtained by MA (6231422 monetary units), while SS also obtains a good result (6272752). The other methods, are slightly worse, but lesser than 2%. If we take into account the average result MA also obtains the best result, while the other methods obtains a higher cost. Figure 5 shows the tendency lines of each algorithm using their best parametric configuration. In this case not all the methods converge to the same result, although they are close to the best result obtained by MA.

Summarizing, all the methods obtain network configurations very similar in terms of cost for both test problems. However, this economic saving become a very important decision variable. For instance, choosing the configuration obtained by MA in Hanoi involves a cost reduction of at least 41330 monetary units in comparison with the best configuration returned by the other methods.

Table 3. Results obtained in Hanoi

	avg. cost	Δ	min. cost	Δ	best config.
SA	6483950	1.02986	6333207	1.01633	Ti=25
ILSSA	6510647	1.03411	6308024	1.01229	Ti=50
MSATS	6538453	1.03852	6352526	1.01943	Ti=50
SS	6688675	1.06238	6272752	1.00663	Ti=25
MA	6295909	1	6231422	1	200 agents, 75% parents

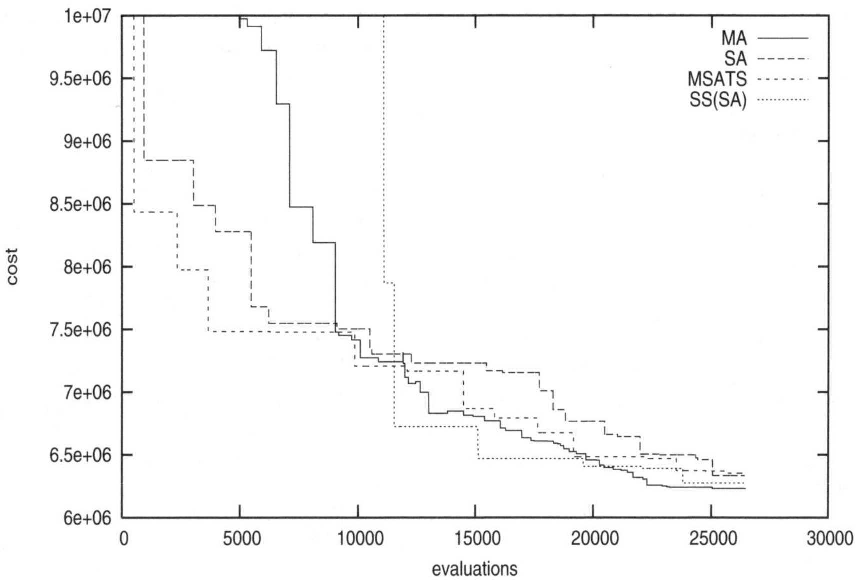

Fig. 5. Comparing MA, SA, MSATS, and SS at Hanoi network

Runtimes are below one minute for all the executions in a workstation of 2 Ghz with 512 Mb of RAM memory.

5 Conclusions and Future Work

This paper analyzes the performance of a new memetic algorithm for the optimal design of looped water distribution systems. It has been developed a new computer model named MENOME, which integrates several heuristic approaches, a hydraulic network solver, a graphical user interface and database management module. In terms of the quality of the solutions, results denote that all the methods here adapted obtain good solutions in the test problems evaluated. In particular, all meta-heuristics work very well in the small network, although the memetic approach work significantly better than the others in the largest network. The cost reduction provided by the new memetic algorithm can be determinant in the decision making process. These conclusions can offer very useful guidelines for solving realistic loop water distribution networks. Future research is focused to solve the problem by considering other objectives, as for instance the reliability of the network.

Acknowledgements

This work was supported by the Spanish MCyT under contracts TIN2005-00447. Authors appreciate the support of the "Structuring the European Research Area" program, RII3-CT-2003-506079, funded by the European Commission.

References

1. Moscato, P., Memetic Algorithms: A Short Introduction, in *F. Glover et al. (eds.)*, New Ideas in Optimization, McGraw-Hill, 1999, 219–234.
2. Holland, J., Adaptation in Natural and Artificial Systems, 1975. MIT Press.
3. Dawkins R., The Selfish Gene, Oxford University Press, New York, 1976.
4. Krasnogor, N. and Gustafson S. Toward Truly "Memetic" Memetic Algorithms: Discussion and Proof of Concepts, in *D. Corne et al. (eds)* in Advances in Nature-Inspired Computation: The PPSN VII Workshops, 2002.
5. Gupta, I., Bassin, J.K., Gupta A., and Khanna P., Optimization of Water Distribution System, *Environmental Software*, Vol.8, 1993, 101-113.
6. Alperovits, E., and Shamir, U., Design of Optimal Water Distribution Systems, *Water Resources Research*, Vol.13, No.6, 1977, 885.
7. Fujiwara, O., and Khang, D. B., A Two-phase Decomposition Method for Optimal Design of Looped Water Distribution Networks, *Water Resources Research*, Vol.26, No.4, 1990, 539-549.
8. Varma, K. V. K., Narasimhan, S., and Bhallamudi, S. M., Optimal Design of Water Distribution Systems Using NLP Method, *Journal of Environmental Engineering, ASCE*, Vol.123, No. 4, 1997, 381.
9. Sherali, H.D., Totlani, R. and Loganathan, G.V., Enhanced Lower Bounds for the Global Optimization of Water Distribution Networks, *Water Resources Research*, Vol.34, No.7, 1998, 1831-1841.
10. Cunha M. D., and Sousa, J., Water Distribution Network Design Optimization: Simulated Annealing Approach, *Journal Of Water Resources Planning And Management, ASCE*, Vol.125, No.4, 1999, 215.
11. Vairavamoorthy, K., and Ali. M., Optimal Design of Water Distribution Systems using Genetic Algorithms, *Computer-Aided Civil and Infraestructure Engineering*, Vol.15, 2000, 372-382.
12. Maier, H.R., Simpson A. R., Zecchin, A. C., Foong W. K., Phang K. Y., Seah H. Y., Tan, C. L., Ant Colony Optimization for Design of Water Distribution Systems, *J. of Water Resources Planning and Management, ASCE*, Vol.129, No. 3, 2003, 200-209.
13. Kirkpatrick, S., Gelatt, C. D., and Vecchi, M. P., Optimization by Simulated Annealing, *Science*, Vol.220, 1983, 671.

14. Gil, C., Ortega, J., Montoya M.G., and Baños R., A Mixed Heuristic for Circuit Partitioning, *Computational Optimization and Applications Journal*, Vol.23, No.3, 2002, 321–340.
15. Ramalhino, H., Martin, O., Stutzle, T., Iterated Local Search, in *F. Glover et al. (eds.)*, Handbook of Metaheuristics, Kluwer, Norwell, MA, 2002, 321–353.
16. Marti, R, and Glover F., Principles of Scatter Search, *European Journal of Operations Research*, Vol.169, No. 2, 2006, 359-372.
17. Talbi, E., A Taxonomy of Hybrid Metaheuristics, *Journal of Heuristics*, Vol. 8, No, 5, 2002, 541–564.
18. Davidor, Y. and Ben-Kiki O. The Interplay Among the Genetic Algorithm Operators: Information Theory Tools Used in a Holistic Way, in *Manner R. et. al (eds)*, Parallel Problem Solving from Nature (PPSN II), 2002, 77-86.
19. Savic, D. A., and Walters G. A., Genetic Algorithms for Least-Cost Design of Water Distribution Networks, *Journal of Water Resources Planning and Management, ASCE*, Vol.123, No. 2, 1997, 67–77.
20. Montesinos, P., Garcia-Guzman A., and Ayuso J.L., Water Distribution Network Optimisation using Modified Genetic Algorithm, *Water Resources Research*, Vol.35, No.11, 1999, 3467-3473.
21. Watters, G.Z. and J. Keller, Trickle Irrigation Tubing Hydraulics, ASAE Paper No. 78-2015, p.18. ASAE, St. Joseph, Michigan, 1978.
22. Rossman, L.A., EPANET 2 user's manual, EPA/600/R-00/057, September 2000.
23. Todini, E. and Pilati, S. A Gradient Method for the Analysis of Pipe Networks, Int. Conf. on Computer Applications for Water Supply and Distribution, UK, 1987.
24. Cotta, C., Scatter Search and Memetic Approaches to the Error Correcting Code Problem, in *Gottlieb et al. (eds.)*, 4th EVOCOP, 2004, 51-61.

Neural Network Models for Air Quality Prediction: A Comparative Study

S.V. Barai[1], A.K. Dikshit[2], and Sameer Sharma[3]

[1] Department of Civil Engineering, Indian Institute of Technology Kharagpur 721 30, India
[2] Center for Environmental Science and Engineering, Indian Institute of Technology Bombay, 400 076, India
[3] TCE Consulting Engineers Ltd., Mumbai 400 009, India

Abstract. The present paper aims to find neural network based air quality predictors, which can work with limited number of data sets and are robust enough to handle data with noise and errors. A number of available variations of neural network models such as Recurrent Network Model (RNM), Change Point detection Model with RNM (CPDM), Sequential Network Construction Model (SNCM), and Self Organizing Feature Maps (SOFM) are implemented for predicting air quality. Developed models are applied to simulate and forecast based on the long-term (annual) and short-term (daily) data. The models, in general, could predict air quality patterns with modest accuracy. However, SOFM model performed extremely well in comparison to other models for predicting long-term (annual) data as well as short-term (daily) data.

Keywords: Air Quality, Change Point Detection, Recurrent Neural Networks, Self Organizing Feature Maps.

1 Introduction

Air pollutants exert a wide range of impacts on biological, physical, and economic systems. Their effects on human health are of particular concern. The decrease in respiratory efficiency and impaired capability to transport oxygen through the blood caused by a high concentration of air pollutants may be hazardous to those having pre-existing respiratory and coronary artery disease (Rao and Rao, 2000). Consequently, it has become a vital task to accurately keep track of the variation of ambient air pollution levels in urban areas.

Natural phenomena are mostly a time series with some degree of randomness. Pollutants in the atmosphere may disperse or concentrate during varied time periods. Previous studies (Giorgio and Piero, 1996) have indicated that the data of ambient air quality are stochastic time series, thereby making it possible to make a short-term forecast on the basis of historical data. Though models may be imperfect, they are the best tool for use in all aspect of air quality planning where prediction is a major component such as for emission control (Melas et al., 2000), accidental release of pollutant, land-use planning, traffic planning (Hadjiiski and Hopke, 2000), planning of measurement programs (Rao and Rao, 2000), analyses of measurements/ trends and episode forecasting (Melas et al., 2000).

A. Saad et al. (Eds.): Soft Computing in Industrial Applications, ASC 39, pp. 290–305, 2007.
springerlink.com

Within the class of statistical methods until now, either the time-series methods, which do not use meteorological inputs, or regression and similar methods, which are mostly based on multivariate linear relationship between meteorological conditions and ambient air pollution concentrations, were commonly used. However, when applying the conventional time-series models to the ambient air pollution forecast, the pollutant level variations are generally not simple autoregressive or moving average models. Analyst must employ statistical graphs of the autocorrelation function and partial autocorrelation function to identify an appropriate time-series model (Chakraborty et al., 1992). In the model identification stage, the resulting model quality frequently relies on individual experience and knowledge of time-series statistic. Furthermore, a time-series model may not be applicable for varied periods of data. A model applicable in one period may require manual adjusting of its model parameters to meet the data characteristics in other time periods. These complexities make applying a time series model to regular air quality forecast an inefficient task. In other words, though the statistical methods do provide reasonable results, these are essentially incapable of capturing complexity and non-linearity of pollution-weather relationships.

The neural networks (Principe and Kuo, 1995) have emerged out to be more flexible, less assumption dependent and adaptive methodology in environment related areas such as rainfall runoff modeling, stream flow forecasting (Thirumalaiah and Deo, 1998), ground water modeling, water management policy, precipitation forecasting , hydrologic forecasting and reservoir operation (Thirumalaiah and Deo, 2000), lake and reservoir modeling (ASCE, 2000a, 2000b), remote sensing and GIS related activities, real time control of water treatment plants, water quality and air quality management (Boznar et al., 1993), adsorbent beds design (Basheer and Najjar, 1996), and hazardous waste management.

The present study investigates the advantage of using neural networks for forecasting the air pollution. The aim is to find better air quality predictors, which can work with low number of data sets and should be robust enough to handle data with noise and errors. The objectives of the study are as follows:

- To implement various available variations of neural network models for predicting air quality.
- To collect suitable data sets for multiple air quality parameters - one containing yearly average pollutant concentrations at a specific location and other containing daily average pollutant concentrations record for significantly long duration.
- To conduct exhaustive simulations using above-developed models with yearly data and hourly data to assess the relative advantage of each model in prediction.
- To perform comparative study to identify suitable air quality prediction model(s) for yearly (long-term) data and that for daily (short-term) data.

2 Implementation of Various Neural Networks Models

This section provides details about selected a few neural networks models applied for air quality prediction.

2.1 Recurrent Network Model (RNM)

For a neural network to be dynamic, it must be given memory. Memory may be divided into "short-term" and "long-term" memory depending upon the retention time. Long-term memory is built into a neural network through supervised learning, whereby the information content of the training data set is stored in the synaptic weights of the networks (Haykin, 2000). However, if the task at hand has a temporal dimension, we need some form of short-term memory to make the network dynamic. The static network accounts for non-linearity and the memory accounts for time. Short-term memory can be implemented in continuous time or in discrete time. Such networks typically use a variant of back-propagation for training. Essentially, there are three ways that a "memory" can be introduced into static neural networks (Connor et al., 1994; Parlos et al., 2000). These are (in increasing order of complexity and capability):

- Tapped Delay Lines Models**:** In these models, the network has past inputs explicitly available (through a tapped delay line) to determine its response at a given point in time. Thus, the temporal pattern is converted to a spatial, which can then be learned through, say, classic back propagation (Haykin, 2000).
- Context Model or Partial Recurrent Models: These models retain the past output of nodes instead of retaining the past raw inputs. For example, the output of the hidden layer neurons of a feed forward network can be used as inputs to the network along with true inputs. These "network derived" inputs are also called context inputs. When the interconnections carrying the context inputs are fixed, classical back propagation can be used for training the network.
- Fully Recurrent Models: These models employ full feedback and interconnection between all nodes (Haykin, 2000). Algorithms to train fully recurrent models are significantly more complex in terms of time and storage requirements.

For the present study, the partial recurrent model has been used.

2.2 Change Point Detection Model (CPDM)

In general, air quality parameters are controlled by various other factors such as emission rate from vehicle and industries etc. This emission change with the introduction or removal of new vehicles, industries and change in atmospheric condition etc. Therefore, we can conjecture that the movement of air quality parameters has a series of change points (Kyong and Han, 2000), which occur because of these changes. The proposed model consists of three stages. The first stage is to detect successive change points in the air quality patterns over a number of years called the *change point detection (CPD) stage*. The second stage is to forecast the change-point group with say back propagation referred as the *change-point-assisted group detection (CPGD) stage*. The final stage is to forecast the output with say back propagation and it is referred as the *output forecasting neural network (OFNN) stage*. The back propagation model is used as a classification tool in CPGD and as a forecasting tool in OFNN. This model obtains intervals divided by change points in the training phase, identifies them as a change point groups in the training phase, and forecast to which group each sample is assigned in the testing phase. In this a series of change point will be detected by the Pettitt test, a nonparametric change-point detection method, as nonparametric statistical property is a suitable match for neural network model that is a kind of nonparametric method (Pettitt, 1979). For the present study, partial recurrent network model has been used in place of classical back propagation model.

2.3 Sequential Network Construction Model (SNCM)

This model introduces an application of the Sequential Network Construction (SCN) to select the size of several popular neural network predictor architectures for various benchmark-training sets. The specific architecture considered here consists of a Finite Impulse Response (FIR) network and the partial recurrent Elman network for adding context units to the output layer (Back and Tsoi, 1991). This model considers an enhancement of a FIR network in which only the weights having relevant time delays are utilized. Bias-variance trade off in relation to the prediction risk estimation by means of Nonlinear Cross Validation is discussed elsewhere (Tomasz and Zurada, 1977).

2.4 Self Organizing Feature Maps (SOFM) Model

These networks are based on competitive learning i.e. the output neurons of the network compete among themselves to be activated or fired, with the result that only one output neuron or one neuron per group is on at any one time (Haykin, 2000). An output neuron that wins the competition is called a winning neuron. One way of inducing a winner-takes-all competition among the output neurons is to use inhibitory connections (negative feedback paths) between them.

In a self organizing map, the neurons are placed at the nodes of a lattice that is usually one- or two-dimensional. The neurons become selectively tuned to various input patterns or a classes of input patterns in the course of a competitive learning. The locations of the neurons so tuned become ordered with respect to each other in such way that a meaningful coordinate system for different input features is created over the lattice. Hence neurons in the lattice are indicative of intrinsic statistical features contained in the input patterns. The spatial location of an output neuron in a topographic map corresponds to a particular domain or feature of data drawn from input space (Kohonen, 1990).

The principal goal of the self organizing map is to transform an incoming signal pattern of arbitrary dimension into a one- or two-dimensional discrete map, and to perform this transformation adaptively in a topologically ordered fashion. Each neuron in the lattice is fully connected to all the source nodes in the input layer. This network represent a feed forward structure with a single computational layer consisting of neurons arranged in rows and columns. The algorithm responsible for the formation of the self organizing map proceeds first by initializing the synaptic weights in the network. This can be done by assigning them small values picked from a random number generator. Once the network has been properly initialized, there are three essential process involved in the formation of the self organizing map, explained elsewhere (Principe and Wang, 1993; Haykin, 2000).

All above-mentioned models were implemented in MATLAB software (Math works, 2000) using Neural Networks Tool box (Demuth and Beale, 1992). More details can be found elsewhere (Sharma, 2002).

3 Data Collection and Properties

Most air quality data are obtained from air quality monitoring stations directly or through remote sensing instruments. Also, an existing model or laboratory experiment can be used to generate data patterns for specific applications. Again there appears to be no fixed method for determining the number of input-output data pairs that will be required. To ensure a good approximation, the number of data pairs used for training should be equal to or greater than the number of parameters (weights) in the network (ASCE, 2000a). For the present study, two types of data sets - one for annual average pollutant concentrations and another for daily average pollutant concentrations have been arranged to apply the model for long-term as well as short-term predictions. The details on data collection, database construction etc. have been discussed in following sections.

3.1 Case Study 1

Data for 115 counties of California State in United States of America has been collected from US EPA website (www.epa.gov). This data is annual average data for 15 years from 1985 to 1999 for seven parameters namely VOC (volatile organic carbon), NO_X (oxides of nitrogen), CO (carbon monoxide), SO_2 (sulphur dioxide), PM10 (particulate matter with size less than 10 microns), PM2.5 (particulate matter with size less than 2.5 microns) and NH_3 (ammonia). All concentrations are in micrograms per cubic meter.

3.2 Case Study 2

The data for three parameters namely RPMA (Respiratory Particulate Matter Average), SO_2 (sulphur dioxide) and NO_2 (nitrogen dioxide) is collected for Delhi State at nine locations. These data are daily average concentrations for last two years from 3/7/2000 to 20/8/2001. This data set has been collected from Tata Energy Research Institute web site www.teri.in. However, only the data for Ashram Chowk has been used for carrying out simulation studies. The dataset size had 110 patterns.

3.3 Statistical Properties of Data Sets

The statistical properties of the data sets are very important for data analysis and pre-processing. They are mean, standard deviation, variance, hypothesis tests etc. Here some properties namely mean, standard deviation, median of the data of two case studies are give below in Tables1 and 2, respectively.

Table 1. Statistical properties of data set – Case Study 1

Parameter	Property		
	Mean	Std. Dev.	Median
VOC	69.91	23.09	66.73
NO_X	3662.1	1762.4	2912
CO	675.69	249.43	539.13
SO_2	551.66	451.46	273.38
PM10	150.07	118.63	78.39
PM2.5	33.21	25.71	40.15
NH3	13.34	16.98	0.08

Table 2. Statistical properties of data set - Case Study 2

Parameter	Property		
	Mean	Std. Dev.	Median
RPMA	172.83	119.88	141
SO_2	9.59	3.7	9
NO_2	77.18	31.59	72

3.4 Preprocessing

The data is processed before used as input to the network. Data can be re-scaled/normalized/standardized according to the requirement and properties of the data sets (Warren, 2002). *Rescaling* a vector means to add or subtract a constant and then multiply or divide by a constant. *Normalizing* a vector most often means dividing by a norm of the vector, for example, to make the Euclidean length of the vector equal to one. In the neural network literature, normalizing also often refers to rescaling by the minimum and range of the vector to make all the elements lie between 0 and 1. *Standardizing* a vector means subtracting a measure of location and dividing by a measure of scale. For example, if the vector contains random values with a Gaussian distribution, one might subtract the mean and divide by the standard deviation, thereby obtaining a 'standard normal' random variable with mean 0 and standard deviation 1.

The data is normalized in three ways. In first method, the whole data set is divided by the maximum value of that parameter. In second method, a normalized value is calculated $N = (w_{max} - w_{present})/(w_{max} - w_{min})$ and then this N is used to renormalize the output value. In third method, all values are mapped between 0 and 1. The maximum value is put equal to one and minimum value to zero and all intermediate values are mapped between 0 and 1.

3.5 Selection of Input and Output Variables

The goal of neural networks is to generalize a relationship of the form

$$Y^n = f(X^m) \quad (1)$$

where X^n is an n-dimensional input vector consisting of variables $x_1, \ldots\ldots x_i, \ldots\ldots x_n$; and Y^m is an m-dimensional output vector consisting of resulting variable of interest $y_1 \ldots\ldots y_j, \ldots\ldots y_m$.

The selection of an appropriate input vector that will allow neural networks to successfully map to the desired output vector is not a trivial task. Unlike physically based models, the set of variables that influence the system are not known a priori. A neural network should not be considered mere a black box, a firm understanding of the system under consideration is an important aspect.

In air quality, the value of x_i can be causal variables such as wind speed, downwind distance, crosswind distance, ambient air temperature, relative humidity, atmospheric stability etc. The values of y_i can be air quality parameters such as ambient concentration of ozone, nitrogen oxides, carbon monoxide, particulate matter etc.

3.6 Presenting the Input to Network

Data can be presented to the network in two ways. In first way, the input given to the network is the year or day number and the normalized values of the parameter for which the network is being trained are kcpt as the target values of the network. This technique has been applied to RNM and CPDM models. In second way, the input to the network is the normalized value of the parameter that is being modeled. A number of the past values of the parameter that is being modeled may also be given as additional input to the network. The number of past values to be given as input varies from the model to model. In SNCM, and SOFM, the second technique has been used with normalized values (I_t) lying between 0.1 and 0.9 as follows:

$$_t = 0.5 - (W_{mean} - W_t) / (W_{max} - W_{min}) * 0.8 \quad (2)$$

where

W_{max} = maximum value in that pattern.
W_{min} = minimum value in that pattern.
$W_{mean} = (W_{max} + W_{min}) / 2$.
W_t = value of any element in that pattern.

After modeling, the normalized output predictions (W_{pr}) produced by model are reverse normalized using:

$$O_t = W_{mean} - (0.5 - W_{pr}) / 0.8 * (W_{max} - W_{min}) \quad (3)$$

where

O_t = Predicted value produced by the model.

3.7 Performance Evaluation of Models for Case Studies

In the Section 2, the background of specific neural network models proposed for present study was discussed. The following four models have been implemented for carrying out simulations to forecast air quality.

- RNM: This is Recurrent Network model and Elaman Networks are used for the simulation.
- CPDM_RNM: This model uses Change point detection technique and Recurrent Elaman Networks for simulation.

- SNCM: This model utilizes the Recurrent Networks for simulation in a different way. It selects the network Architecture for time series modeling by adding neurons one by one and training the network again and again.
- SOFM: This model exploits the properties of Self-organizing Feature Maps and utilized for non-linear time series modeling.

3.8 Network Error Calculation

The performance of various models for different air quality parameters was estimated in terms of mean percentage error (PE) defined as follows over a specified number of iterations or for achieving the goal of specified sum squared error (SSE) by the network, whichever reach earlier:

$$PE = (target - output) / target * 100 \qquad (4)$$

The model having minimum PE is expected to be the best model for forecasting.

4 Results and Discussion

This section provides the numerical experimentation carried out using various neural networks for two datasets.

4.1 Results of Case Study 1

The parameters of various models used for modeling annual time series data are presented in Tables 3 to 6. The model performance was evaluated using following approache:

Training on nine data points of time series and predicting remaining part of time series six data points. The performance of various models for different air quality parameters was estimated in terms of mean percentage error (PE). The model having minimum PE is expected to be the best model for forecasting. The performance of various models have been compared in Table 7 and discussed in forthcoming section.

Table 3. The RNM Parameters (Case Study 1)

Air-quality Parameter	VOC	NOx/CO/ SO_2/PM10	PM2.5/NH_3
Model Parameters↓			
NN Architecture	1-6-6-1	1-5-5-1	1-5-5-1
Activation Function	Tansig/purelin	Tansig/purelin	Tansig/purelin
Learning Rate	0.03	0.03	0.04
No. of Epochs	5000	5000	5000
SSE	1.0e-5.	1.0e-5.	1.0e-5.
Input parameters	Yr. No	Yr. No	Yr. No
Output Parameters	VOC	NOx/ CO/ SO_2/PM10	PM2.5//NH_3

Table 4. The CPDM_RNM Parameters (Case Study 1)

Air-quality Parameter	VOC	NOx/ CO/ SO_2/PM10	PM2.5/NH_3
Model Parameters↓			
NN Architecture	1-6-6-1	1-5-5-1	1-5-5-1
Activation Function	Tansig/purelin	Tansig/purelin	Tansig/purelin
Learning Rate	0.03	0.03	0.04
No. of Epochs	1000	1000	1000
SSE	1.0e-6.	1.0e-6.	1.0e-7.
Input parameters	Yr. No	Yr. No	Yr. No
Output Parameters	VOC	NOx/ CO/ SO_2/PM10	PM2.5/ NH_3

Table 5. The SNCM Parameters (Case Study 1)

Air-quality Parameter	VOC/NOx/ CO/ SO_2/PM10/ PM2.5/NH_3
Model Parameters↓	
NN Architecture	1-3-3-1 to 1-8-8-1.
Activation Function	Tansig/purelin
Learning Rate	0.03
No. of Epochs	5000.
SSE	0.1e-7.
Input parameters	Yr. No
Output Parameters	VOC/ NOx/ CO/ SO_2/PM10/ PM2.5/NH_3

Table 6. The SOFM Model Parameters (Case Study 1)

Air-quality Parameter	VOC/NOx/ CO/ SO_2/PM10/ PM2.5//NH_3
Model Parameters↓	
NN Architecture	1-4-1
Learning Rate	1
No. of Epochs	5000.
Input parameters	VOC/NOx/ CO/ SO_2/PM10/ PM2.5//NH_3
Output Parameters	VOC/NOx/ CO/ SO_2/PM10/ PM2.5//NH_3

Table 7. Neural networks models performance (Case Study 1)

Air-quality Parameter	VOC	NOx	CO	SO_2	PM10	PM2.5	NH3
Model ↓							
RNM	201	88	81	69	36	20.5	26.9
CPDM_RNM	46.87	42.01	45.905	27.72	14.56	7.71	21.49
SNCM	43.24	47.77	42.587	45.986	47.63	36.752	41.84
SOFM	**23.81**	**16.34**	**7.806**	**15.722**	**12.73**	**7.87**	**14.73**

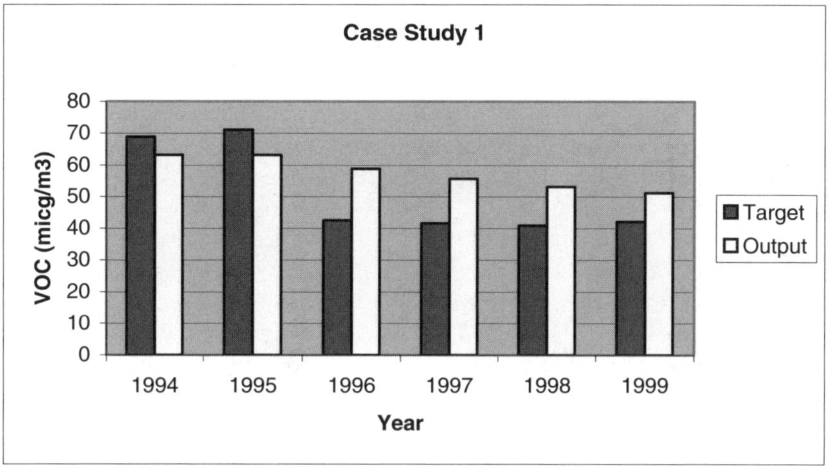

Fig. 1. VOC emissions with SOFM

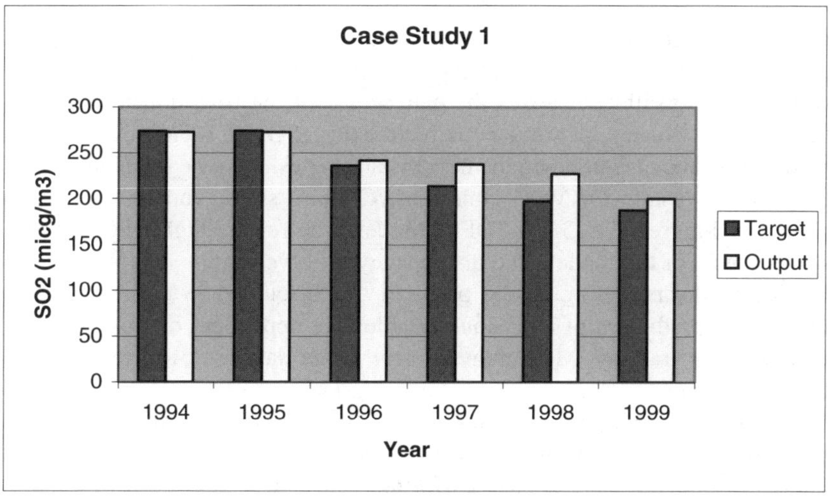

Fig. 2. SO_2 emissions with SOFM

4.1.1 Observations

From the Table 7, given in above section, following observations can be made.

- Time-series prediction method demonstrates the more realistic problems in the field. Here one would like to forecast about air quality parameters based on the past history. The models in general could predict with some accuracy. However, among all the models implemented, Self-organizing Feature Map (SOFM) based

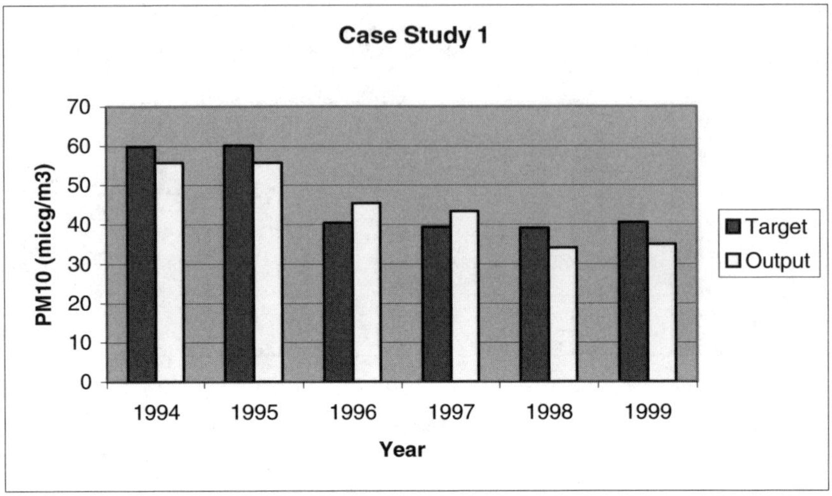

Fig. 3. PM10 emissions with SOFM

model has performed extremely well in comparison to other models. This performance could be attributed to the basic underlying characteristic of the algorithm. The algorithm localizes the data points of the time domain and classifies them in the cluster of similar characteristics. SOFM identifies the parameter values from the cluster when the new instance is asked to be predicted,

- The typical results for VOC emissions, CO emissions, and NO_x emissions are shown in Figures 1 to 3 for SOFM Model. The results highlight the very good performance of the model. The discrepancy observed in the model prediction can be due to the modeling of the problem. As discussed in earlier about SOFM model, the prediction of emission variables are dependent on past history of the data used for training. Measurement error in the data used for training may lead to prediction error.
- The case study demonstrated an example of an annual average emission (long-term) data prediction using various neural networks models for a very limited dataset. Models in general have performed reasonably well even with the limited historical data. It is expected that availability of more annual average emission data can improve the performance of models studied.

4.2 Results of Case Study 2

The parameters of various models for modeling daily average data are shown in Tables 8 to 11. The performance of various models was evaluated for following scenarios: Training on 80 data points of time series and predicting remaining part of time series 30 data points.

Table 8. The RNM Parameters (Case Study 2)

Air-quality Parameter	RPMA/ SO_2/NO_2
NN Architecture	1-10-10-1
Activation Function	Tansig/purelin
Learning Rate	0.04
No. of Epochs	5000
SSE	0.1e-4
Input parameters	Day No.
Output Parameters	RPMA/ SO_2 /NO_2

Table 9. The CPDM_RNM Parameters (Case Study 2)

Air-quality Parameter	RPMA/ SO_2 /NO_2
NN Architecture	1-4-4-1
Activation Function	Tansig/purelin
Learning Rate	0.03
No. of Epochs	5000
SSE	0.1e-4
Input parameters	Day No.
Output Parameters	RPMA/ SO_2 /NO_2

Table 10. The SNCM Parameters (Case Study 2)

Air-quality Parameter	RPMA/ SO_2 /NO_2
NN Architecture	1-3-3-1 to 1-8-8-1.
Activation Function	Tansig/purelin
Learning Rate	0.03
No. of Epochs	1000.
SSE	0.1e-7.
Input parameters	Day No.
Output Parameters	RPMA/ SO_2 /NO_2

Table 11. The SOFM Parameters (Case Study 2)

Air-quality Parameter	RPMA	SO_2	NO_2
NN Architecture	1-5-1	1-5-1	1-5-1
Learning Rate	1	1	1
No. of Epochs	5000.	5000.	5000.
Input parameters	RPMA	SO_2	NO_2
Output Parameters	RPMA	SO_2	NO_2

The models' performance in terms of PE has been shown in Table 12 for all three air quality parameters. The salient features of different models have been discussed and compared in next section.

Table 12. Neural networks models performance (Case Study 2)

Air-quality Parameter	RPMA	SO2	NO2
RNM	56.76	48.63	43.5
CPDM_RNM	45.36	41.83	38.9
SNCM	33.45	37.79	35.87
SOFM	**25.6**	**30.73**	**28.94**

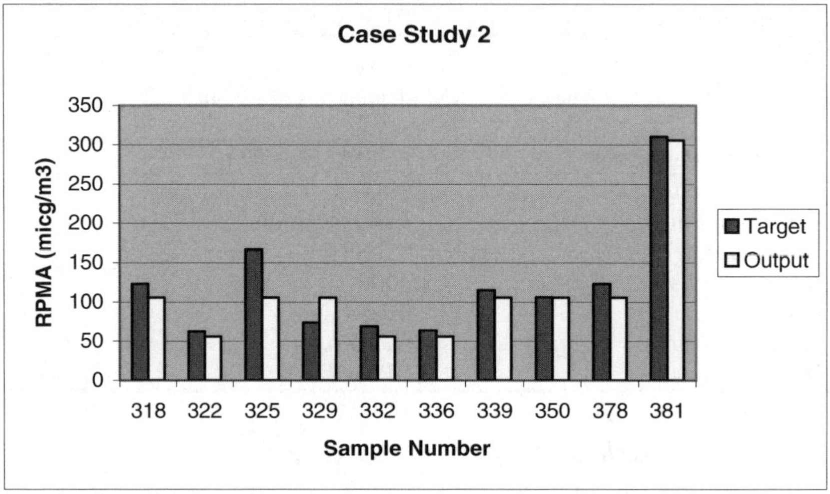

Fig. 4. RPMA emissions with SOFM

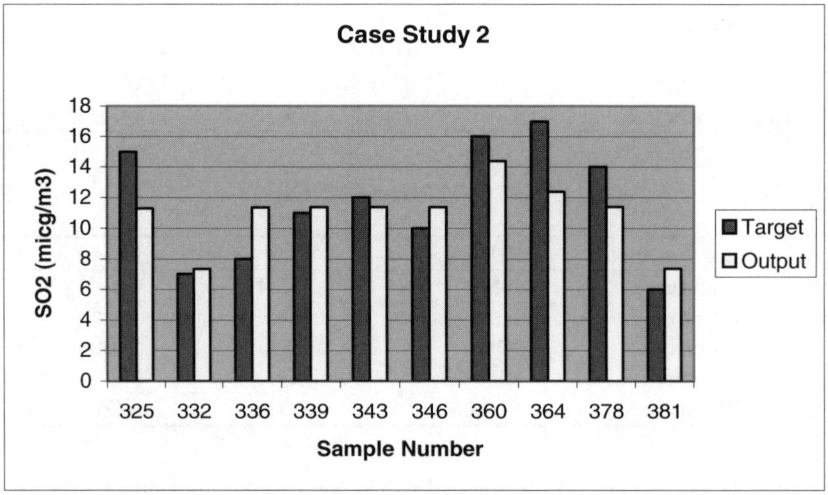

Fig. 5. SO_2 emissions with SOFM

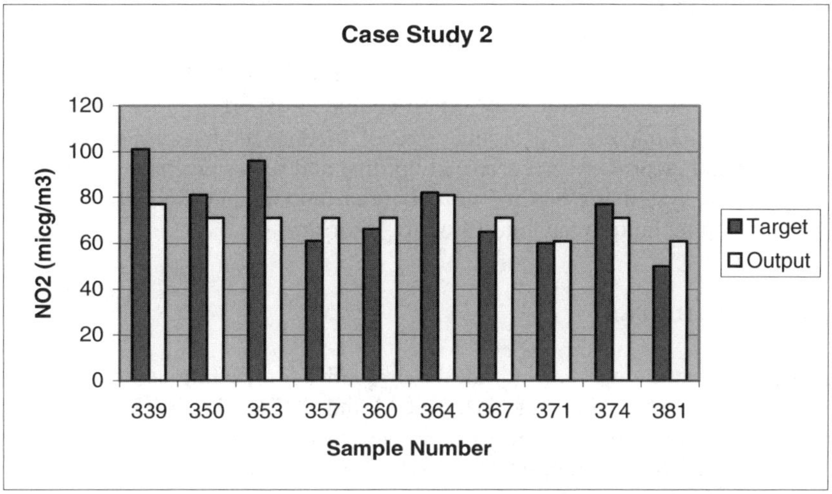

Fig. 6. NO_2 emissions with SOFM

4.2.1 Observations

The following observation can be made on the basis of results of various simulations shown in Table 12.

- The models studied for this case, in general could predict with modest accuracy. However, among all the models implemented, SOFM based model has performed extremely well in comparison to other models. This may be attributed to the fact that network learns the history much more. Similar trend was observed during earlier case study too.
- As discussed above, the typical results for RPMA emissions, SO_2 emissions, and NO_2 emissions are shown in Figures 4 to 6. The results shown in Figures 4 to 6 depicts reasonably good match between model predictions with target prediction.

The case study demonstrated an example of a daily average emission data prediction using various neural networks model for a very large size dataset. Models in general have performed reasonably well even though data was showing randomness in time domain.

5 Future Projections

The predictions of studied models are based on the limited history of air quality. However, the model prediction can be improved by carrying out investigation incorporating following aspects:

- Models can have as inputs data from multiple sources, such as historical air quality measurements, meteorological data etc.
- Models can have along with emissions data, episode levels definition, historical measurements of surface and upper air meteorological data.

- Models should be able to give predictions for the following three different time windows: 1 day, 1 week and 1 month predictions.
- Models should have an easy to use interface and should be able to present the results in an understandable way to non-computer experts.
- The model parameters and architecture of models in this project were arrived with trial and error. One can arrive at optimal and better performance model after carrying out systematic studies on networks models and their parameters using optimization techniques such as Genetic Algorithm.

6 Closing Remarks

In this paper, the study was carried out on air quality forecasting using various neural network models: RNM, CPDM_RNM, SNCM and SOFM. The study was focused at preliminary investigation of single variable based time series prediction. The investigation was carried out for long-term as well as short-term air quality data set. Self-Organizing Features Maps (SOFM) used for time series prediction came up as the best tool for time series forecasting. These were found to be very useful for large training datasets. The results shown here are indications that the neural network techniques can be useful tool in the hands of practitioners of air quality management and prediction. In that case, practitioners need not know even about the development of the model. The models studied in this study are easily implemented, and they can deliver prediction in real time, unlike other modeling techniques. The models can very well easily deal with input noise and uncertainty.

References

ASCE Task Committee on Application of Artificial Neural Networks in Hydrology (2000a), "Artificial neural networks in hydrology. I: preliminary concepts", *J. Hydrologic Engineering*, ASCE, 5(2), 115-123.

ASCE Task Committee on Application of Artificial Neural Networks in Hydrology (2000b), "Artificial neural networks in hydrology. II: preliminary concepts", *J. Hydrologic Engineering*, ASCE, 5(2), 124-137.

Back, A.D., and Tsoi, A.C. (1991), "FIR and IIR synapses, a new neural network architecture for time series modeling", *Neural Computation*, 3, 375-385.

Basheer, A., and Najjar, M. (1996), "Predicting dynamic response of adsorption columns with Neural Nets", *J. Computing in Civil Engineering*, ASCE, 10(1), 211-220.

Boznar, M., Lesjak, M., and Mlakar, P (1993), "A neural network-based method for short term prediction of ambient SO_2 concentrations in highly polluted industrial areas of complex terrain", *Atmos. Environ.*, 27B, 221-230.

Chakraborty, K., Mehrotra, K., Mohan, C.K., and Ranka, S. (1992), "Forecasting the behaviour of multivariate time series using neural networks", *Neural Networks*, 2, 53-58.

Connor, J.T, Martin, R.D., and Atlas, L.E. (1994), "Recurrent neural networks and robust time series prediction", *IEEE Trans. Neural Networks*, 5(2).

Demuth, H., and Beale, M. (1992), "*Neural Networks Toolbox Manual*", Math Works Inc.

Giorgio, F., and Piero, M. (1996), "Mathematical models for planning and controlling air quality", *Proceedings of IIASA Workshop*, 17.

Hadjiiski, L., and Hopke, P. (2000), "Application of artificial neural networks to modeling and prediction of ambient ozone concentrations", *J. Air and Waste Manage. Assoc.*, 50, 894-901.

Haykin, S. (2000), "*Neural Networks*", Second Edition, Addison Wesely Longman.

Kohonen, T. (1990), "The self-organizing map", *Proceedings of the IEEE*, 78, 1464-1480.

Kyong, J.O., and Han, I. (2000), "Using change-point detection to support artificial neural networks for interest rate forecasting", *Expert System with Applications,* 19, 105-115.

Math Works (2000), "*MATLAB User's Guide*", Math Works Inc.

Melas, D., Kioutsioukis, I., and Ziomas, I.C. (2000),"Neural network models for predicting peak photochemical pollutant levels", *J. Air and Waste Manage. Assoc.,* 50, 495-501.

Parlos, A.G., Rais, O.T., and Atiya, A.F. (2000), "Multi-step-ahead prediction using dynamic recurrent neural networks", *Neural Networks,* 13, 765-786.

Pettitt, A.N. (1979), "A non-parametric approach to the change-point problem", *Applied Statistics*, 28(2), 126-135.

Principe, J., and Wang, L. (1993), "Non-linear time series modeling with self-organization feature maps", *Neural Networks,* 3.

Principe, J., and Kuo, J.M. (1995), "Dynamic modeling of chaotic time series with neural networks", *Proc. of Neural Infor. Proc. Sys.*, 7.

Rao, M.N., and Rao, H.V. (2000), "*Air Pollution*", Tata McGraw-Hill, New Delhi.

Sharma, S. (2002), "*Neuro Air Quality Predictors*", M.Tech. Thesis, IIT Kharagpur.

Thirumalaiah, K., and Deo, M.C. (1998), "River stage forecasting using artificial neural networks", *Journal of Hydrologic Engineering*, ASCE, 3(1), 26-32.

Thirumalaiah, K., and Deo, M.C. (2000), "Hydrological forecasting using neural networks", *Journal of Hydrologic Engineering*, ASCE, 5(2). 180-189.

Tomasz, J.C., and Zurada, J.M. (1977), "Sequential Network Construction for time series prediction", In *Proceedings of the IEEE International Joint Conference on Neural Networks,* 2034–2039, Houston, Texas, USA, June 9–12.

Warren, S. S. (2002), "*FAQ in Neural Networks*", ftp://ftp.sas.com/pub/neural/FAQ.html

Recessive Trait Cross over Approach of GAs Population Inheritance for Evolutionary Optimization

Amr Madkour, Alamgir Hossain, and Keshav Dahal

Modeling Optimization Scheduling And Intelligent Control (MOSAIC) Research Centre
Department of Computing, University of Bradford, Bradford, BD7 1DP, UK
{A.A.M.Madkour, m.a.hossain1, k.p.dahal}@bradford.ac.uk

Abstract. This research presents an investigation into a new population inheritance approach using a concept taken from the recessive trait idea for evolutionary optimization. Evolutionary human inheritance recessive trait idea is used to enhance the effectiveness of the traditional genetic algorithms. The capability of the modified approach is explored by two examples *(i)* a mathematical function of two variables, and *(ii)* an active vibration control of a flexible beam system. Finally, a comparative performance for convergence is presented and discussed to demonstrate the merits of the modified genetic algorithms approach over the traditional ones.

Index Terms: Genetic algorithms, PEAKS function, System identification, Flexible beam, Active vibration control.

1 Introduction

Over the last decade, genetic algorithms (GAs) have been extensively used as search and optimization tools in various problem domains, including the science, commerce, and engineering. The primary reasons for their success are their broad applicability, ease of use, and global perspective.

The concept of the GAs was first conceived by John Holland of the University of Michigan [1]. Thereafter he and his student have contributed much to develop this field. Most of the initial research work has been reported in various literatures [1-8]. However, there now exist several textbooks on GAs; many journals are now dedicated to promote research in this field. In addition, most GAs applications can also be found in various domain-specific conferences and journals [2].

This investigation demonstrates a modified approach for population inheritance to reduce the randomization "lucky" of the traditional GA crossover operator (TCGA). The GA with this Recessive trait Crossover is referred to here as (RCGA). The merits and capability of the modified approach is investigated by two examples *(i)* a mathematical function of two variables, and *(ii)* control parameter optimization of an active vibration control (AVC) system for a flexible beam.

The mathematical function 'PEAKS' in Matlab is chosen because it is a simple function with only two variables and contains a good number of local minimum that can trap the minimal seeker. On the other hand, an active vibration control algorithm for a flexible beam system is considered for control parameter optimization. It is

A. Saad et al. (Eds.): Soft Computing in Industrial Applications, ASC 39, pp. 306–315, 2007.
springerlink.com

worth noting that many researchers have already been used similar platform to investigate system identification and active vibration control system using TCGA [9-15]. For both examples, comparative performances of the TCGA and RCGA for convergence are measured and verified. Finally, these performances are presented and discussed to demonstrate the merits and capabilities of the RCGA over the Traditional Uniform Crossover Genetic Algorithm TUCGA.

2 Genetic Algorithms

GAs simultaneously evaluates many points in the parameter space and converges towards the global solution. GAs differ from other search techniques by the use of concepts taken from natural genetics and evolution theory.

The theory of evolution originated with Darwin in the nineteenth century, however, the idea that species mutate over time has been around for a long time in one form or another [16, 17]. Darwin suggested that in the universal struggle for life, nature "selects" those individuals who are best suited (fittest) for the struggle, and these individuals in turn reproduce more than those who are less fit, thus changing the composition of the population.

In fact, when a baby is conceived it is supplied with two copies of every chromosome: one copy from the mother and the other one from the father. Each parent donates of his/ her own chromosomes. The information from all of those genes takes together to makes up the blueprint or plan for the human body, its functions and its properties [18].

There are three methods of human inheritance, dominant, recessive and sex linked [19]. The sex-linked properties expressing depend on the person's sex. For dominant properties, only one genetic trait is needed for this property to be expressed. However, if a genetic trait is recessive, a person needs to inherit two copies of the gene for the trait to be expressed. Thus, both parents have to be carriers of a recessive trait in order to express a child for that trait. If both parents are carriers, there is a 25% chance with each child to show the recessive trait and it is becomes 100% if the both have that recessive trait.

Fig. 1 illustrates the inheritance of a recessive property. Let us assume that two brown eyes parents careering of a recessive trait of blue eyes -blue eyes carrier- are married, there is a 25% chance with each child to be blue eyes, a 25% chance with each child to be pure brown eyes, and a 50% chance with each child to be blue eyes carrier like there parents. From this second generation; if one of the carrier children married the blue eyes one there is a 50% chance with each child to be blue eyes, a 0% chance with each child to be pure brown eyes, and a 50% chance with each child to be blue eyes carrier. From the third generation of this relationship, if the two blue eyes children married, 100% of there children will be blue eyes.

From this example, we can understand why the blue eyes, yellowy heir and white skin are very common at the western societies, in contrast, black hair and eyes and dark skin are common at the African societies.

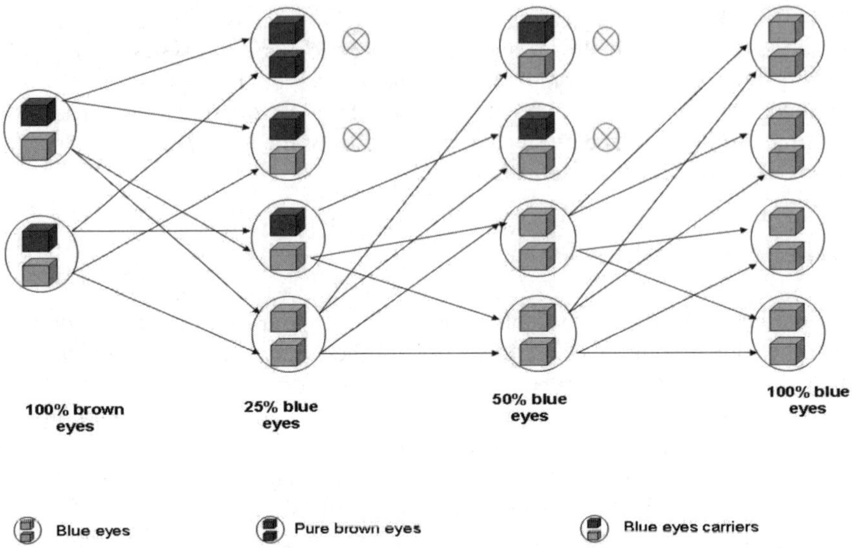

Fig. 1. Inheritance of blue eyes recessive property

3 Genetic Algorithms as a Computation Work

According to Goldberg [3], GAs are different from more normal optimization and search procedures in four ways:

- GAs work with a coding of the parameter set, not the parameters themselves.
- GAs search from a population of points, not a single point.
- GAs use payoff (objective function) information, not derivatives or other auxiliary knowledge.
- GAs use probabilistic transition rules, not deterministic rules.

3.1 Traditional Crossovers Genetic Algorithms (TCGA)

The TCGA evolution can be summarized as follows: create a population of individuals, evaluate their fitness, generate a new population by applying genetic operators, and repeat this process for a number of times.

To start the optimization process, TCGA generates randomly or with other means a population of individuals, say N individuals. Generally, each individual in the population consists of encoded strings representing a solution. Each solution has a fitness value evaluated by the some objective function and constraint satisfaction. The individuals with higher fitness value are usually selected and sent to the mating pool. Different selection methods such as roulette wheel selection and stochastic universal sampling can be used for this operation. Solutions having higher fitness values are most likely to survive for the next generation. Crossover operator is used on these strings to obtain the new solutions that inherit the good and bad properties of their parent solutions. The crossover operator works on randomly selected pairs of selected solutions from mating pool with certain crossover rate. The crossover rate is defined as the probability of applying crossover to a pair of selected solutions. There are many ways of defining this operator such as single point, double point, multipoint and uniform crossover. These traditional crossover operators are discussed in [6].

3.2 Recessive Trait Crossover Genetic Algorithms (RCGA)

The TCGA works on randomly pairs of selected solutions from mating pool with certain crossover rate. This operation exchanges the genes between the two random selected solutions. Using the concepts taken from the recessive property inheritance and Darwin theory of evolution the RCGA produces children by selecting the common genes between both parents, and choosing the remaining genes randomly according to the fact of the complementary of all of the chromosome parts makes its survival fitness. The main difference between the two algorithms is the way of how the new population is inherited from the previous generations.

Now let us assume that two parents have the eight genes chromosome as shown in Table 1. It is worth noting that those parents have common genes at (1, 3, 5, and 6). Over the evolution process the survival fitness of two parents depend on those common genes. So we will keep these common genes without any change when children are reproduced and try to make better children by introducing the different genes using the four possible binary combinations randomly. This is the only random operation in this crossover approach. The new solutions are as shown in Table 2.

Table 1. An example for two parents have 8 genes chromosome

Gene NO.	1	2	3	4	5	6	7	8
Parent 1	0	1	1	1	0	1	0	1
Parent 2	0	0	1	0	0	1	1	0

Table 2. The reproducing in table 1 using RCGA

Gene NO.	1	2	3	4	5	6	7	8
Child 1	0	1	1	1	0	1	0	0
Child 2	0	1	1	0	0	1	1	0
Child 3	0	0	1	0	0	1	0	1
Child 4	0	0	1	1	0	1	1	1

Referring to recessive trail behavior, the selection of the married parents is very important as both of them should be at lest fitness trait carrier. This selection operation will be done by sorting the old populations according to their fitness and then reproducing the first parent with the second to generate four new populations, as shown in the above example, and so on.

The RCGA can be written as:

1. Create a random population of *N* individuals.
2. Evaluate their fitness.

3. Sort the individuals in the population according to their fitness.
4. Choose the best *N/2* individuals to generate the new *N* population.
5. Generate the new *N* individuals by reproducing the nearest two sorted parents keeping the common genes and change the different genes using the four possible binary combinations randomly (every two parents generate four children).
6. Apply mutation operation with a probability.
7. Repeat steps from 2 to 6 for the best fitness value.

4 Numerical Examples

To demonstrate the effectiveness of the RCGA as compared to the TUCGA, two numerical examples are presented below. The parameter sensitivity analysis is done for the first example using a set of mutation rate and set of population size for a certain number of experiments. Each experiment started with the same random initial population set for both algorithms.

4.1 PEAKS Function

PEAKS is a function of two variables, obtained by translating and scaling Gaussian distributions evaluated as [13].

$$f(x,y) = 3(1-x)^2 e^{-x^2-(y+1)^2} - 10\left(\frac{x}{5} - x^3 - y^5\right) e^{-x^2-y^2} - \frac{1}{3} e^{-(x+1)^2-y^2} \quad (1)$$

The surface plot of this function is shown in Fig 2, it can be obtained directly by using MATLAB peaks function [21]. This function was chosen because it contains a good number of local minimum that can trap the minimal seeker.

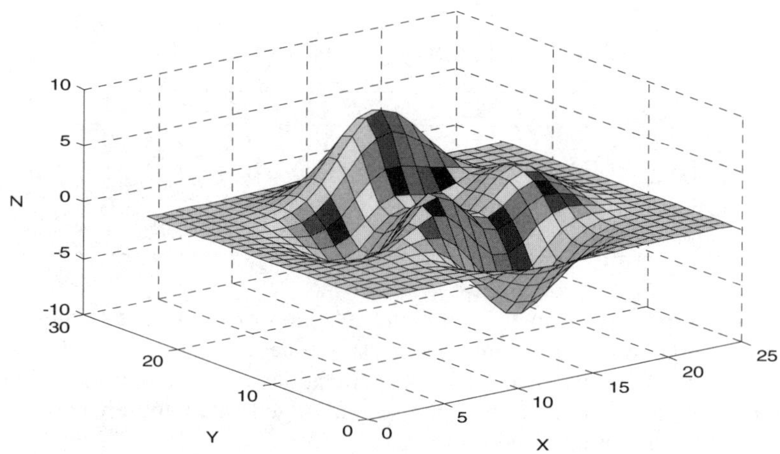

Fig. 2. Matlab peaks function

Using the minimum value of the peaks as a fitness function, one hundred experiments for every item of both, mutation rate and population size were done to compare the two algorithms. Those experiments were done by using mutation rate set defined as [0% %5 10% 15% 20% 50%] and population size set defined as [20 60 100 200 300 500].

The average results obtained after running the two algorithms for 100 generations using the same initial populations of 16-bit representation for each variable of the one hundred experiments is shown in table 3, where 'T' represents TUCGA and 'R' represents RCG.

Table 3. The average of the peaks function values obtained after 100 generations

		Population size					
Mutation		20	60	100	200	300	500
0%	R	-5.35204	-6.31998	-6.48472	-6.53505	-6.54830	-6.55063
	T	-5.03608	-6.38537	-6.49890	-6.54637	-6.55017	-6.55107
5%	R	-6.42770	-6.52496	-6.54628	-6.55113	-6.55113	-6.55113
	T	-6.37187	-6.49793	-6.54395	-6.54953	-6.55113	-6.55113
10%	R	-6.41143	-6.52071	-6.55107	-6.55113	-6.55113	-6.55113
	T	-6.36684	-6.53314	-6.54641	-6.55113	-6.55113	-6.55113
15%	R	-6.48776	-6.54639	-6.55113	-6.55113	-6.55113	-6.55113
	T	-6.45994	-6.54775	-6.55111	-6.55113	-6.55113	-6.55113
20%	R	-6.52503	-6.55110	-6.55111	-6.55113	-6.55113	-6.55113
	T	-6.52573	-6.54930	-6.55105	-6.55110	-6.55111	-6.55112
50%	R	-6.48626	-6.52943	-6.53825	-6.54337	-6.54678	-654856
	T	-6.47995	-6.52497	-6.53935	-6.54352	-6.54628	-6.54848

Table 4. Execution time (sec) average taken for 100 generations

		Population size					
Mutation		20	60	100	200	300	500
0%	R	82.12	52.41	44.51	20.42	19.78	16.74
	T	Fail	55.65	47.59	25.65	23.48	20.25
5%	R	55.69	45.18	25.74	17.3	12.42	10.76
	T	60.50	57.42	47.90	22.62	21.62	17.56
10%	R	49.20	36.34	21.9	17.8	13.66	12.32
	T	53.58	49.44	37.30	28.96	24.90	23.79
15%	R	47.36	33.46	29.38	22.56	18.76	16.14
	T	79.56	63.76	50.40	35.88	32.96	29.62
20%	R	72.37	58.12	42.12	35.04	30.22	23.12
	T	98.23	87.60	76.40	51.68	50.96	39.64
50%	R	Fail	Fail	100	99.52	99.02	98.08
	T	Fail	Fail	Fail	Fail	Fail	Fail

Table 4 shows the generations number average of the one hundred experiments to obtained the peaks function minimum value of (-6.551) using the same initial populations of 16-bit representation for each variable for the two algorithms.

To explore further, we executed both algorithms for 100 generations starting from the same initial population set of 60 members with 16-bit representation and 10% mutation rate. The performance for both the algorithms is shown in Fig 3. The minimum value of the Matlab PEAKS function (equation 1) as a fitness function and its corresponding two variables (x, y) is show in Table 5. It is noted that the RCGA successfully obtained the global minimum of the function after 10 generations. In contrast, the TUCGA achieved very close value of the global minimum after 15 generations using the same initial populations.

Table 5. The result after 100 iterations with 16- bit representation, 10% mutation rate, and population size of 60

Algorithm	F(x, y)	x	y
TUCGA	-6.3259	0.2579	-1.5000
RCGA	-6.5511	0.2283	-1.6255

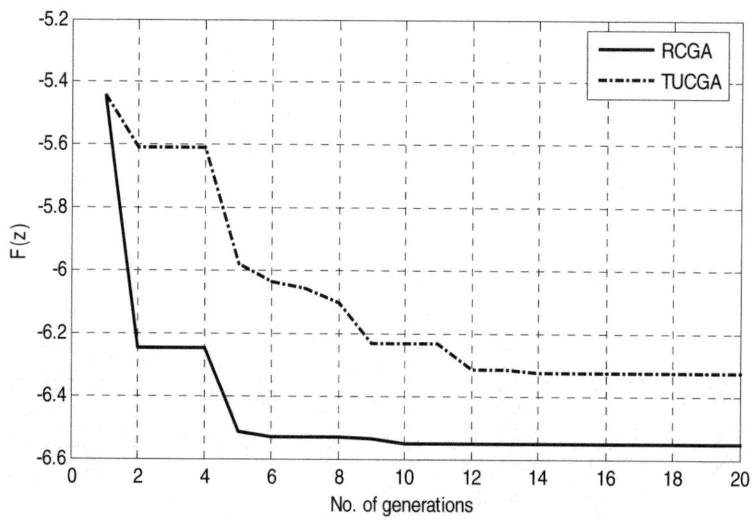

Fig. 3. Minimum value evaluation of the peaks function got by the two algorithms

4.2 The Flexible Beam System

To demonstrate the effectiveness further in real application, we consider a cantilever beam system with a force $F(x,t)$ applied at a distance x from its fixed (clamped) end at time t. This results a deflection $y(x,t)$ of the beam from its stationary position at the point where the force has been applied as shown in Fig. 4. This flexible beam was used by many researchers as a platform to investigates there algorithms for system identification and active vibration control [9-15].

TUCGA and RCGA are used to estimate the parameters of the AVC system. The algorithms for control parameters estimation and similar work using TCGA is reported earlier [11].

Fig. 5 depicts the auto-power spectral density before cancellation and the auto-power spectral density after cancellation in implementing the AVC system using TUCGA, and RCGA. Fig. 6 shows the time-domain performance in implementing the AVC system using TUCGA and RCGA, where the solid lines represent fluctuation of the beam at the end point before cancellation.

It is noted that the convergence achieved using the RCGA is consistently better than the TUCGA. The proposed RCGA using a new population inheritance approach through the recessive trait crossover offers for propagating good building blocks of genes to subsequent generations. These features RCGA, we believe, offered a significant impact in providing better results and convergence.

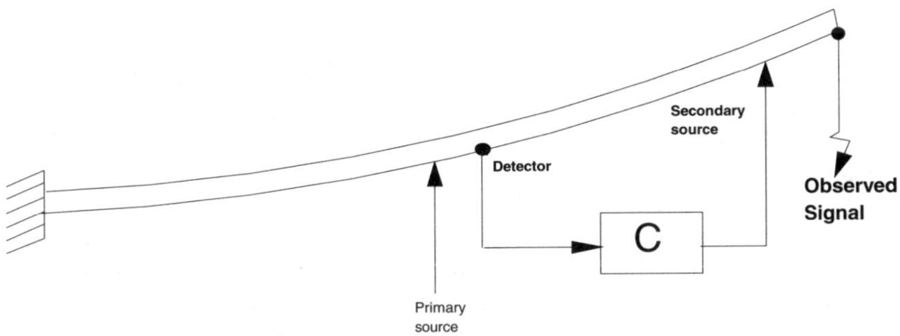

Fig. 4. Active vibration control structure

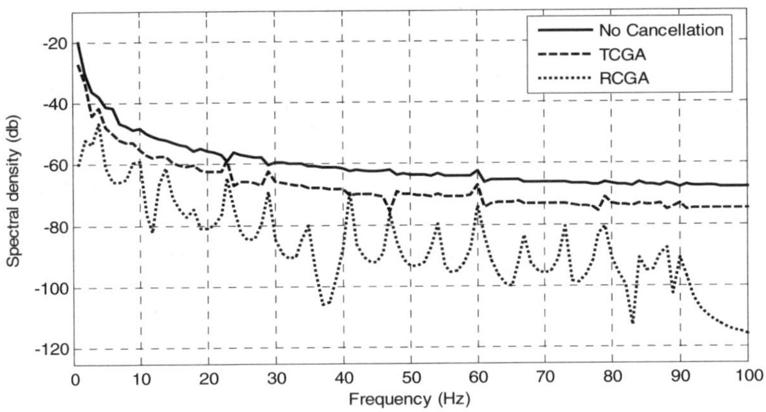

Fig. 5. Performance of the TUGA and RCGA in auto-power spectral density

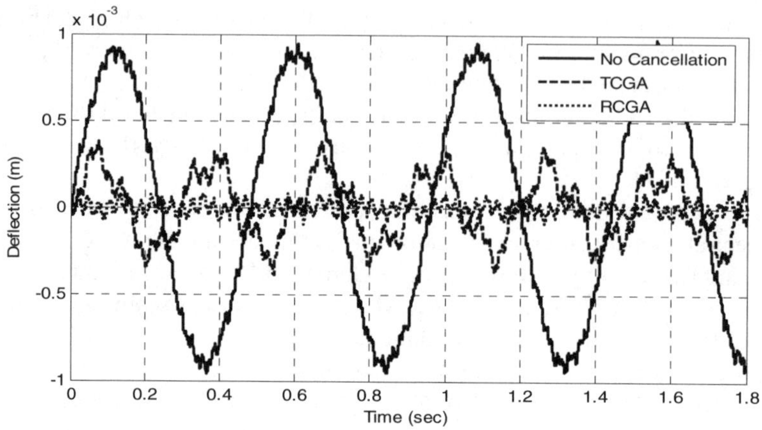

Fig. 6. Beam fluctuation at the end point

5 Conclusion

This paper has presented the investigation into a modified GAs population inheritance using a concept taken from the recessive trait idea. A comparative performance of the modified and traditional approaches has been presented and discussed through a set of experiments.

Two numerical examples have been provided to demonstrate the merits and capabilities of the modified cross over approach. In both cases, it is noted that using same initial populations, bit representation, and mutation rate, the RCGA offered better convergence, higher accuracy and faster solution for each problem. Farther more, the RCGA is very sample and easy to use for any numerical optimization problem not only as a minimum or maximum seeker but also for any fitness function.

References

[1] J. H. Holland, 'Adaptation in Natural and Artificial Systems' The University of Michigan Press, 1975.

[2] K. Deb, 'Multi-objective optimization using evolutionary algorithms', John wiley & sons LTD, 81-169,2001.

[3] D. E. Goldberg, 'Genetic Algorithms for Search, Optimization, and Machine Learning', Reading, MA: Addison-Wesley, 1989.

[4] M. Gen, R. Cheng, 'Genetic Algorithms and Engineering Design', New York: Wiley, 1997.

[5] Z. Michalewicz, 'Genetic Algorithms + Data Structures = Evolution Programs' , Berlin: Springer-Verlag, 1992.

[6] M., Mitchell, 'Introduction to Genetic Algorithms' Ann Arbor, MI: MIT Press 1996

[7] M. D. Vose, ' Simple Genetic Algorithm: Foundation and Theory' , Ann Arbor, MI: MIT Press 1999

[8] K.F.Man, K.S.Tang and S.Kwong, 'Genetic Algorithms: Concept and Designs', Springer, London.,1999
[9] P. K. Kourmoulis, 'Parallel processing in the simulation and control of flexible beam structure systems', PhD thesis, Dept. of Automatic Control & Systems Engineering, The University of Sheffield, 1990.
[10] M. A Hossain. 'Digital signal processing and parallel processing for real-time adaptive noise and vibration control', Ph.D. thesis, Department of Automatic Control and System Engineering, The University of Sheffield, UK.,1995.
[11] M. A. Hossain,and M. O. Tokhi 'Evolutionary adaptive active vibration control, Proc Inst. Mechanical Eng'., 211(1), pp. 183-193, 1997.
[12] M. O Tokhi,. M. A. Hossain, and M. H. Shaheed, 'Parallel Computing for Real-time Signal Processing and Control,' Springer, London,2002.
[13] A. M. Madkour, M. A, Hossain, K. P. Dahal, and H. Yu, 'Real-time System Identification using Intelligent Algorithms', Proceedings of IEEE SMC UK-RI Chapter Conference 2004 on Intelligent Cybernetic Systems, pp. 236-241,2004.
[14] M. A, Hossain, A. A. M. Madkour, K. P. Dahal, and H. Yu,. Intelligent Active Vibration Control for a Flexible Beam System', Proceedings of IEEE SMC UK-RI Chapter Conference 2004 on2004 on Intelligent Cybernetic Systems, pp. 236-241,2004.
[15] S. Z. Mohd Hashim, M. O. Tokhi, and I. Z. Mat Darus, 'Genetic Adaptive Active Vibration Control of Flexible Structures', Proceedings of IEEE SMC UK-RI Chapter Conference 2004 on2004 on Intelligent Cybernetic Systems, pp. 166-171,2004.
[16] G. Himmelfarb, 'Darwin and the Darwinian Revolution', Doubleaday & Company Inc. New York, 1959
[17] Fogel, D. B., Evolutionary Computation, Toward a New Philosophy of Machine Intelligence, IEEE Press, Piscataway, New Jersey, 1995.
[18] http://www.muscle.ca/content/index.php?id=315, Produced by Muscular Dystrophy Canada,Updated: December 2002
[19] E. Richard and M. D Robert.,"Nelson Essentials of Paediatrics", 3ed edition,W.B. Saunders Company, ISBN 0-7216-7229-9 , 1998.
[20] J. S. R. Jang, C. T. Sun, and E. Mizutani,'Neuro-Fuzzy and Soft Computing: A Computational Approach to Learning and Machine Intelligence'Prentice-Hall Inc.,USA,ISBN 0-13-261066-3,1997
[21] http://www.mathworks.com/access/helpdesk/help/techdoc/ref/peaks.html, Produced by the Mathworks Inc., December 2005.

Automated Prediction of Solar Flares Using Neural Networks and Sunspots Associations

T. Colak and R. Qahwaji

Department of Electronic Imaging and Media Communications, University of Bradford
Richmond Road, Bradford BD7 1DP, England, UK
t.colak@bradford.ac.uk, r.s.r.qahwaji@brad.ac.uk

Abstract. An automated neural network-based system for predicting solar flares from their associated sunspots and simulated solar cycle is introduced. A sunspot is the cooler region of the Sun's photosphere which, thus, appears dark on the Sun's disc, and a solar flare is sudden, short lived, burst of energy on the Sun's surface, lasting from minutes to hours. The system explores the publicly available solar catalogues from the National Geophysical Data Center to associate sunspots and flares. Size, shape and spot density of relevant sunspots are used as input values, in addition to the values found by the solar activity model introduced by Hathaway. Two outputs are provided: The first is a flare/ no flare prediction, while the second is type of the solar flare prediction (X or M type flare). Our system provides 91.7% correct prediction for the possible occurrences and, 88.3% correct prediction for the type of the solar flares.

Keywords: Neural Networks, Solar Physics.

1 Introduction

The term "space weather" refers to adverse conditions on the Sun that may affect space-borne or ground-based technological systems and can endanger human health or life [1].The importance of space weather is increasing day after day because of the way solar activities affect life on Earth and it will continue to increase as we rely more and more on different communication and power systems. The established effects of space weather activities on our daily lives can be summarized as follows:

Ground based systems: Induced electric fields and currents can disrupt the normal operation of high voltage power transmission grids, pipelines, telecommunications cables, metallic oil and gas pipelines and railway signaling [2]. The great geomagnetic storm of March 13, 1989 closed down the entire Hydro Quebec system [3]. [4] predicted that a credible electric power outage could result in a direct loss to the US Gross Domestic Product of $3 - $6 billion. [5] predicted that timely forecasts could save the US power industry $365M per year.

Communications systems: Wireless communications systems suffer from interruption of service like frequency jamming and dropped communications due to radio bursts caused by solar microwave emissions [6]. Solar activities can produce X-rays that disrupt point-to-point high frequency radio communications and radio noise that interferes with communications and radar systems [7].

A. Saad et al. (Eds.): Soft Computing in Industrial Applications, ASC 39, pp. 316–324, 2007.
springerlink.com

Space based systems: Adverse space weather conditions can cause anomalies and system failures and increased drag on the movement of satellites and spacecraft leading to slow-downs, changes in orbits and shorter life-times of missions. Other radiation hazards include direct collision damage and/or electrical defects, caused by charged particles [8]. On 19 May 1998, the PanAmSat Corporation's Galaxy 4 satellite experienced a failure in its altitude-control system, leading to the suspension of paging service for 45 million people [9].

There have been noticeable developments recently in solar imaging and the automated detection of various solar features, by: [10], [11], [12], [13], [14] and [15]. Despite the recent advances in solar imaging, machine learning and data mining have not been widely applied to solar data. [16] described a method for the automatic detection of solar flares from optical H-alpha images using the multi-layer perceptron (MLP) with back-propagation training rule. In [17], the classification performance for features extracted from solar flares is compared using Radial Basis Functions (RBF), Support Vector Machines (SVM) and MLP methods. Each flare is represented using nine features. However, these features provide no information about the position, size and verification of solar flares. Neural Networks (NNs) were used in [18] for filament recognition in solar images. However, machine learning algorithms are still not applied properly for the automated prediction of solar flares and space weather activities. This is a very challenging task because of the following reasons:

- There are an increasing number of space missions and ground based observatories providing continuous observation of the Sun at many different wavelengths. We are becoming "data rich" but without automated data analysis and knowledge extraction techniques, we continue to be "knowledge poor".
- A long standing problem in solar physics is establishing a correlation between the occurrence of solar activity (e.g., solar flares and coronal mass ejections (CMEs)) and solar features (sunspots, active regions and filaments) observed in various wavelengths.
- An efficient prediction system requires the successful integration of solar physics, machine learning and maybe solar imaging.
- There is no machine learning algorithm that is known to provide the "best" learning performance especially in the solar domain. In most cases, empirical studies, in a manner similar to [17], must be carried out to compare the performances of these algorithms before the final decision on which learning algorithm to use can be made.

A first attempt in addressing these challenges is reported in the recent work of [19] where different neural network (NN) topologies were studied to determine the best NN topology to process sunspots and associate them with solar flares. The findings of this work are used in this paper to address the challenges highlighted above. In general, we aim to investigate the degree of correlation between sunspot classes and the occurrence of solar flares that can affect our life on Earth using neural networks and timing information that represent the solar activity.

This paper is organized as follows: the data used in this paper is described in Section 2. The NN topology is discussed in Section 3. Section 4 is devoted to the practical implementation and the evaluation of the performance. Finally, the concluding remarks are given in Section 5.

2 Data

2.1 Sunspots and Flare Catalogues

A sunspot is the cooler region of the Sun's photosphere which, thus, appears dark on the Sun's disc, and a solar flare is sudden, short lived, burst of energy on the Sun's surface, lasting from minutes to hours. Solar flare research has shown that flares are mostly related to sunspots and active regions [20], [21], [22]. There are many publicly available catalogues which includes the information about flares and sunspots occurred in the past.

The sunspot group catalogue and solar flare catalogue are used from the National Geophysical Data Center (NGDC) [23] for this research. Flare catalogue includes data for all type of detected class flares. Flares are classified according to their x-ray brightness in the wavelength range from 1 to 8 Angstroms. *C*, *M*, and *X* class flares can effect earth. *C*-class flares are moderate flares with few noticeable consequences on Earth (i.e., minor geomagnetic storms). *M*-class flares are large; they generally cause brief radio blackouts that affect Earth's Polar Regions by causing major geomagnetic storms. *X*-class flares can trigger planet-wide radio blackouts and long-lasting radiation storms. This catalogue supplies information about dates, starting and ending times, location, the National Oceanic and Atmospheric Administration (NOAA) number of the corresponding active region and x-ray classification of detected flares. NOAA numbers active regions consecutively since January 5, 1972 as they are observed on the Sun. An active region must be observed by two different observatories before it is given a number or a flare must be observed to occur in it [24]. Not all the flares have their related NOAA number, so in this study only flares with associated NOAA numbers are included.

NGDC hold records of sunspot groups reported from several observatories from all around the world, supplying their location, time, physical properties and classification data. Two classification systems exist for sunspots: McIntosh and Mt. Wilson. McIntosh classification depends on the size, shape and spot density of sunspots, while the Mt. Wilson classification [25] is based on the distribution of magnetic polarities within spot groups [26]. The McIntosh classification is the standard for the international exchange of solar geophysical data. It is a modified version of the Zurich classification system developed by Waldmeir. The general form of the McIntosh classification is *Zpc* where, *Z* is the modified Zurich class, *p* is the type of spot, and *c* is the degree of compactness in the interior of the group. Mt. Wilson classification consists of letters taken from the Greek alphabet from alpha to delta and their different combination.

2.2 Solar Cycle Prediction Data

Besides the flare and sunspot group associations, 'monthly average sunspot number' generated using Equation 1 from [27] as a part of the NN inputs. The comparison of real and generated 'monthly average sunspot number' for a given date is given in Fig. 1. The values generated by Equation 1 are more suitable for our research instead of historical data because with this Equation we can generate predicted 'monthly average sunspot numbers' and use it in our trained NN system. In Equation 1, parameter a

represents the amplitude and is related to the rise of the cycle minimum, b is related to the time in months from minimum to maximum; c gives asymmetry of the cycle; and to denotes the starting time. More info about Equation 1 can be found in [27].

$$f(t) = a(t - t_0)^3/\{\exp[(t - t_0)^2/b^2] - c\} \tag{1}$$

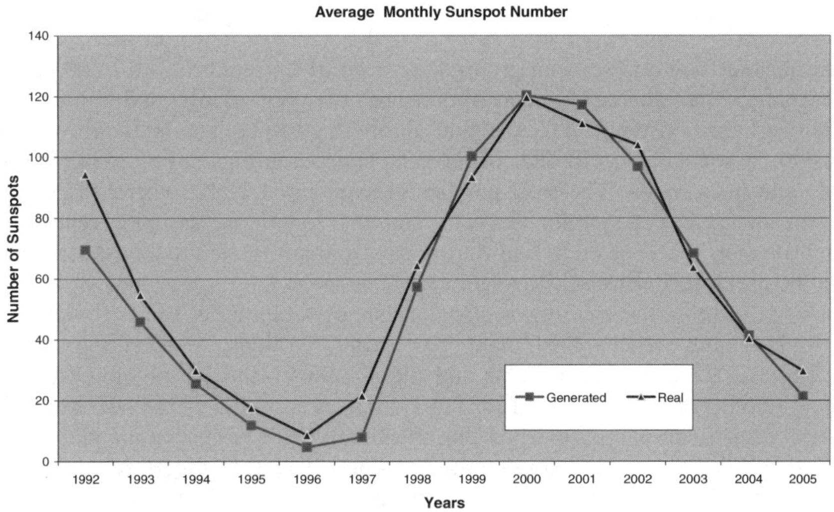

Fig. 1. The comparison of real and generated 'monthly average sunspot number' between 1992 and 2005

3 The Topology and Design of NN

The NN has proven to be a very good tool for solving many real-life problems. The efficient implementation of the NN requires training sessions that depends on the training vector and on the topology of the NN. The NN manages to converge if the training data are adequate to create the appropriate discriminations between the different output classes. The topology of the NN also plays an important role in the training process. If the network topology is too small then most properly the network will not be able to converge. On the contrary, if the network is too large and the training examples are presented many times, the network focuses on the singular statistical characteristics of the training set and looses its generalization ability. A network should be large enough to learn and small enough to generalize [28], [29]. The training is improved if the NN is optimized. A NN is optimized if the best hidden layer nodes and the optimum learning time are reached [30]. In general, two hidden layers are sufficient to perform any classification task although real-life problems are often much simpler and can be solved with a single hidden layer [31]. However, once trained the NN provides fast response.

In [19], several NN topologies were evaluated and it was found that the cascade forward NN with backpropagation training provides the best performance in terms of convergence time, optimum network structure and recognition performance. In cascade forward NN, the first layer has weights coming from the input and each subsequent layer has weights coming from the input and all previous layers.

4 Practical Implementation

4.1 The Training Process

The associated flare and sunspot group data, from 01/01/1992 to 31/12/2005 are used for training. The degree of correspondence between flares and sunspots was determined based on their NOAA region number and time. The software we created manages to associate 1425 M and X flares and sunspot groups out of 29343 flares and 110241 sunspot groups. The total number of samples is 2882, where 1425 samples represent sunspots that produced flares and the remaining samples represent the distinct sunspots that existed in non-flaring days and not related to any sunspot groups within the previous flaring sunspot samples.

For each sample, the training vector consists of 6 elements and is divided into 2 parts: input and target. The input part has 4 values representing McIntosh classification (3 values) of sunspots and the simulated number of monthly average sunspots number generated based on the Hathaway's model [27]. The 3 values for McIntosh classification are modified Zurich class, type of largest spot and the sunspot distribution. The target part consists of 2 values. The first target value is used to predict whether the sunspot is going to produce a flare or not. The other target value is used to determine whether the predicted flare is an X or M class flare.

4.2 Evaluating the Performance

The NN training and testing was carried out based on the statistical Jackknife technique, which is usually implemented to provide a correct statistical evaluation for the performance of the classifier, when implemented on a limited number of samples. This technique divides the total number of samples into 2 sets: a training set and a testing set. Practically, a random number generator decides which samples are used for the training of the NN and which are kept for testing it. The classification error depends mainly on the training and testing samples. For a finite number of samples, the error counting procedure can be used to estimate the performance of the classifier [32]. In each experiment, 80% of the samples were randomly selected and used for training while the remaining 20% were used for testing. Hence, the number of samples used in the training is 2305, while 577 samples are used in the testing of the NN.

As illustrated in [19], the cascade forward backpropagation trained NN provides the optimum performance for our case. However, before applying Jackknife technique, the number of hidden nodes was found empirically. We have started with 1 hidden node and continuously were increasing the number of hidden nodes until 35 hidden nodes were reached. Every time a new number of hidden nodes were used, the error rate and the recognition rate were recorded. After carrying out all the empirical experiments it was found that optimum performance was reached with 9 hidden nodes.

After the hidden nodes number was set to 9, ten training and testing experiments, based on the Jackknife technique were carried out to evaluate the NN performance, as illustrated in Table 1. On average, our system can provide 91.7% correct prediction for the possible occurrence of a solar flare and it can predict the class of this flare correctly in 88.3% of all cases.

Table 1. Experiments and results with Jackknife technique. (CFP= Correct Flare Prediction, CFTP= Correct Flare Type Prediction).

Experiment No	Convergence Error	% CFP in Total Flares	% CFTP in Total Flares
1	0.02850	91.50780	88.90815
2	0.03010	93.93414	90.29463
3	0.03400	90.64125	87.69497
4	0.02990	91.68111	90.12132
5	0.02754	89.60139	86.82842
6	0.02755	92.37435	86.82842
7	0.02740	90.29463	86.82842
8	0.04290	92.20104	88.04159
9	0.03070	91.85442	87.86828
10	0.02980	93.41421	89.94801
Average	0.03084	91.75043	88.33622

5 Conclusions

In this paper, an automated NN-based system that provides efficient prediction of solar activities that can affect life on Earth is presented. The system processes two publicly available solar catalogues from the National Geophysical Data Center and compares the reported occurrences of M and X solar flares with the relevant sunspots that were classified earlier and exist in the same NOAA region. To increase the accuracy of prediction, a mathematical model, based on the work of [27], is implemented to simulate the solar activity during the times of flares occurrences. The simulated activity and classified sunspots are converted to the appropriate numerical formats and fed to cascade forward backpropagation NN, to predict whether a significant flare will occur and whether it is going to be an X or M flare.

Also, with this work, the results found for correct flare prediction in [19] is improved to 91.7% from 85% and correct flare type prediction is improved to 88.3% from 78% by using a simpler topology and also number of inputs in the network is reduced to 4 from 8. By all means this study outperforms the previous work in [19] and can be used for automated flare prediction from sunspot groups.

Our practical findings in this paper show that there is a direct relation between the eruptions of flares and certain McIntosh classes of sunspots such as Ekc, Fki and Fkc. which are in accordance with [33], [34], [35], but it is the first time to verify this relation using machine learning.

We believe that the quality of this work can be enhanced if evolutionary learning is used in conjunction with the learning algorithms presented here. This paper is a first step toward building a real-time flares prediction model. Hence, it is important for our learning algorithms to be able to improve its learning and generalisation capabilities by continuous learning from the new sunspots data, which is available on daily basis. This can be done if automatic learning algorithms that require no user intervention are applied once the prediction model is built. For our future work, we intend on exploring the feasibility of using evolutionary learning, which is slow but very versatile, for our application. The feasibility of designing a hybrid system that combines evolutionary learning with cascade correlation neural networks in a manner similar to Nessy algorithm [36] will be explored. We will also explore the feasibility of integrating SVM in such system. In addition, we intend to apply more testing criteria for the newly developed learning algorithms.

In this work, we have tested our system using the Jack-knife technique, which provides a correct statistical evaluation for the performance of the classifier, when implemented on a limited number of samples. This test has proven that a nonlinear relation exists between sunspots data and the occurrence of flares. However, for our future work we will extend our evaluation criteria by training the model on certain periods of time and testing it on the remaining periods.

Acknowledgments. This work is supported by an EPSRC Grant (GR/T17588/01), which is entitled "Image Processing and Machine Learning Techniques for Short-Term Prediction of Solar Activity".

References

1. Koskinen, H., E. Tanskanen, R. Pirjola, A. Pulkkinen, C. Dyer, D. Rodgers, and P. Cannon, Space Weather Effects Catalogue, in ESA Space Weather Programme Feasibility Studies, FMI, QinetiQ, RAL Consortium. (2001).
2. Clark, T.D.G., A Review of the Effects of Space Weather on Ground Based Technology, in Space Weather Workshop: Looking towards a European Space Weather Programme: Noordwijk, The Netherlands. (2001).
3. Erinmez, I.A., J.G. Kappenman, and W.A. Radasky, "Management of the geomagnetically induced current risks on the national grid company's electric power transmission system", Journal of Atmospheric and Solar-Terrestrial Physics, 64(5-6): (2002), 743.
4. Barnes, P. and J.V. Dyke, On the Vulnerability of Electric Power to Geomagnetic Storms, Oak Ridge National Laboratory: Tennesse. (1990).
5. Weiher, R. and T. Teisberg, "Economic Valuation of Geomagnetic Storm Forecasts in the North American Electic Industry", Journal of Policy Analysis and Management, 19(2): (2000).
6. Gary, D.E., L.J. Lanzerotti, G.M. Nita, and D.J. Thomson. Effects of Solar Radio Bursts on Wireless Systems. in NATO ARW on Effects of Space Weather on Technology Infrastructure. 25 to 29 March 2003. Rhodes, Greece: Springer, (Year).
7. Shea, M.A. and D.F. Smart, Space weather: The effects on operations in space, in Solar-Terrestrial Relations: Predicting The Effects On The Near- Earth Environment. p. 29-38, (1998).

8. Guy, F. and D.C. Johnson, Space Environmental Impacts on DoD Operations, AIR FORCE SPACE COMMAND. (2003).
9. Worden, S.P., "The Air Force and Future Space Directions: Are We Good Stewards?" Aerospace Power Journal, 15(1): (2001), 50-57.
10. Gao, J.L., H.M. Wang, and M.C. Zhou, "Development of an automatic filament disappearance detection system", Solar Physics, 205(1): (2002), 93-103.
11. Turmon, M., J.M. Pap, and S. Mukhtar, "Statistical pattern recognition for labeling solar active regions: application to SOHO/MDI imagery", Astrophysical Journal, 568(1): (2002), 396-407.
12. Shih, F.Y. and A.J. Kowalski, "Automatic Extraction of Filaments in H-alpha images", Solar Physics, 218(1-2): (2003), 99 – 122.
13. Benkhalil, A., V. Zharkova, S. Ipson, and S. Zharkov. Automatic Identification of Active Regions (Plages) in the Full-Disk Solar Images Using Local Thresholding and Region Growing Techniques. in AISB'03 Symposium on Biologically-inspired Machine Vision, Theory and Application. 2003. University of Wales, Aberystwyth (Year).
14. Lefebvre, S. and J.P. Rozelot, "A new method to detect active features at the solar limb", Solar Physics, 219(1): (2004), 25-37.
15. Qahwaji, R. and T. Colak, "Automatic Detection And Verification of Solar Features", International Journal of Imaging Systems & Technology, 15(4): (2006), 199 - 210.
16. Borda, R.A.F., P.D. Mininni, C.H. Mandrini, D.O. Gomez, O.H. Bauer, and M.G. Rovira, "Automatic solar flare detection using neural network techniques", Solar Physics, 206(2): (2002), 347-357.
17. Qu, M., F.Y. Shih, J. Jing, and H.M. Wang, "Automatic solar flare detection using MLP, RBF, and SVM", Solar Physics, 217(1): (2003), 157-172.
18. Zharkova, V. and V. Schetinin, "A Neural Network Technique for Recognition of filaments in solar images", Proceedings of the seventh International Conference on Knowledge-Based Intelligent Information & Engineering Systems KES'03: (2003), 148-154.
19. Qahwaji, R. and T. Colak, Neural Network-based Prediction of Solar Activities, in CITSA2006: Orlando. (2006).
20. Liu, C., N. Deng, Y. Liu, D. Falconer, P.R. Goode, C. Denker, and H.M. Wang, "Rapid change of delta spot structure associated with seven major flares", Astrophysical Journal, 622(1): (2005), 722-736.
21. Zirin, H. and M.A. Liggett, "Delta-Spots And Great Flares", Solar Physics, 113(1-2): (1987), 267-283.
22. Shi, Z.X. and J.X. Wang, "Delta-Sunspots And X-Class Flares", Solar Physics, 149(1): (1994), 105-118.
23. National Geophysical Data Center. Available from: ftp:// ftp.ngdc.noaa.gov/ STP/SOLAR_DATA/ Date Accessed: 2006
24. Holman, G. and S. Benedict, Questions and Answers, NASA. (2006).
25. Hale, G.E., F. Ellerman, S.B. Nicholson, and A.H. Joy, "The Magnetic Polarity of Sun-Spots", Astrophysical Journal, 49: (1919), 153.
26. Greatrix, G.R. and G.H. Curtis, "Magnetic Classification Of Sunspot Groups", Observatory, 93(994): (1973), 114-116.
27. Hathaway, D., R.M. Wilson, and E.J. Reichmann, "The Shape of the Sunspot Cycle", Solar Physics, 151: (1994), 177.
28. Horn, D., "Neural Computation Methods and Applications: Summary Talk of the AI", Journal of Nuclear Instruments and Methods in Physics research, 389: (1997), 381-387.

29. Tzafestas, E., A. Nikolaidou, and S. Tzafestas, "Performance Evaluation and Dynamic Node Generation Criteria for ‘ Principal Component Analysis’ Neural Network", Mathematics and Computer in Simulation, 51: (2000), 145-156.
30. Ma, Q., A.Yan, Z. Hu, Z. Li, and B. Fan, "Principal Component Analysis and Arti-ficial Neural Networks Applied to the Classification of Chinese Pottery Neolithic age", Analytica Chimica Acta Journal, 406: (2000), 247-256.
31. Kim, J., A. Mowat, P. Poole, and N. Kasabov, "Linear And Non-Linear Pattern Recognition Models For Classification Of Fruit From Visible-Near Infrared Spectra", Chemometrics And Intelligent Laboratory Systems, 51: (2000),201-216.
32. Fukunaga, K., Introduction to Statistical Pattern Recognition," Academic Press, New York, 1990. New York: Academic Press (1990).
33. McIntosh, P.S., "The Classification Of Sunspot Groups", Solar Physics, 125(2): (1990), 251-267.
34. Warwick, C.S., " Sunspot Configurations and Proton Flares ", Astrophysical Journal, 145: (1966), 215.
35. Sakurai, K., "On the magnetic configuration of sunspot groups which produce solar proton flares", Planetary and Space Science, 18(1): (1970), 33.
36. Köppen, M., M. Teunis, and B. Nickolay. Nessy - An Evolutionary Learning Neural Network. in SOCO 97. 1997. Nimes, France(Year).

Keyword Index

Author Index

Printing: Mercedes-Druck, Berlin
Binding: Stein+Lehmann, Berlin